Advanced Mathematics

高 等 数 学

主　编　付　敏　李圆媛　阮正顺

武汉理工大学出版社
·武　汉·

内 容 简 介

本书专为共建“一带一路”国家留学生设计，旨在提供系统化、实用性强的高等数学学习资源。本书内容涵盖微积分核心模块，结合工程、经济与管理学科中的实际案例，帮助学生将数学工具应用于专业领域。全书采用英文编写，语言简洁，配有丰富的图表、习题，兼顾了不同文化背景学生的学习需求。本书融入中国传统文化特色，包含《易经》的哲学思想，并且强化数学建模能力，从基础理论到高阶应用循序渐进，适应不同基础的学习者。本书适用于理工科、经管类共建“一带一路”国家本科留学生，需强化数学基础的研究生，以及职业技术人员。本书致力于推动“一带一路”学术交流，为国际化人才培养提供坚实的数学支持。

This textbook is specifically designed for international students from Belt and Road countries. It provides systematic and practical learning resources for advanced mathematics, covering core topics, such as calculus. By integrating real-world examples from engineering, economics, and management disciplines, it helps students apply mathematical tools to their professional fields. The concrete English content features clear explanations, abundant diagrams, exercises, catering to the diverse learning needs of students from different cultural backgrounds. This textbook integrates distinctive elements of traditional Chinese culture, including philosophical concepts from the *I Ching* (*Book of Changes*), while emphasizing the development of mathematical modeling skills. Designed from foundational theories to advanced applications—it adapts to learners with diverse starting points. This textbook is suitable for undergraduate students in STEM (Science, Technology, Engineering, Mathematics), economics, and management disciplines under Belt and Road countries, as well as graduate students seeking to strengthen their mathematical foundations and professionals in technical fields. This textbook is dedicated to promoting academic exchanges within the BRI (Belt and Road Initiative) framework and providing robust mathematical support for cultivating global talent.

图书在版编目(CIP)数据

高等数学 = Advanced Mathematics ：英文 / 付敏，李圆媛，阮正顺主编. -- 武汉 ：武汉理工大学出版社，2025. 5. -- ISBN 978-7-5629-7427-7

Ⅰ. O13

中国国家版本馆 CIP 数据核字第 2025VW3642 号

项目负责人：王　思(027-87290908)　　责任编辑：王　思
责任校对：张明华　　版面设计：正风图文
出版发行：武汉理工大学出版社　　地　址：武汉市洪山区珞狮路 122 号
邮　编：430070　　网　址：http://www.wutp.com.cn
经　销：各地新华书店　　印　刷：武汉中远印务有限公司
开　本：787 mm×1092 mm　1/16　　印　张：17.75
字　数：591 千字
版　次：2025 年 5 月第 1 版　　印　次：2025 年 5 月第 1 次印刷
定　价：48.00 元

Preface

Under the backdrop of Belt and Road Initiative driving global connectivity and educational collaboration, the internationalization of higher education has become a pivotal direction for cultivating talent in the new era. In response to this demand and to empower students from Belt and Road countries with robust mathematical foundations and scientific thinking, we have compiled *Advanced Mathematics*, aiming to construct an academic bridge integrating mathematical theory with interdisciplinary practice.

This textbook aligns with the China's Emerging Engineering Education principles, systematically reconstructing the calculus knowledge framework. Focuses on core theories of functions, limits, and single-variable calculus, emphasizing geometric intuition and economic interpretations of definitions and theorems.

Each chapter features a "Chinese tradition" column, showcasing mathematical applications in transnational collaboration. Drawing on expertise from China's National Exemplary Courses for international students, the text incorporates specialized adaptations for non-native learners.

The contents of this textbook are written by the authors as follows: Associate Professor Fu Min for Chapter 1 to Chapter 3, Professor Li Yuanyuan for Chapter 4 and Chapter 5, Associate Professor Ruan Zhengshun for Chapter 6 and Chapter 7. Additionally, the textbook development received collective guidance from faculty members at Wuhan Institute of Technology's School of Mathematics and Physics, with multiple teachers participating in exercise verification and teaching trials. The compilation was also assisted by Yin Weilin, Liu Yue, and Yang Ziyu, students majoring in Information and Computing Science at Wuhan Institute of Technology, and we extend our gratitude to all of them.

This publication benefits from the academic and financial support of Wuhan Institute of Technology and Wuhan Smart Wing Automatic (Yongneng Zhixiang) company. We extend gratitude to them.

May this textbook serve as a master key for Belt and Road scholars to unlock scientific frontiers, fostering civilization dialogue through the dialectics of differentiation and integration. Due to the limited level of the editor, we sincerely request colleagues and readers to criticize and correct the omissions in this textbook.

Authors
Apr. 2025

Catalog

Chapter 1 Functions and Limits

The Daoist Zhuangzi said "A one-foot pole is halved daily, it will never be exhausted, even after ten thousand generations."

In ancient China, "the hammer of one foot, the day takes half of it, eternal life" (*Zhuangzi* · *Tianxia*) embodies the simple limit thought. This coincides with "infinity approaches but does not equal" in modern limit theory. For example, the daily itinerary of a caravan on the Silk Road can be seen as an item in a sequence of numbers that can be viewed over time, the limit of the total distance is the exact value of the total distance.

Imagine a pole of length 1 foot. Each day, we cut it in half and keep one half. The remaining length forms a sequence:

$$L^n = \frac{1}{2^n}, \quad \text{where } n = 0, 1, 2, \cdots.$$

Day 0: $L_0 = 1$ (whole pole)

Day 1: $L_1 = \frac{1}{2}$

Day 2: $L_2 = \frac{1}{4}$

...

Day n: $L_n = \frac{1}{2^n}$

The limit of the sequence, as n approaches infinity ($n \to \infty$), the remaining length mathematically approaches zero:

$$\lim_{n \to \infty} \frac{1}{2^n} = 0.$$

The pole's length gets infinitely close to zero but never truly vanishes. This mirrors the paradox of infinite divisibility: physically, you can always cut the remaining piece in half, but mathematically, the "endpoint" of this process is zero.

Function is one of the mathematical concepts in natural sciences. It is the basic concept of mathematics, and also the basic research object of calculus. And the main method of studying the function is the limit. This chapter introduces the concept of functions, the properties of functions and the operation of functions, the concept, properties and calculation methods of function limit. There are four possible ways to represent a function: verbally (by a description in words), numerically (by a table of values), visually (by a graph), algebraically (by an explicit formula). We focus on the algebraic explanation stared from mapping.

1.1 Mapping and Function

1.1.1 Mapping

Definition 1.1.1 A,B are two non-empty sets, if there is a correspondence such that each element in set A, by having uniquely identified elements corresponding to it in B, is called the map, denoted as

$$f:A\to B, a\mapsto b \text{ or } b=f(a),$$

b is called the image, a is called the original image, A is called the domain, noted as D_f, B is called as the set of all the elements of the map of the value domain, noted as R_f.

Note:

(1) Each element has an image, and the image is all in it;

(2) The image of each element is unique;

(3) The mapped value domain is a subset.

Each element in the value domain has the original image, but the original image can not be unique.

Example 1.1.1 $f:\mathbf{R}\to\mathbf{R}, x\in\mathbf{R}, f(x)=|x|$, where $D_f=\mathbf{R}$, and $R_f=\{y\mid y\geqslant 0\}$.

Obviously, the original image of each element is not unique.

Example 1.1.2 $f:[0,1]\to[1,2], x\in[0,1], f(x)=x+1$, where $D_f=[0,1]$, and $R_f=[1,2]$, the original image of each element is unique.

Example 1.1.3 $f:\mathbf{R}\to[-1,1], x\in\mathbf{R}, f(x)=\cos x$, where $D_f=\mathbf{R}$, and $R_f=[-1,1]$, each element in the value domain is not unique.

A mapping is a correspondence between elements of sets, which can be many-to-one or one-to-one. According to the correspondence, mappings can be classified. From classification, it can be seen that Example 1.1.1 is neither an injection nor a surjection; Example 1.1.2 is a bijection; Example 1.1.3 is not an injection but a surjection.

1.1.2 Function

Definition 1.1.2 Let f be the map from the set A to the set B, i.e. $\forall x\in A$. According to the map, there are unique $y\in B$, if B is the number of sets, the map is called a function, denoted as f.

$$f:A\to B, x\mapsto y=f(x).$$

When B is a real number set, it is called real function; when it is a complex set, it is called complex function.

Note:

(1) About the domain

The domain of a function can be several sets or not several sets. When it is real set, it is called real variable function; when it is complex set, it is called complex variable function. Given a function, the domain of the function is given sometimes, a given function does not explicitly indicate its domain, then the domain of the function is self-evident, that is, the domain is a set of numbers that make the function meaningful.

For example, a given function $f(x)=\sqrt{1-x}$ does not indicate its domain, which is the set of numbers that make the function meaningful, i.e. $A:\{x\mid x\leqslant 1\}$.

In functions with practical meaning, the domain of the function is constrained by practical meaning. For example, the area of a circle with a radius can be taken as any real number from the abstract mathematical formula $S=\pi r^2$, but from its practical sense, the radius of a circle cannot take a negative value, that is, the domain is an interval $[0,+\infty)$.

(2) About the function f

In function definition, the function f is abstract, only within specific functions is the function f concrete. Specific function representations include data or table, analytical expression, and graphical method. As long as for any x in the set A, it can be clearly indicated that there is a unique corresponding y, a mapping f is specifically represented, that is, a function f defined on A is given, and the method used to provide this correspondence rule is irrelevant.

(3) About the range

The range R_f of the function f is generally a subset of the set E, and in specific function expressions, it is determined by the elements in the set A, that is

$$R_f=f(A)=\{y\mid y=f(x),x\in A\}.$$

Example 1.1.4 The Floor Function: $\forall x\in\mathbf{R}$ (where $\mathbf{R}$ is the set of real numbers), defined as

$$f(x)=\lfloor x\rfloor=n\ (n\leqslant x<n+1,n\text{ is integer}),$$

that is, $\lfloor x\rfloor$ is the greatest integer less than or equal to x.

As shown in Figure 1-1, obviously there is

$$\lfloor x\rfloor\leqslant x<\lfloor x\rfloor+1.$$

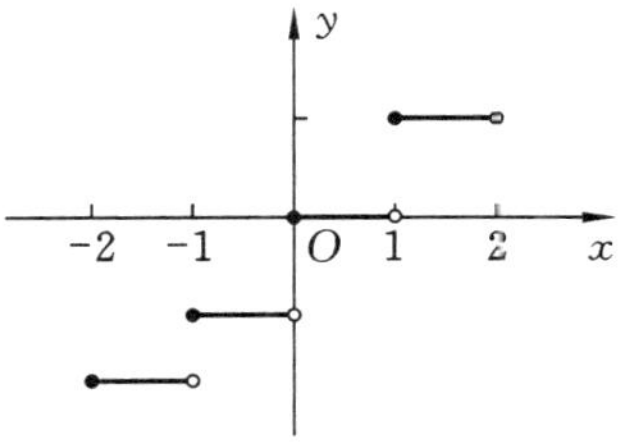

Figure 1-1

Example 1.1.5 The Sign Function:

$$\operatorname{sgn}(x)=\begin{cases}1, & x>0\\ 0, & x=0.\\ -1, & x<0\end{cases}$$

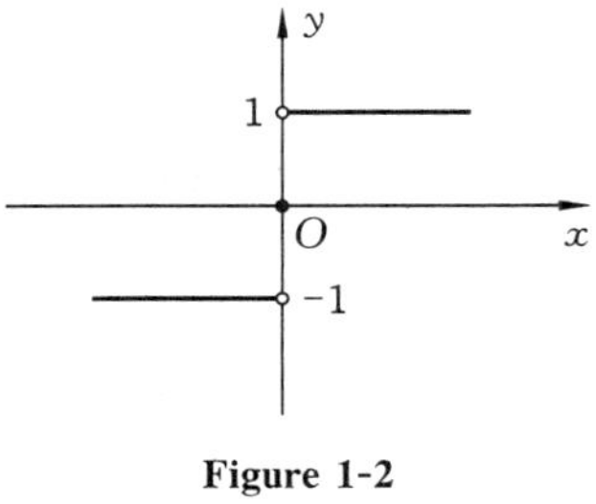

Figure 1-2

$\forall x\in\mathbf{R}$, where $|x|=x\operatorname{sgn}x$ holds, thus $\operatorname{sgn}x$ serves as a sign. This function is called the sign function, as shown in Figure 1-2. If a function is defined piece-wise by different analytical expressions on different intervals of its domain (individual intervals can also shrink to a single point), it is referred to as a piece-wise function, or simply a piece-wise function. Example 1.1.5 is an example of such a function.

1.1.3 Elementary Functions

In the development of mathematics, six functions have been gradually selected as the simplest, most fundamental, and most commonly used: constant function, exponential function, logarithmic function, power function, trigonometric function, and inverse trigonometric function. These six functions are collectively referred to as the elementary functions.

(1) Constant Function

$$y=C \text{ or } f(x)=C \quad (C \text{ is a constant}, x\in \mathbf{R}).$$

Domain $D_f=\mathbf{R}$, Range $R_f=\{C\}$.

(2) Exponential Function

$$y=a^x \quad (a>0, a\neq 1, x\in \mathbf{R}).$$

Domain $D_f=\mathbf{R}$, Range $R_f=(0,+\infty)$.

(3) Logarithmic Function

$$y=\log_a x \quad (a>0, a\neq 1, x>0).$$

Domain $D_f=(0,+\infty)$, Range $D_f=\mathbf{R}$.

(4) Power Function

$$y=x^\alpha \quad (\alpha\in \mathbf{R}).$$

The behavior of the power function $y=x^\alpha$ is related to the power exponent α. Typically, the domain of the power function $y=x^\alpha$ is $(0,+\infty)$.

(5) Trigonometric Function

The domains of $y=\sin x$ and $y=\cos x$ are $\mathbf{R}$, and their ranges are the closed interval $[-1,1]$.

The domains of $y=\tan x$ and $y=\sec x$ are

$$\left\{x \,\middle|\, x\in\left(k\pi-\frac{\pi}{2}, k\pi+\frac{\pi}{2}\right), k\in \mathbf{Z}\right\}.$$

The ranges are $\mathbf{R}$ and $(-\infty,-1]\cup[1,+\infty)$, respectively.

The domains of $y=\cot x$ and $y=\csc x$ are

$$\{x \mid x\in(k\pi,(k+1)\pi), k\in \mathbf{Z}\}.$$

The ranges are $\mathbf{R}$ and $(-\infty,-1]\cup[1,+\infty)$, respectively.

(6) Inverse Trigonometric Function

The domains of $y=\arcsin x$ and $y=\arccos x$ are the closed interval $[-1,1]$, and their ranges are $\left[-\frac{\pi}{2},\frac{\pi}{2}\right]$ and $[0,\pi]$, respectively.

The domain of $y=\arctan x$ and $y=\operatorname{arccot} x$ is $\mathbf{R}$, with their respective ranges being $\left(-\frac{\pi}{2},\frac{\pi}{2}\right)$ and $(0,\pi)$.

In calculus, many functions are constructed from basic elementary functions. Therefore, mastering the properties of basic elementary functions is very helpful for learning calculus.

Exercise 1.1

1. Which of the following mappings are injective, surjective, or bijective? Why?

(1) $f:\mathbf{N}\to\mathbf{N}, f(x)=x+2$;

(2) $f:\mathbf{Z}\to\mathbf{Z}, f(x)=x+2$;

(3) $f:\mathbf{R}\to\mathbf{R}, f(x)=x^2+1$.

2. Given $f(x)=\dfrac{1}{x}$, compute $\dfrac{f(x_0+h)-f(x_0)}{h}$ $(x_0\neq 0, x_0+h\neq 0)$.

3. Are the functions $f(x)$ and $g(x)$ the same in the following questions? Why?

(1) $f(x)=1, g(x)=\csc^2 x-\cot^2 x$;

(2) $f(x)=\ln x^2, g(x)=2\ln x$;

(3) $f(x)=\sqrt[3]{x^4-x^3}, g(x)=x\sqrt[3]{x-1}$.

4. Find the domain of the following functions:

(1) $y=\arccos\dfrac{2x}{1+x}$;

(2) $y=\sqrt{\sin\sqrt{x}}$;

(3) $y=\sqrt{x-1}+\sqrt{1-x}+1$;

(4) $f(x)=\sqrt{2+x+x^2}$;

(5) $f(x)=\sqrt{\dfrac{1-x}{1+x}}$;

(6) $f(x)=\ln(2x+1)+\sqrt{4-3x}$;

(7) $f(x)=\sqrt{\cos 2x}$;

(8) $f(x)=\log_3(\log_2 x)$;

(9) $f(x)=\sqrt{\sin x}+\sqrt{16-x^2}$;

(10) $f(x)=\sqrt{\log_4 \sin x}$.

5. Given $f(x)=\begin{cases}|\sin x|, & |x|<\dfrac{\pi}{3}\\ 0, & |x|\geqslant\dfrac{\pi}{3}\end{cases}$, find $f\left(\dfrac{\pi}{4}\right)$, $f\left(-\dfrac{\pi}{4}\right)$, $f(-2)$, and draw the graph of the function $y=f(x)$.

6. Express the area and perimeter of an equilateral triangle as a function of its side length x.

7. Remove a sector with arc length x from a circular disk of radius $\mathbf{R}$, and join the two remaining sides to form a cone.

(1) Express the circumference of the cone's base in terms of x;

(2) Express the radius of the cone's base in terms of x;

(3) Express the volume of the cone in terms of x.

8. From a rectangular sheet of metal with length a and width b, cut squares of side length x from each corner, and fold up the sides to form an open box. Express the volume of the box V in terms of x.

9. Given the height of a right cylinder is h and the radius is r, taking the radius r as the independent variable, express the surface area of the cylinder S in terms of r.

1.2 Properties of Function

1.2.1 Boundness

Definition 1.2.1 Let the function $f(x)$ be defined on the set of numbers A. If the range of the function $f(x)$ has an upper bound (or a lower bound, or is bounded), then the function $f(x)$ is said to have an upper bound (or a lower bound, or is bounded) on A. Otherwise, the function $f(x)$ is said to have no upper bound (or no lower bound, or is unbounded) on A.

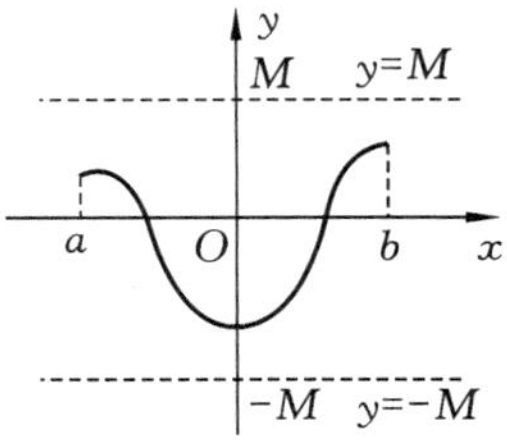

Figure 1-3

The geometric meaning of the function $f(x)$ being bounded on the interval $[a, b]$ is that there exist two lines $y=M$ and $y=-M$, and the graph of the function $y=f(x)$ lies between these two lines, as shown in Figure 1-3.

Example 1.2.1 The arc-tangent function $y=\arctan x$ and the arc-cotangent function $y=\operatorname{arccot} x$ are bounded on their domain $\mathbf{R}$, as shown in Figure 1-4 and Figure 1-5.

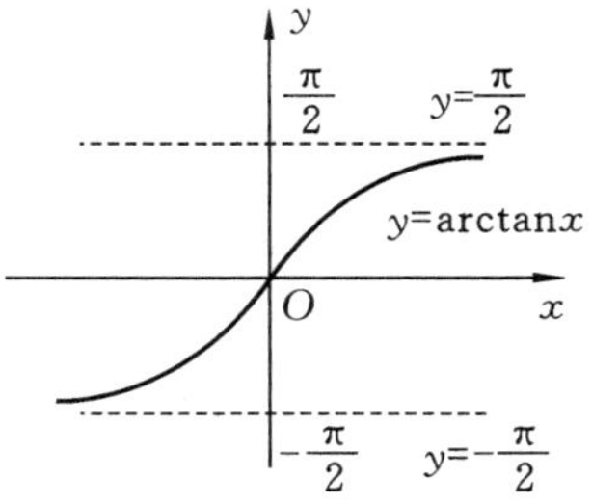

Figure 1-4

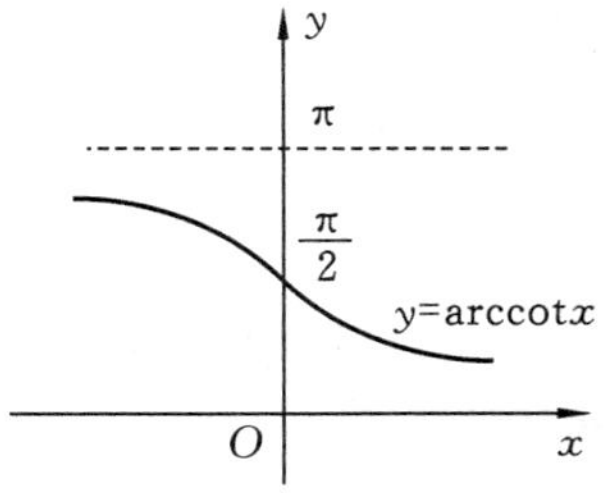

Figure 1-5

Theorem 1.2.1 Let $f(x)$, $g(x)$ all be bounded functions on A, then $f(x)\pm g(x)$, $f(x)g(x)$ is also a bounded function on A.

Proof Since $f(x)$, $g(x)$ are bounded on A, there exists a constant $M_1>0$, $M_2>0$, $\forall x\in A$, such that

$$|f(x)|\leqslant M_1,\ |g(x)|\leqslant M_2,$$

then

$$|f(x)\pm g(x)|\leqslant|f(x)|+|g(x)|\leqslant M_1+M_2,$$
$$|f(x)g(x)|=|f(x)||g(x)|\leqslant M_1M_2.$$

This shows that $f(x)\pm g(x)$, $f(x)g(x)$ is bounded on A.

Note: Theorem 1.2.1 can be generalized as: The sum and product of a finite number of bounded functions remain bounded functions.

1.2.2 Monotonicity

Definition 1.2.2 Let the function $f(x)$ be defined on A. If $\forall x_1, x_2 \in A$, and $x_1 < x_2$, then

$$f(x_1) \leqslant f(x_2) \text{ (or } f(x_1) \geqslant f(x_2)\text{)},$$

the function $f(x)$ is said to be monotonically increasing (or monotonically decreasing) on A.

If the above inequality is changed to

$$f(x_1) < f(x_2) \text{ (or } f(x_1) > f(x_2)\text{)},$$

the function $f(x)$ is said to be strictly increasing (or strictly decreasing) on A.

Geometrically, $f(x)$ being monotonically increasing (or monotonically decreasing) means: The graph of $y = f(x)$ gradually rises (or falls) along the positive direction of the x axis.

1.2.3 Symmetry

Definition 1.2.3 Let the function $f(x)$ be defined on the set of numbers A. If $\forall x \in A$, and $-x \in A$, then

$$f(-x) = -f(x) \text{ (or } f(-x) = f(x)\text{)},$$

the function $f(x)$ is called an odd function (or even function) on A.

Geometrically, the graph of an odd function is symmetric about the origin, and the graph of an even function is symmetric about the y axis.

Theorem 1.2.2

(1) The sum of two odd functions is an odd function, and the sum of two even functions is an even function;

(2) The product of two odd functions or two even functions is an even function;

(3) The product of an odd function and an even function is an odd function.

Proof We only prove (3).

Let $f(x)$, $g(x)$ be an odd function and an even function on A, i.e., $\forall x \in A$, then $f(-x) = -f(x)$, $g(-x) = g(x)$, thus

$$f(-x)g(-x) = [-f(x)]g(x) = -[f(x)g(x)].$$

Therefore, $f(x)g(x)$ is an odd function. Similarly, (1) and (2) can also be proved.

1.2.4 Periodicity

Definition 1.2.4 Let the domain of function $f(x)$ be A. If there exists a positive constant $T \neq 0$, $\forall x \in A$, and $x \pm T \in A$, then

$$f(x+T) = f(x).$$

If a function $f(x)$ is periodic, then T is called a period of $f(x)$.

If T is a period of $f(x)$, then nT $(n=1,2,\cdots)$ are all periods of $f(x)$, hence $f(x)$ has infinitely many periods.

If a function $f(x)$ has the smallest positive period, this smallest positive period is commonly referred to as the period of the function, also known as the fundamental period.

For example, $f(x)=\sin x$ and $f(x)=\cos x$ are both periodic functions with period 2π over the domain **R**.

Exercise 1.2

1. Determine the boundness of function $y=\begin{cases}\frac{1}{x}, & 0<x\leqslant 1\\ 0, & x=0\end{cases}$ over the given interval.

2. Determine the monotonicity of the following functions over the given intervals:

(1) $f(x)=x^2+1$, $(-\infty,0)$, $(0,+\infty)$;

(2) $f(x)=\sin x-1$, $\left(-\frac{\pi}{2},\frac{\pi}{2}\right)$;

(3) $f(x)=3\ln x^2$, $(-\infty,0)$, $(0,+\infty)$.

3. Determine the parity of the following functions:

(1) $f(x)=\sin x+x$;　　(2) $f(x)=\ln\left(x+\sqrt{1+x^2}\right)$;

(3) $f(x)=\sin x^2$;　　(4) $f(x)=\ln\frac{1-x}{1+x}$;

(5) $f(x)=\frac{1}{2}(a^x+a^{-x})$;　　(6) $f(x)=x^2\sin\frac{1}{x}$;

(7) $f(x)=x^2\operatorname{sgn}x$;　　(8) $f(x)=x\sin x$.

4. Prove:

(1) The sum and product of two even functions are even functions;

(2) The sum of two odd functions is odd, and their product is even;

(3) The product of an odd function and an even function is an odd function.

5. Determine the periodicity of the following functions. If periodic, specify their periods.

(1) $f(x)=\sin(x+2)$;　　(2) $f(x)=x\cos x$;

(3) $f(x)=\sin x^2$;　　(4) $f(x)=\sin^2 x$;

(5) $f(x)=\lfloor x\rfloor$;　　(6) $f(x)=x-[x]$.

6. Prove: If the function $f(x)$ is defined on **R**, then

$$F_1(x)=f(x)+f(-x),\ F_2(x)=f(x)-f(-x).$$

It can be expressed as the sum of an even function and an odd function, and any function defined on **R** can be represented as the sum of an even function and an odd function.

7. Prove the monotonicity of the following functions over the given intervals:

(1) $f(x)=x^3$ $(x\in(-\infty,+\infty))$;　　(2) $f(x)=\ln x$ $(x>0)$.

1.3 Form a New Function from Known Functions

There are three basic methods to form a new function from known functions: arithmetic operations, composite function and reverse function.

1.3.1 Arithmetic Operations of Functions

Definition 1.3.1 Let the domains of two functions f and g be D_f and D_g respectively, and let $D_f \cap D_g \neq \varnothing$. Then, the sum $f+g$, difference $f-g$, product $f \cdot g$, and quotient $\frac{f}{g}$ of functions f and g are defined as

$$(f+g)(x)=f(x)+g(x) \quad (x \in D_f \cap D_g);$$
$$(f-g)(x)=f(x)-g(x) \quad (x \in D_f \cap D_g);$$
$$(f \cdot g)(x)=f(x)g(x) \quad (x \in D_f \cap D_g);$$
$$\left(\frac{f}{g}\right)(x)=\frac{f(x)}{g(x)} \quad (x \in D_f \cap D_g-\{x \mid g(x)=0\}).$$

For example, the domain of $f(x)=\sqrt{x}$ is $A=[0,\infty)$ and the domain of $g(x)=\sqrt{2-x}$ is $B=(-\infty,2]$, so the domain of $(f+g)(x)=\sqrt{x}+\sqrt{2-x}$ is $A \cap B=[0,2]$. And if $f(x)=x^2$ and $g(x)=x-1$, then the domain of the rational function $\left(\frac{f}{g}\right)(x)=\frac{x^2}{x-1}$ is $\{x \mid x \neq 1\}$, or $(-\infty,1) \cup (1,\infty)$.

In engineering technology, a class of so-called hyperbolic functions is commonly encountered, which are constructed through arithmetic operations of exponential functions e^x and e^{-x}. The definitions are as follows:

Hyperbolic sine $\operatorname{sh}x=\frac{e^x-e^{-x}}{2} \quad (x \in \mathbf{R})$;

Hyperbolic cosine $\operatorname{ch}x=\frac{e^x+e^{-x}}{2} \quad (x \in \mathbf{R})$;

Hyperbolic tangent $\operatorname{th}x=\frac{\operatorname{sh}x}{\operatorname{ch}x}=\frac{e^x-e^{-x}}{e^x+e^{-x}} \quad (x \in \mathbf{R})$;

Hyperbolic cotangent $\operatorname{coth}x=\frac{\operatorname{ch}x}{\operatorname{sh}x}=\frac{e^x+e^{-x}}{e^x-e^{-x}} \quad (x \in \mathbf{R}-\{0\})$.

1.3.2 Composite Function

Composite functions are also a type of functional operation. This operation is performed by two or more functions using the method of "correspondence rule transmission".

Definition 1.3.2 Let the domain of function $y=f(u)$ be D_f, and the domain of function $u=g(x)$ be D_g, with its range being $R_g \subset D_f$. Then, the function determined by

$$y=f[g(x)] \quad (x\in D_g)$$

is called the composite function of f and g, denoted as $f\circ g$, that is,

$$(f\circ g)(x)=f[g(x)],$$

where: the variable u is referred to as the intermediate variable.

The conditions in Definition 1.3.2 $R_g\subset D_f$ are crucial, otherwise, they may be meaningless for certain $x\in D_g$, $f[g(x)]$. If the notation $f[g(x)]$ is used under conditions where $R_g\subset D_f$ is not met, it implies that the domain of the composite function $f[g(x)]$ will be a subset of D_g (when $x\in D$, $g(x)\in D_f$).

For example, suppose that $y=f(u)=\sqrt{u}$ and $u=g(x)=x^2+1$. Since y is a function of u and u is, in turn, a function of x, it follows that y is ultimately a function of x. We compute this by substitution:

$$y=f(u)=f[g(x)]=f(x^2+1)=\sqrt{x^2+1}.$$

The procedure is called composition, because the new function is composed of the two given functions f and g.

Another example, taking $f(x)=\sqrt{x}$, $g(x)=1-x^2$, although $g(x)$ is defined within $(-\infty,+\infty)$, it is only defined when $|x|\leqslant 1$, at which point $f[g(x)]=\sqrt{1-x^2}$ is defined, and the domain of $f\circ g$ is $D=[-1,1]$.

Definition 1.3.2 can be extended to the case of multiple function compositions.

For example, let $f(x)=\sqrt{x}$, $g(x)=1-x^2$, $h(x)=\sin x$, then

$$(f\cdot g\cdot h)(x)=\sqrt{1-\sin^2 x}=|\cos x|.$$

1.3.3 Inverse Function

Suppose the domain of function $y=f(x)$ is D_f, and its range is R_f. According to the mapping f, $\forall x\in D_f$, there is a unique $y\in R_f$. Conversely, for $\forall y\in R_f$, is there a unique $x\in D_f$ such that $y=f(x)$ holds? This is the issue to be discussed below.

Definition 1.3.3 Suppose the domain of function $y=f(x)$ is D_f, and its range is R_f. If $y=f(x)$ is a one-to-one map from D_f to R_f, then for $\forall y\in R_f$, there is a unique $x\in D_f$, such that $f(x)=y$, i.e., a function is defined on R_f, called the inverse function of $y=f(x)$, denoted as f^{-1}, that is

$$x=f^{-1}(y) \quad (y\in R_f).$$

Obviously, if $x=f^{-1}(y)$ is the inverse function of $y=f(x)$, then $y=f(x)$ is also the inverse function of $x=f^{-1}(y)$.

Note:

(1) The range of f is the domain of f^{-1}, and the domain of f is the range of f^{-1}.

(2) From $y=f(x)\Leftrightarrow x=f^{-1}(y)$, it is known that the graph of $y=f(x)$ is the same as the graph of $x=f^{-1}(y)$.

(3) $f^{-1}[f(x)]=f^{-1}(y)=x$ $(x\in D_f)$, $f[f^{-1}(y)]=f(x)=y$ $(y\in R_f)$.

The inverse function $x=f^{-1}(y)$, y of a function $y=f(x)$ is the independent variable. However, by convention, the independent variable is represented by x, so after swapping x and y in the inverse function $x=f^{-1}(y)$, it is rewritten as $y=f^{-1}(x)$.

For example, the inverse function $x=\log_a y$ of the exponential function $y=a^x$ $(0<a\neq1)$, after swapping x and y, is rewritten as $y=\log_a x$. When x and y are swapped in the inverse function $x=f^{-1}(y)$, the graph of the function $y=f(x)$ and its inverse function $y=f^{-1}(x)$ will not be the same.

Clearly, if the point $M(a,b)$ lies on the graph of the function $y=f(x)$, then the point $M'(b,a)$ must lie on the graph of the inverse function $y=f^{-1}(x)$. Conversely, the same holds true. Therefore, the graph of the function $y=f(x)$ and its inverse function $y=f^{-1}(x)$ are symmetric about the line $y=x$, as shown in Figure 1-6.

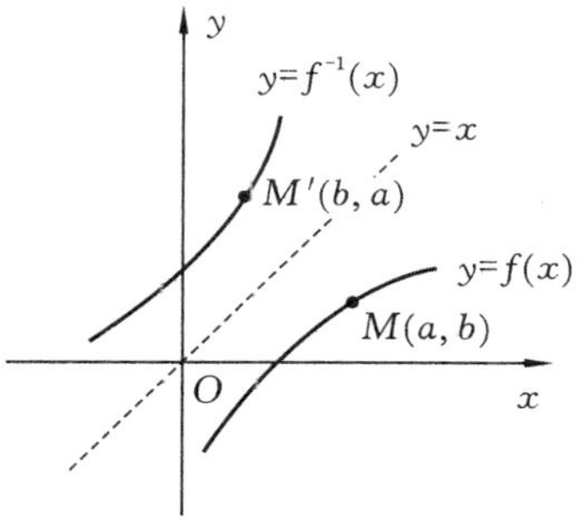

Figure 1-6

Example 1.3.1 Find the inverse function of the following:

(1) $y=\sqrt{x+1}$; (2) $y=\dfrac{1-x}{1+x}$.

Solution

(1) The domain of $y=\sqrt{x+1}$ is $[-1,+\infty)$, and its range is $[0,+\infty)$, thus

$$x=y^2-1 \quad (x\in[-1,\infty), y\in[0,+\infty)).$$

By swapping x and y, the inverse function is obtained

$$y=x^2-1 \quad (x\in[0,+\infty)).$$

(2) The domain of $y=\dfrac{1-x}{1+x}$ is $\mathbf{R}-\{-1\}$, and its range is $\mathbf{R}-\{-1\}$, solving for

$$x=\frac{1-y}{1+y} \quad (x\in\mathbf{R}-\{1\}, y\in\mathbf{R}-\{-1\}).$$

By swapping x and y, the inverse function is obtained $y=\dfrac{1-x}{1+x}$ $(x\in\mathbf{R}-\{-1\})$.

1.3.4 Primary Function

Definition 1.3.4 A function that is constructed by performing a finite number of arithmetic operations and finite compositions on elementary functions and can be expressed by a single formula is called an primary function.

For example, $f(x)=x^2\cos\dfrac{1}{x}+e^x$, $g(x)=\ln\left(x+\sqrt{1+x^2}\right)$ are all primary functions, but there are also some that are not. For instance, the floor function $y=\lfloor x\rfloor$, the sign function $y=\operatorname{sgn}x$ are not primary functions.

Exercise 1.3

1. Find the following composite functions $f[g(x)]$ and determine their domains:

(1) $f(x)=x^3, g(x)=x+2$;

(2) $f(x)=\sqrt{x}, g(x)=x^2$;

(3) $f(x)=\sqrt{x^2+1}, g(x)=\tan x$;

(4) $f(x)=a^x, g(x)=\log_a x$;

(5) $f(x)=\sin x, g(x)=\arcsin x$.

2. Which elementary functions are composed to form the following functions?

(1) $y=3^{2\sin x}$; (2) $y=3^{\sin 2x}$;

(3) $y=\tanh 2x$; (4) $y=(\arccos 2^x)^2$.

3. Let $f(x)=\dfrac{1}{1+x}, g(x)=1+x^2$, find $f[f(x)], g[g(x)], f[g(x)], g[f(x)]$, $g[f(1)], f[g(1)], f\{f[f(x)]\}$.

4. Let $f(x+1)=\begin{cases} x^2, & 0\leqslant x\leqslant 1 \\ 2x, & 1<x\leqslant 2 \end{cases}$. Find $f(x)$.

5. Prove: If the domain and range of function $f(x), g(x), h(x)$ are both D, $\forall x\in D$, with $f(x)\leqslant g(x)\leqslant h(x)$, and $f(x), g(x), h(x)$ are all monotonically increasing, then $\forall x\in D$, there exists $f[f(x)]\leqslant g[g(x)]\leqslant h[h(x)]$.

6. Prove: If $f(x)$ and $g(x)$ are both odd functions, then $f[g(x)]$ and $g[f(x)]$ are also odd functions.

7. Find the inverse functions of the following:

(1) $y=\sqrt[3]{x+1}$; (2) $y=\dfrac{1}{2}(e^x-e^{-x})$;

(3) $y=1+\ln(x+2)$; (4) $y=\begin{cases} x, & x\in[0,1] \text{ is rational} \\ 1-x, & x\in[0,1] \text{ is irrational} \end{cases}$.

8. Over what range do the following equations hold? Why?

(1) $y=\sin(\arcsin y)$; (2) $y=\tan(\arctan y)$;

(3) $x=a^{\log_a x}\quad (a>0, a\neq 1)$.

9. Prove: If $f(x)$ is an odd function over $\mathbf{R}$ and has an inverse, then its inverse is also an odd function.

10. Let x, y be any given real number, denote $\max\{x,y\}$ (or $\min\{x,y\}$) as the largest (or smallest) among x, y. Prove:

$$\max\{x,y\}=\frac{x+y+|x-y|}{2}, \min\{x,y\}=\frac{x+y-|x-y|}{2}.$$

11. A fertilizer plant produces a product 1000 t, priced at 130 yuan/t, with orders within 700 t being charged at the full price, and any excess over 700 t being charged at a 90% discount of the full price. Find the functional relationship between the order quantity x and the order amount y.

1.4 The Limit of a Function

1.4.1 The Definition of a Limit

Let's investigate the behavior of the function f defined by $f(x)=x^2-x+2$ for values of x near 2. Table 1-1 gives values of $f(x)$ for values of x close to 2, but not equal to 2.

Table 1-1

X	$f(x)$	X	$f(x)$
1.0	2.000000	3.0	8.000000
1.5	2.750000	2.5	5.750000
1.8	3.440000	2.2	4.640000
1.9	3.710000	2.1	4.310000
1.95	3.852500	2.05	4.152500
1.99	3.970100	2.01	4.030100
1.995	3.985025	2.005	4.015025
1.999	3.997001	2.001	4.003001

From the table we see that when x is close to 2 (on either side of 2), $f(x)$ is close to 4. In fact, it appears that we can make the values of $f(x)$ as close as we like to 4 by taking x sufficiently close to 2. We express this by saying "the limit of the function $f(x)=x^2-x+2$ as x approaches 2 is equal to 4". The notation for this is

$$\lim_{x\to 2}(x^2-x+2)=4.$$

In general, we use the following notation.

Definition 1.4.1 Suppose $f(x)$ is defined when x is near the number a (this means that f is defined on some open interval that contains a, except possibly at a itself). Then we write

$$\lim_{x\to a}f(x)=L,$$

and say "the limit of $f(x)$, as x approaches a, equals L".

If we can make the values of $f(x)$ arbitrarily close to L by taking x to be sufficiently close to a (on either side of a), but not equal to a.

Roughly speaking, this says that the values of $f(x)$ approach L as x approaches a. In other words, the values of $f(x)$ tend to get closer and closer to the number L as x gets closer and closer to the number a (from either side of a), but $x\neq a$ (a more precise definition will be given in Section 1.4.2).

An alternative notation for

$$\lim_{x \to a} f(x) = L$$

is

$$f(x) \to L \text{ as } x \to a,$$

which is usually read "$f(x)$ approaches L as x approaches a".

Note: The phrase "but $x \neq a$" in the definition of limit. This means that in finding the limit of $f(x)$ as x approaches a, we never consider $x = a$. In fact, $f(x)$ need not even be defined when $x = a$. The only thing that matters is how f is defined near a.

1.4.2 The Precise Definition of a Limit

When discussing the limit of function $f(x)$ at a finite point x_0, it involves a neighborhood of point x_0. Therefore, we first introduce the concept of a neighborhood of a point.

Let $\delta > 0$, x_0 be a point on the real axis, referred to as

$$U(x_0, \delta) = (x_0 - \delta, x_0 + \delta) = \{x \mid |x - x_0| < \delta\},$$

the δ-neighborhood of point x_0, where point x_0 is the center of the neighborhood, and δ is called the radius of the neighborhood.

$$U(x_0, \delta) = (x_0 - \delta, x_0) \cup (x_0, x_0 + \delta) = \{x \mid 0 < |x - x_0| < \delta\},$$

the δ-deleted neighborhood of point x_0.

$$U_-(x_0, \delta) = (x_0 - \delta, x_0) = \{x \mid 0 < x_0 - x < \delta\},$$

the δ-left neighborhood of point x_0.

$$U_+(x_0, \delta) = (x_0, x_0 + \delta) = \{x \mid 0 < x - x_0 < \delta\},$$

the δ-right neighborhood of point x_0.

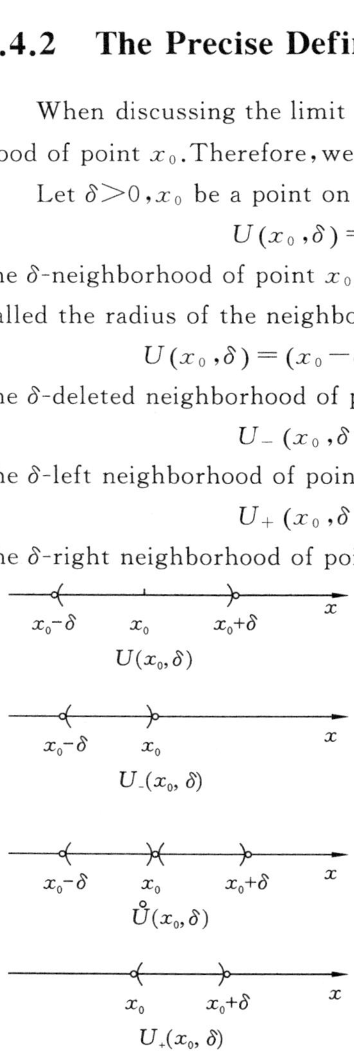

Figure 1-7

When δ is not emphasized, $U(x_0, \delta)$, $U(x_0, \delta)$, $U_-(x_0, \delta)$, $U_+(x_0, \delta)$ is abbreviated to $U(x_0)$, $U(x_0)$, $U_-(x_0)$ and $U_+(x_0)$, respectively, and referred to as the neighborhood, deleted neighborhood, left neighborhood, and right neighborhood of point x_0, as shown in Figure 1-7.

An open interval containing point x_0 is also commonly referred to as the neighborhood of point x_0.

Definition 1.4.2 Let function $f(x)$ be defined in some deleted neighborhood of point x_0. If $\forall \varepsilon > 0$, $\exists \delta > 0$, when $0 < |x - x_0| < \delta$, then

$$|f(x) - A| < \varepsilon.$$

The constant A is called the limit of $f(x)$ as x approaches x_0, denoted as

$$\lim_{x \to x_0} f(x) = A \text{ or } f(x) \to A \ (x \to x_0).$$

Note:

(1) The "$0 < |x - x_0| < \delta$" in the definition refers to the deleted neighborhood

$U(x_0,\delta)$, and the function $f(x)$ may or may not be defined at point x_0.

(2) When function $f(x)$ is defined at point x_0, the limit of function $f(x)$ at point x_0 is unrelated to the function value $f(x_0)$ at point x_0. The limit value $\lim\limits_{x\to x_0} f(x)$ may or may not equal the function value $f(x_0)$.

(3) The geometric meaning of $\lim\limits_{x\to x_0} f(x)=A$: $\forall\varepsilon>0$, $\exists\delta>0$, when $0<|x-x_0|<\delta$, there is $|f(x)-A|<\varepsilon$ or $A-\varepsilon<f(x)<A+\varepsilon$. These points fall within the horizontal strip region between the two lines $y=A-\varepsilon$ and $y=A+\varepsilon$, as shown in Figure 1-8.

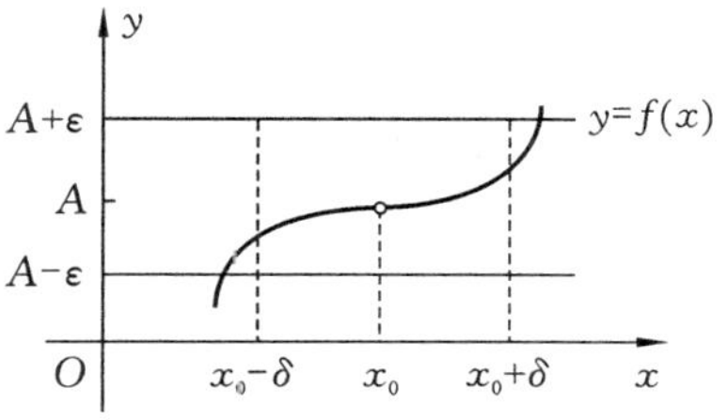

Figure 1-8

Example 1.4.1 Prove: $\lim\limits_{x\to 2}(2x-1)=3$.

Proof $\forall\varepsilon>0$, solve the inequality

$$|(2x-1)-3|=2|x-2|<\varepsilon,$$

obtain $|x-2|<\frac{\varepsilon}{2}$. Taking $\delta=\frac{\varepsilon}{2}$, then when $0<|x-2|<\delta$,

$$|(2x-1)-3|<\varepsilon.$$

That is,

$$\lim_{x\to 2}(2x-1)=3.$$

Example 1.4.2 Prove: $\lim\limits_{x\to x_0} x^2=x_0^2$.

Proof Restrict $|x-x_0|<1$, that is, $|x|<|x_0|+1$, $\forall\varepsilon>0$, solve the inequality

$$|x^2-x_0^2|=|x+x_0|\,|x-x_0|\leqslant(|x|+|x_0|)|x-x_0|<(2|x_0|+1)|x-x_0|<\varepsilon.$$

Obtain $|x-x_0|<\frac{\varepsilon}{2|x_0|+1}$. Taking $\delta=\min\left\{1,\frac{\varepsilon}{2|x_0|+1}\right\}$, then when $0<|x-x_0|<\delta$, there is $|x^2-x_0^2|<\varepsilon$, that is,

$$\lim_{x\to x_0} x^2=x_0^2.$$

1.4.3 One-Sided Limit

The upper limit definition considers the case where both sides approach the point x_0, but in practice, sometimes it is only necessary to consider the limit as one approaches the point x_0 from the left or right. Hence, there are corresponding left-sided and right-sided limits.

Definition 1.4.3

(1) Let $f(x)$ be defined in some left neighborhood of point x_0. If $\forall\varepsilon>0$, $\exists\delta>0$, as $0<x_0-x<\delta$, then

$$|f(x)-A|<\varepsilon.$$

The constant A is called the left-sided limit of $f(x)$ at point x_0, denoted as

$$\lim_{x\to x_0^-} f(x)=A \text{ or } f(x_0-0)=A.$$

(2) Let $f(x)$ be defined in some right neighborhood of point x_0. If $\forall\varepsilon>0$, $\exists\delta>0$, as $0<x-x_0<\delta$, then

$$|f(x)-A|<\varepsilon.$$

The constant A is called the right-sided limit of $f(x)$ at point x_0, denoted as

$$\lim_{x\to x_0^+} f(x)=A \text{ or } f(x_0+0)=A.$$

Theorem 1.4.1 $\lim\limits_{x\to x_0} f(x)=A \Leftrightarrow \lim\limits_{x\to x_0^-} f(x)=A$, $\lim\limits_{x\to x_0^+} f(x)=A$.

Proof Omitted.

Example 1.4.3 Let $f(x)=\begin{cases}2x-1, & x<0\\ a^x, & x>0\end{cases}$ $(a>0, a\neq 1)$.

Prove: $f(0-0)=-1$, $f(0+0)=1$.

Proof When $x<0$, $f(x)=2x-1$. $\forall\varepsilon>0$, solve the inequality

$$|(2x-1)-(-1)|=2|x|=2(-x)<\varepsilon,$$

obtain $0-x<\dfrac{\varepsilon}{2}$. Taking $\delta=\dfrac{\varepsilon}{2}$, then when $0<0-x<\delta$, there is

$$|(2x-1)-(-1)|<\varepsilon,$$

that is, $\lim\limits_{x\to 0^-} f(x)=-1$ or $f(0-0)=-1$.

When $x>0$, $f(x)=a^x$. If $a>1$, $\forall\varepsilon>0$, solve the inequality

$$|a^x-1|=a^x-1<\varepsilon,$$

obtain $a^x<1+\varepsilon$, that is, $x<\log_a(1+\varepsilon)$. Taking $\delta=\log_a(1+\varepsilon)$, then when $0<x-0<\delta$, there is $|a^x-1|<\varepsilon$, that is

$$\lim_{x\to 0^+} f(x)=1 \text{ or } f(0+0)=1.$$

When $0<a<1$, similarly, it can be proven that $f(0+0)=1$.

1.4.4 Infinite Limits

To find $\lim\limits_{x\to 0}\dfrac{1}{x^2}$ if it exists, we find as x becomes close to 0, x^2 also becomes close to 0, and $\dfrac{1}{x^2}$ becomes very large. In fact, it appears from the graph of the function $f(x)=\dfrac{1}{x^2}$ values of $f(x)$ can be made arbitrarily large by taking x close enough to 0. Thus the values of $f(x)$ do not approach a number, so $\lim\limits_{x\to 0}\dfrac{1}{x^2}$ does not exist.

To indicate the kind of behavior, we use the notation

$$\lim_{x\to 0}\frac{1}{x^2}=\infty,$$

this does not mean that we are regarding ∞ as a number, nor does it mean that the limit exists. It simply expresses the particular way in which the limit does not exist: $1/x^2$ can be made as large as we like by taking x close enough to 0.

In general, we write symbolically

$$\lim_{x\to a} f(x)=\infty$$

to indicate that the values of $f(x)$ tend to become larger and larger (or "increase without bound") as x becomes closer and closer to a.

Definition 1.4.4 Let function $f(x)$ be defined in $(a,+\infty)$, A be a constant, if for $\forall\varepsilon>0$, $\exists X>0(>a)$, $\forall x>X$, there exists

$$|f(x)-A|<\varepsilon,$$

then the constant A is called the limit of $f(x)$ as $x\to+\infty$. It is denoted as

$$\lim_{x\to+\infty} f(x)=A \text{ or } f(x)\to A \quad (x\to+\infty).$$

The geometric meaning of the limit $\lim_{x\to+\infty} f(x)=A$: $\forall\varepsilon>0$, $\exists X>0$, when $x>X$, the curve $y=f(x)$ lies between the horizontal lines $y=A-\varepsilon$ and $y=A+\varepsilon$. As shown in Figure 1-9.

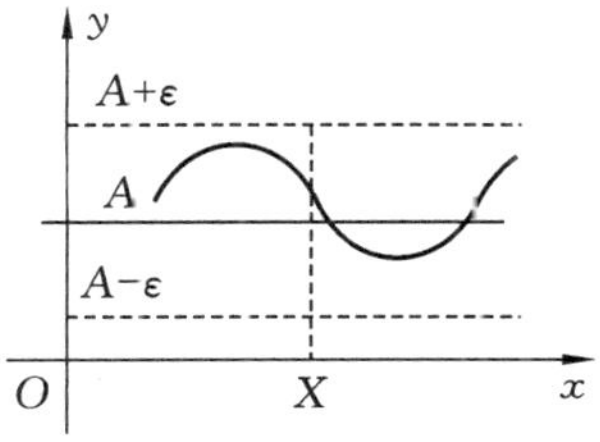

Figure 1-9

It is easy to see that the limit definition of the function $f(x)(x\to+\infty)$ is very similar to the limit definition of the sequence $\{a_n\}$. This is because their independent variables both change in an infinitely increasing manner. However, their independent variables differ in their modes of change. The independent variable x of the function $f(x)$ takes all real numbers in the interval $(a,+\infty)$ continuously and infinitely increasing, while the independent variable n of the sequence $\{a_n\}$ takes natural numbers discretely and infinitely increasing.

There are also two other cases similar to the mode of change of $x\to+\infty$: $x\to-\infty$ and $x\to\infty$. The limit definition of the corresponding function $f(x)$ is as follows.

Definition 1.4.5 Let the function $f(x)$ be defined on the interval $(-\infty,a)$, and A be a constant. If for $\forall\varepsilon>0$, $\exists X>0(-X<a)$, $\forall x<-X$, there exists

$$|f(x)-A|<\varepsilon,$$

then the constant A is called the limit of $f(x)$ as $x\to-\infty$. It is denoted as

$$\lim_{x\to-\infty} f(x)=A \text{ or } f(x)\to A \quad (x\to-\infty).$$

Definition 1.4.6 Let the function $f(x)$ be defined as $|x|>a$ $(a>0)$, and A be a constant. If for $\forall\varepsilon>0$, $\exists X>0(|X|>a)$, $\forall|x|>X$, there exists

$$|f(x)-A|<\varepsilon.$$

It is said that the constant A is the limit of $f(x)$ as $x\to\infty$ approaches, denoted as

$$\lim_{x\to\infty} f(x)=A \text{ or } f(x)\to A \quad (x\to\infty).$$

It can be seen that:

$$\lim_{x\to\infty} f(x)=A \Leftrightarrow \lim_{x\to-\infty} f(x)=A,\ \lim_{x\to+\infty} f(x)=A.$$

Example 1.4.4 Prove: $\lim_{x\to-\infty} a^x=0(a>1)$.

Proof $\forall\varepsilon>0(0<\varepsilon<1)$, solve the inequality

$$|a^x-0|=a^x<\varepsilon.$$

Taking the logarithm to the base a on both sides of the inequality, we have $x<\log_a\varepsilon$. Taking $X=|\log_a\varepsilon|$, then when $x<-X=\log_a\varepsilon$, there is $|a^x-0|<\varepsilon$. Thus

$$\lim_{x\to-\infty} a^x=0.$$

Example 1.4.5 Prove: $\lim\limits_{x\to\infty}\dfrac{2x^3-3}{x^2+1}=2$.

Proof $\forall\varepsilon>0$, solve the inequality

$$\left|\frac{2x^2-3}{x^2+1}-2\right|=\left|\frac{-5}{x^2+1}\right|<\frac{5}{|x|^2}<\varepsilon,$$

we get $|x|>\sqrt{\dfrac{5}{\varepsilon}}$. Taking $X=\sqrt{\dfrac{5}{\varepsilon}}$, then when $|x|>X$, there is $\left|\dfrac{2x^2-3}{x^2+1}-2\right|<\varepsilon$. Thus

$$\lim_{x\to\infty}\frac{2x^3-3}{x^2+1}=2.$$

Exercise 1.4

1. Prove the following equations using the definition of limits:

(1) $\lim\limits_{x\to1}\dfrac{x^2-1}{x-1}=2$; (2) $\lim\limits_{x\to+\infty}(\sqrt{x+1}-\sqrt{x})=0$;

(3) $\lim\limits_{x\to\infty}\dfrac{1}{x}\sin\dfrac{1}{x}=0$; (4) $\lim\limits_{x\to0}x\sin\dfrac{1}{x}=0$;

(5) $\lim\limits_{x\to5^+}\sqrt{x-5}=0$.

2. Prove: $\lim\limits_{x\to+\infty}f(x)=\lim\limits_{x\to-\infty}f(x)=A\Leftrightarrow\lim\limits_{x\to\infty}f(x)=A$.

3. Find the left and right limits of $f(x)=\dfrac{x}{|x|}$, $g(x)=\dfrac{|x|}{x}$ as $x\to0$, and explain whether the limit exists as $x\to0$.

4. Write the following limit statements using "ε-δ" language or "ε-X" language:

(1) $\lim\limits_{x\to-\infty}f(x)=A$, $\lim\limits_{x\to-\infty}f(x)\neq A$;

(2) $\lim\limits_{x\to\infty}f(x)=A$, $\lim\limits_{x\to\infty}f(x)\neq A$;

(3) $\lim\limits_{x\to x_0^+}f(x)=A$, $\lim\limits_{x\to x_0^+}f(x)\neq A$;

(4) $\lim\limits_{x\to x_0^-}f(x)=A$, $\lim\limits_{x\to x_0^-}f(x)\neq A$.

1.5 Properties of Function Limits

1.5.1 Properties of Function Limits

There are six forms in the limits of functions:

$$\lim_{x\to\infty}f(x),\ \lim_{x\to-\infty}f(x),\ \lim_{x\to+\infty}f(x),$$

$$\lim_{x\to x_0} f(x), \lim_{x\to x_0^-} f(x), \lim_{x\to x_0^+} f(x).$$

Each form of function limit has similar properties and operational rules, here, only the $\lim\limits_{x\to x_0} f(x)$ form is explained.

Theorem 1.5.1 (Uniqueness) If $\lim f(x)$ exists, then its limit is unique.

Proof Suppose $\lim\limits_{x\to x_0} f(x)=A$, $\lim\limits_{x\to x_0} f(x)=B$, without loss of generality, let $B>A$. For $\varepsilon_0=\frac{B-A}{2}>0$, from $\lim\limits_{x\to x_0} f(x)=A$, $\exists\ \delta_1>0$, when $0<|x-x_0|<\delta_1$, there is $|f(x)-A|<\varepsilon_0$, that is

$$\frac{3A-B}{2}<f(x)<\frac{A+B}{2}.$$

From $\lim\limits_{x\to x_0} f(x)=B$, $\exists\ \delta_2>0$, when $0<|x-x_0|<\delta_2$, there is $|f(x)-B|<\varepsilon_0$, that is

$$\frac{A+B}{2}<f(x)<\frac{3B-A}{2}.$$

Taking $\delta=\min\{\delta_1,\delta_2\}$, then when $0<|x-x_0|<\delta$, there is

$$\frac{A+B}{2}<f(x)<\frac{A+B}{2}.$$

This is impossible. Therefore, $A=B$.

Theorem 1.5.2 (Preservation of Sign) Suppose $\lim\limits_{x\to x_0} f(x)=A>0\ (<0)$, then there exists a deleted neighborhood $\mathring{U}(x_0)$ of point x_0, such that when $x\in U(x_0)$, there is $f(x)>0\ (<0)$.

Proof For $\varepsilon_0=\frac{A}{2}>0$, from $\lim\limits_{x\to x_0} f(x)=A$, $\exists\ \delta>0$, when $0<|x-x_0|<\delta$, there is

$$|f(x)-A|<\varepsilon \text{ or } \frac{A}{2}<f(x)<\frac{3A}{2},$$

thus, when $x\in U(x_0,\delta)$, $f(x)>\frac{A}{2}>0$.

The same can be proved when $A<0$.

Corollary 1.5.1 (Preservation of Order) Suppose $\lim\limits_{x\to x_0} f(x)=A$, $\lim\limits_{x\to x_0} g(x)=B$, and $A<B$. Then there exists a deleted neighborhood $U(x_0)$ of point x_0, such that when $x\in U(x_0)$, there is

$$f(x)<g(x).$$

Corollary 1.5.2 Let $\lim\limits_{x\to x_0} f(x)=A$, $\lim\limits_{x\to x_0} g(x)=B$, if there exists a deleted neighborhood $\mathring{U}(x_0)$ of point x_0, such that when $x\in U(x_0)$, there is $f(x)\leqslant g(x)$, then $A\leqslant B$.

Proof By contradiction.

If $A > B$, by Corollary 1.5.1, there exists a deleted neighborhood $U_1(x_0)$ of point x_0, such that when $x \in U_1(x_0)$, there is $f(x) > g(x)$. Consequently, when $x \in U_1(x_0) \cap U(x_0)$, there is $f(x) > g(x)$, which contradicts $f(x) \leqslant g(x)$.

Theorem 1.5.3 (Local Boundedness) If $\lim_{x \to x_0} f(x) = A$, then there exists a deleted neighborhood $U(x_0)$ of point x_0, within which $f(x)$ is bounded.

Proof Given $\lim_{x \to x_0} f(x) = A$, $\exists\, \varepsilon_0 > 0$, $\exists\, \delta > 0$, when $0 < |x - x_0| < \delta$, there is $|f(x) - A| < \varepsilon_0$, that is

$$|f(x)| < |A| + \varepsilon_0,$$

thus, $\exists\, M = |A| + \varepsilon_0$, when $x \in U(x_0, \delta)$, $|f(x)| < M$, that is, $f(x)$ is bounded within $U(x_0, \delta)$.

1.5.2 Relationship Between Function Limit and Sequence Limit

Function limits and sequence limits are defined separately, but essentially they can be converted into each other.

Theorem 1.5.4 If $\lim_{x \to x_0} f(x) = A$, then for any sequence $\{x_n\}$, $x_n \in D_f (x_n \neq x_0)$, when $x_n \to x_0 (n \to \infty)$, there exists

$$\lim_{n \to \infty} f(x_n) = A.$$

Theorem 1.5.5 Serves as a bridge between function limits and sequence limits. The limit of function $f(x)$ at point x_0 can be transformed into the limit of a sequence of function values, and it is also convenient to use the limit of a sequence of function values to determine the non-existence of a function limit at a certain point. The proof is omitted.

Example 1.5.1 Proves that $\lim_{x \to 0} \sin \frac{1}{x}$ does not exist.

Proof Take sequences $\left(x_n = \frac{1}{2n\pi + \frac{\pi}{2}} \right)$ and $\left(x'_n = \frac{1}{2n\pi - \frac{\pi}{2}} \right)$, it is obvious that $x_n \to 0$, $x'_n \to 0 (n \to \infty)$.

Consequently

$$\lim_{n \to \infty} \sin \frac{1}{x_n} = \lim_{n \to \infty} \sin \left(2n\pi + \frac{\pi}{2} \right) = 1,$$

$$\lim_{n \to \infty} \sin \frac{1}{x'_n} = \lim_{n \to \infty} \sin \left(2n\pi - \frac{\pi}{2} \right) = -1.$$

Therefore, by Theorem 1.5.1, it is known that $\lim_{x \to 0} \sin \frac{1}{x}$ does not exist.

1.6 Calculating Limits Using the Limit Laws

1.6.1 Limit Laws

Theorem 1.6.1 If $\lim f(x)=A$, $\lim g(x)=B$, then

(1) $\lim[f(x)\pm g(x)]=\lim f(x)\pm \lim g(x)=A\pm B$;

(2) $\lim[f(x)g(x)]=\lim f(x)\cdot \lim g(x)=AB$;

(3) $\lim \dfrac{f(x)}{g(x)}=\dfrac{\lim f(x)}{\lim g(x)}=\dfrac{A}{B}$, where $B\neq 0$.

These laws can be stated verbally as follcws: The limit of a sum is the sum of the limits. The limit of a difference is the difference of the limits. The limit of a product is the product of the limits. The limit of a quotient is the quotient of the limits (provided that the limit of the denominator is not 0).

Proof Here we only prove (2) and (3).

Let $\lim\limits_{x\to x_0} f(x)=A$, $\lim\limits_{x\to x_0} g(x)=B$, according to the definition of function limit, $\forall \varepsilon>0$, $\exists\, \delta_1>0$, when $0<|x-x_0|<\delta_1$, there is $|f(x)-A|<\varepsilon$; $\exists\, \delta_2>0$, and when $0<|x-x_0|<\delta_2$, there is $|g(x)-B|<\varepsilon$.

(2) Given $\lim\limits_{x\to x_0} f(x)=A$, $\exists\, M>0$, $\exists\, \delta_0>0$, when $0<|x-x_0|<\delta$, there is $|f(x)|\leqslant M$. Taking $\delta=\min\{\delta_0,\delta_1,\delta_2\}$, then when $0<|x-x_0|<\delta$, there is

$$|f(x)-A|<\varepsilon,\ |g(x)-B|<\varepsilon,\ |f(x)|\leqslant M.$$

Thus, when $0<|x-x_0|<\delta$, there is

$$\begin{aligned}|f(x)g(x)-AB|&=|f(x)g(x)-Bf(x)+Bf(x)-AB|\\&\leqslant|f(x)|\,|g(x)-B|+|B|\,|f(x)-A|<M\varepsilon+|c|\varepsilon=(M+|c|)\varepsilon.\end{aligned}$$

That is

$$\lim_{x\to x_0}[f(x)g(x)]=AB=\lim_{x\to x_0} f(x)\lim_{x\to x_0} g(x).$$

(3) Given $\lim\limits_{x\to x_0} g(x)=B\neq 0$, $\exists\, \varepsilon_0=\dfrac{|B|}{2}>0$, $\exists\, \delta_0>0$, when $0<|x-x_0|<\delta$, there is $|g(x)-B|<\dfrac{|B|}{2}$, that is $|g(x)|>|B|-\dfrac{|E|}{2}=\dfrac{|B|}{2}$, hence

$$\frac{1}{|g(x)|}<\frac{2}{|B|}.$$

Taking $\delta=\min\{\delta_0,\delta_1,\delta_2\}$, then when $0<|x-x_0|<\delta$, there is

$$|f(x)-A|<\varepsilon,\ |g(x)-B|<\varepsilon,\ \frac{1}{|g(x)|}<\frac{2}{|B|}.$$

Thus, when $0<|x-x_0|<\delta$, there is

$$\left|\frac{f(x)}{g(x)}-\frac{A}{B}\right|=\frac{|Bf(x)-Ag(x)|}{|Bg(x)|}=\frac{|Bf(x)-AB+AB-Ag(x)|}{|Bg(x)|}$$

$$\leqslant\frac{1}{|B||g(x)|}(|B||f(x)-A|+|A||g(x)-B|)<\frac{2}{|B|^2}(|B|+|A|)\varepsilon,$$

that is

$$\lim_{x\to x_0}\frac{f(x)}{g(x)}=\frac{A}{B}=\frac{\lim\limits_{x\to x_0}f(x)}{\lim\limits_{x\to x_0}g(x)}.$$

Note: The (1)、(2) in the theorem can be generalized to the case of a finite number of functions.

Corollary 1.6.1 If $\lim f(x)$ exists, then $\lim[Cf(x)]=C\lim f(x)$, C is a constant.

Corollary 1.6.2 If $\lim f(x)$ exists, then $\lim[f(x)]^n=[\lim f(x)]^n$, n is a positive integer.

Regarding sequences, similar rules for the arithmetic of limits can also be applied.

Theorem 1.6.2 Let sequences $\{x_n\}$ and $\{y_n\}$ converge, and $\lim\limits_{n\to\infty}x_n=A$, $\lim\limits_{n\to\infty}y_n=B$, then

(1) $\lim\limits_{n\to\infty}(x_n\pm y_n)=A\pm B$;

(2) $\lim\limits_{n\to\infty}x_ny_n=AB$;

(3) $\lim\limits_{n\to\infty}\frac{x_n}{y_n}=\frac{A}{B}$, where $y_n\neq0\,(n=1,2,\cdots)$ and $B\neq0$.

Example 1.6.1 Calculate:

(1) $\lim\limits_{x\to x_0}(a_nx^n+a_{n-1}x^{n-1}+\cdots+a_1x+a_0)$;

(2) $\lim\limits_{x\to\infty}\left(a_n\frac{1}{x^n}+a_{n-1}\frac{1}{x^{n-1}}+\cdots+a_1\frac{1}{x}\right)$.

Solution

(1)
$$\begin{aligned}&\lim_{x\to x_0}(a_nx^n+a_{n-1}x^{n-1}+\cdots+a_1x+a_0)\\&=\lim_{x\to x_0}(a_nx^n)+\lim_{x\to x_0}(a_{n-1}x^{n-1})+\cdots+\lim_{x\to x_0}(a_1x)+\lim_{x\to x_0}a_0\\&=a_n\lim_{x\to x_0}x^n+a_{n-1}\lim_{x\to x_0}x^{n-1}+\cdots+a_1\lim_{x\to x_0}x+a_0\\&=a_nx_0^n+a_{n-1}x_0^{n-1}+\cdots+a_1x_0+a_0.\end{aligned}$$

(2)
$$\begin{aligned}&\lim_{x\to\infty}\left(a_n\frac{1}{x^n}+a_{n-1}\frac{1}{x^{n-1}}+\cdots+a_1\frac{1}{x}\right)\\&=\lim_{x\to\infty}\left(a_n\frac{1}{x^n}\right)+\lim_{x\to\infty}\left(a_{n-1}\frac{1}{x^{n-1}}\right)+\cdots+\lim_{x\to\infty}\left(a_1\frac{1}{x}\right)\\&=a_n\lim_{x\to\infty}\frac{1}{x^n}+a_{n-1}\lim_{x\to\infty}\frac{1}{x^{n-1}}+\cdots+a_1\lim_{x\to\infty}\frac{1}{x}\\&=a_n\cdot0+a_{n-1}\cdot0+\cdots+a_1\cdot0=0.\end{aligned}$$

Example 1.6.2 Find the following limits:

(1) $\lim\limits_{x\to 2}\dfrac{x-1}{2x-1}$; (2) $\lim\limits_{x\to 1}\dfrac{x^2-1}{x-1}$; (3) $\lim\limits_{x\to 3}\dfrac{x-3}{\sqrt{x}-\sqrt{3}}$.

Solution

(1) Since $\lim\limits_{x\to 2}(2x-1)=2\times 2-1=3\neq 0$, therefore

$$\lim_{x\to 2}\frac{x-1}{2x-1}=\frac{\lim\limits_{x\to 2}(x-1)}{\lim\limits_{x\to 2}(2x-1)}=\frac{1}{3}.$$

(2) Since $\lim\limits_{x\to 1}(x-1)=0$, $\lim\limits_{x\to 1}(x^2-1)=0$, the limit laws cannot be applied directly. Typically, factorize and cancel out the common factors, thereby

$$\lim_{x\to 1}\frac{x^2-1}{x-1}=\lim_{x\to 1}\frac{(x+1)(x-1)}{x-1}=\lim_{x\to 1}(x+1)=1+1=2.$$

(3)

$$\lim_{x\to 3}\frac{x-3}{\sqrt{x}-\sqrt{3}}=\lim_{x\to 3}\frac{(x-3)(\sqrt{x}+\sqrt{3})}{(\sqrt{x}-\sqrt{3})(\sqrt{x}+\sqrt{3})}$$
$$=\lim_{x\to 3}\frac{(x-3)(\sqrt{x}+\sqrt{3})}{x-3}=\lim_{x\to 3}(\sqrt{x}+\sqrt{3})=2\sqrt{3}.$$

Theorem 1.6.3 (Limit of Composite Functions) Let the function $f[g(x)]$ be composed of $f(u)$, $u=g(x)$. If $\lim\limits_{x\to x_0}g(x)=u_0$ within a deleted neighborhood of point x_0, where $g(x)\neq u_0$ and $\lim\limits_{u\to u_0}f(u)=A$, then

$$\lim_{x\to x_0}f[g(x)]=A.$$

Proof Omitted.

Example 1.6.3 Find the following limits:

(1) $\lim(\sin x-\sin^2 x+1)\left(x\to\dfrac{\pi}{2}\right)$; (2) $\lim\limits_{x\to 0}\dfrac{\sqrt[n]{1+x}-1}{x}$ (n is a natural number).

Solution

(1) Make the substitution $u=\sin x$, when $x\to\dfrac{\pi}{2}$, $u\to 1$, therefore

$$\lim_{x\to\frac{\pi}{2}}(\sin x-\sin^2 x+1)=\lim_{u\to 1}(u-u^2+1)=1.$$

(2) When $n=1$, $\lim\limits_{x\to 0}\dfrac{\sqrt[n]{1+x}-1}{x}=\lim\limits_{x\to 0}\dfrac{1+x-1}{x}=1$.

When $n\geqslant 2$, let $u=\sqrt[n]{1+x}$, then $x=u^n-1$, when $x\to 0$, $u\to 1$, therefore

$$\lim_{x\to 0}\frac{\sqrt[n]{1+x}-1}{x}=\lim_{u\to 1}\frac{u-1}{u^n-1}=\lim_{u\to 1}\frac{u-1}{(u-1)(u^{n-1}+u^{n-2}+\cdots+u+1)}$$
$$=\lim_{u\to 1}\frac{1}{u^{n-1}+u^{n-2}+\cdots+u+1}=\frac{1}{n}.$$

1.6.2 Two Important Limits

Theorem 1.6.4 (Squeeze Theorem) Suppose within a deleted neighborhood of point x_0.

(1) $f(x)\leqslant h(x)\leqslant g(x)$;

(2) $\lim\limits_{x\to x_0} f(x)=\lim\limits_{x\to x_0} g(x)=A$.

Then $\lim\limits_{x\to x_0} h(x)$ exists, and $\lim\limits_{x\to x_0} h(x)=A$.

The proof is omitted.

When the limit process $x\to x_0$ in the theorem is replaced with any other form of limit process, the conclusion of the theorem still holds.

Ⅰ (Important Limit) $\lim\limits_{x\to 0}\frac{\sin x}{x}=1$.

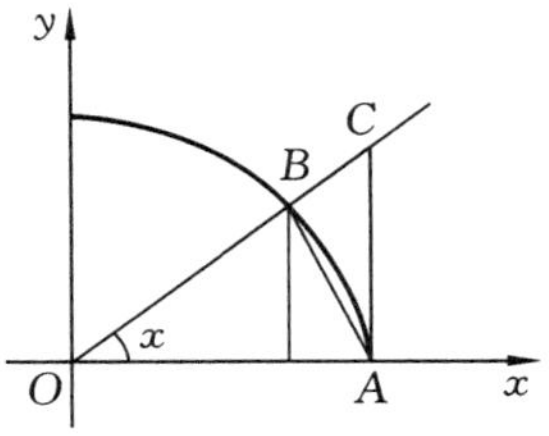

Figure 1-10

Proof First, prove that $\lim\limits_{x\to 0^+}\frac{\sin x}{x}=1$.

As shown in Figure 1-10, $\overset{\frown}{AB}$ is the arc of a circle with center at point O and radius 1. A tangent to the arc $\overset{\frown}{AB}$ drawn from point A intersects the extension of OB at point C. Let $\angle AOB=x$ (in radians), $0<x<\frac{\pi}{2}$, then the area $\triangle AOB$ < area sector AOB < area $\triangle AOC$, that is, $\frac{1}{2}\sin x<\frac{1}{2}x<\frac{1}{2}\tan x$ or $\cos x<\frac{\sin x}{x}<1$.

From $\sin x<x$, we get $\sin\frac{x}{2}<\frac{x}{2}$. Therefore,

$$0<1-\cos x=2\sin^2\frac{x}{2}\leqslant 2\left(\frac{x}{2}\right)^2=\frac{x^2}{2}.$$

Since $\lim\limits_{x\to 0^+}\frac{x^2}{2}=0$, it follows that $\lim\limits_{x\to 0^+}(1-\cos x)=0$, hence,

$$\lim_{x\to 0^+}\cos x=1;\quad \lim_{x\to 0^+}\frac{\sin x}{x}=1.$$

Next, when $x<0$, let $u=-x$, then when $x\to 0^-$, $u\to 0^+$, thus,

$$\lim_{x\to 0^-}\frac{\sin x}{x}=\lim_{u\to 0^+}\frac{\sin(-u)}{-u}=\lim_{u\to 0^+}\frac{\sin u}{u}=1.$$

In conclusion, we obtain

$$\lim_{x\to 0}\frac{\sin x}{x}=1.$$

Example 1.6.4 Find the following limits:

(1) $\lim\limits_{x\to 0}\frac{\sin\alpha x}{\sin\beta x}$ $(\alpha\beta\neq 0)$; (2) $\lim\limits_{x\to 0}\frac{\arcsin x}{x}$; (3) $\lim\limits_{x\to 0}\frac{1-\cos x}{x^2}$.

Solution

(1) $\lim\limits_{x\to 0}\frac{\sin\alpha x}{\sin\beta x}=\lim\limits_{x\to 0}\left(\frac{\sin\alpha x}{\alpha x}\frac{\beta x}{\sin\beta x}\frac{\alpha}{\beta}\right)=\frac{\alpha}{\beta}$.

(2) Let $\arcsin x=t$, then $x=\sin t$, and when $x\to 0$, $t\to 0$, then:

$$\lim_{x\to 0}\frac{\arcsin x}{x}=\lim_{t\to 0}\frac{t}{\sin t}=1.$$

(3) $\displaystyle\lim_{x\to 0}\frac{1-\cos x}{x^2}=\lim_{x\to 0}\frac{2\sin^2\frac{x}{2}}{x^2}=\lim_{x\to 0}\frac{1}{2}\left(\frac{\sin\frac{x}{2}}{\frac{x}{2}}\right)^2=\frac{1}{2}.$

Ⅱ (Important Limit) $\displaystyle\lim_{x\to\infty}\left(1+\frac{1}{x}\right)^x=\mathrm{e}.$

Proof First, we prove $\displaystyle\lim_{x\to+\infty}\left(1+\frac{1}{x}\right)^x=\mathrm{e}.$

$\forall x\geqslant 1$, taking $n=[x]$, from $[x]\leqslant x<[x]+1$ we know,

$$n\leqslant x<n+1 \text{ or } \frac{1}{n+1}<\frac{1}{x}\leqslant\frac{1}{n},$$

hence

$$1+\frac{1}{n+1}<1+\frac{1}{x}\leqslant 1+\frac{1}{n}.$$

Each inequality above has both ends greater than 1,

$$\left(1+\frac{1}{n+1}\right)^n<\left(1+\frac{1}{x}\right)^x<\left(1+\frac{1}{n}\right)^{n+1},$$

and

$$\lim_{n\to\infty}\left(1+\frac{1}{n+1}\right)^n=\lim_{n\to\infty}\frac{\left(1+\frac{1}{n+1}\right)^{n+1}}{1+\frac{1}{n+1}}=\mathrm{e},$$

$$\lim_{n\to\infty}\left(1+\frac{1}{n}\right)^{n+1}=\lim_{n\to\infty}\left(1+\frac{1}{n}\right)^n\left(1+\frac{1}{n}\right)=\mathrm{e}.$$

When $x\to+\infty$, there is $n\to\infty$,

$$\lim_{x\to+\infty}\left(1+\frac{1}{x}\right)^x=\mathrm{e}.$$

Secondly, when $x\to-\infty$, let $u=-x$, then when $x\to-\infty$, $u\to+\infty$, thus

$$\lim_{x\to-\infty}\left(1+\frac{1}{x}\right)^x=\lim_{u\to+\infty}\left(1-\frac{1}{u}\right)^{-u}=\lim_{u\to+\infty}\left(\frac{u}{u-1}\right)^u=\lim_{u\to+\infty}\left(1+\frac{1}{u-1}\right)^u$$

$$=\lim_{u\to+\infty}\left(1+\frac{1}{u-1}\right)^{u-1}\left(1+\frac{1}{u-1}\right)=\mathrm{e}.$$

Therefore

$$\lim_{x\to\infty}\left(1+\frac{1}{x}\right)^x=\mathrm{e}.$$

An equivalent form of the important limit Ⅱ:

$$\lim_{x\to 0}(1+x)^{\frac{1}{x}}=\mathrm{e}.$$

Example 1.6.5 Find the following limits:

(1) $\lim\limits_{x\to\infty}\left(\dfrac{x-1}{x+1}\right)^x$; (2) $\lim\limits_{x\to 0}(1-x)^{\frac{2}{x}}$; (3) $\lim\limits_{x\to\frac{\pi}{2}}(\csc^2 x)^{\tan^2 x}$.

Solution

(1) $\lim\limits_{x\to\infty}\left(\dfrac{x-1}{x+1}\right)^x=\lim\limits_{x\to\infty}\left(\dfrac{1-\frac{1}{x}}{1+\frac{1}{x}}\right)^x=\lim\limits_{x\to\infty}\dfrac{\left(1-\frac{1}{x}\right)^x}{\left(1+\frac{1}{x}\right)^x}=\lim\limits_{x\to\infty}\dfrac{1}{\left(1+\frac{1}{x}\right)^x\left(1-\frac{1}{x}\right)^{-x}}=\dfrac{1}{e^2}.$

(2) $\lim\limits_{x\to 0}(1-x)^{\frac{2}{x}}=\lim\limits_{x\to 0}(1-x)^{-\frac{1}{x}(-2)}=e^{-2}.$

(3) $\lim\limits_{x\to\frac{\pi}{2}}(\csc^2 x)^{\tan^2 x}=\lim\limits_{x\to\frac{\pi}{2}}(1+\cot^2 x)^{\frac{1}{\cot^2 x}}=e.$

Exercise 1.6

1. Find the following limits:

(1) $\lim\limits_{x\to -3}(6x^2-2x+1)$;

(2) $\lim\limits_{x\to 4}x(2x-1)(3x+2)$;

(3) $\lim\limits_{x\to 5}\dfrac{x-5}{x^2-25}$;

(4) $\lim\limits_{x\to -5}\dfrac{x^2+3x-10}{x+5}$;

(5) $\lim\limits_{x\to 1}\dfrac{x-1}{\sqrt{x+3}-2}$;

(6) $\lim\limits_{x\to 0}\dfrac{x^2}{\sin^2\frac{x}{3}}$;

(7) $\lim\limits_{x\to 0}\dfrac{\sin(\alpha+x)-\sin(\alpha-x)}{x}$;

(8) $\lim\limits_{x\to 1}\dfrac{x+x^2+\cdots+x^n-n}{x-1}$;

(9) $\lim\limits_{x\to\pi^+}\dfrac{\sqrt{1+\cos x}}{\sin x}$;

(10) $\lim\limits_{x\to+\infty}(\sin\sqrt{x+1}-\sin\sqrt{x})$;

(11) $\lim\limits_{x\to\infty}\left(\dfrac{x-2}{x}\right)^{x+2}$;

(12) $\lim\limits_{x\to\frac{\pi}{2}}(1+\cos x)^{3\sec x}$.

2. Prove: If $\lim\limits_{x\to a}f(x)=A$, then

(1) $\lim\limits_{x\to a}|f(x)|=|A|$; (2) $\lim\limits_{x\to a}f^2(x)=A^2$; (3) $\lim\limits_{x\to a}\sqrt{f(x)}=\sqrt{A}$ $(A>0)$.

3. Determine the parameters in the following limit expressions a, b:

(1) $\lim\limits_{x\to\infty}\left(\dfrac{x^2+1}{x+1}-ax-b\right)=\dfrac{1}{2}$;

(2) $\lim\limits_{x\to 1}\dfrac{\sin^2(x-1)}{x^2+ax+b}=1$.

4. Prove that if $\lim\limits_{x\to+\infty}f(x)=A$, $\lim\limits_{x\to+\infty}g(x)=B$, $\lim\limits_{x\to+\infty}h(x)=C$, then,

$$\lim_{x\to+\infty}[f(x)-g(x)+h(x)]=A-B+C.$$

5. Prove that if $\lim\limits_{x\to a^+}f(x)=A\neq 0$, then $\lim\limits_{x\to a^+}\dfrac{1}{f(x)}=\dfrac{1}{A}$.

1.7 Infinitesimals and Infinite Limits

1.7.1 Infinitesimals

Definition 1.7.1 If $\lim\alpha=0$, then α is called an infinitesimal in this limit process, or simply an infinitesimal.

Note:

(1) An infinitesimal is a function whose limit is zero in a certain limit process, it is related to the limit process. For example, when $x\to 0$, x^2, $\sin x$, $\tan x$ are infinitesimals, but when $x\to 1$, x^2, $\sin x$, $\tan x$ are not infinitesimals.

(2) An infinitesimal is not a "very small number" (except for the constant zero). An infinitesimal is a quantity whose absolute value can be made arbitrarily smaller than any given positive number ε in a limit process, whereas a "very small number" (except for the constant zero) cannot be made smaller than any given positive number ε.

There is the following relationship between limits and infinitesimals.

Theorem 1.7.1 $\lim f(x)=A \Leftrightarrow f(x)=A+\alpha$, where $\lim\alpha=0$.

Proof Let $\alpha=f(x)-A$. Suppose $\lim\limits_{x\to x_0} f(x)=A$, i.e., $\forall\varepsilon>0$, $\exists\,\delta>0$, when $0<|x-x_0|<\delta$, there is

$$|f(x)-A|<\varepsilon.$$

This is $|\alpha|<\varepsilon$, thus $\lim\limits_{x\to x_0}\alpha=0$.

Next, prove the sufficiency.

Suppose $\lim\limits_{x\to x_0}\alpha=0$, i.e., $\forall\varepsilon>0$, $\exists\,\delta>0$, when $0<|x-x_0|<\delta$, there is

$$|\alpha|<\varepsilon.$$

This is $|f(x)-A|<\varepsilon$, thus $\lim\limits_{x\to x_0} f(x)=A$.

Theorem 1.7.2 In the same limit process, ① the sum of a finite number of infinitesimals is infinitesimal; ② the product of a finite number of infinitesimals is infinitesimal; ③ the product of a bounded function and an infinitesimal is infinitesimal. In particular, the product of a constant and an infinitesimal is infinitesimal.

Proof We only prove the case of the sum of two infinitesimals in ①, leaving the rest for the reader to prove.

Suppose $\lim\limits_{x\to x_0}\alpha=0$, $\lim\limits_{x\to x_0}\beta=0$, then $\forall\varepsilon>0$, $\exists\,\delta_1>0$, when $0<|x-x_0|<\delta_1$, there is $|\alpha|<\frac{\varepsilon}{2}$; $\exists\,\delta_2>0$, when $0<|x-x_0|<\delta_2$, there is $|\beta|<\frac{\varepsilon}{2}$. Taking $\delta=\min\{\delta_1,\delta_2\}$, then when $0<|x-x_0|<\delta$, there is

$$|(\alpha+\beta)-0|=|\alpha+\beta|\leqslant|\alpha|-|\beta|<\frac{\varepsilon}{2}+\frac{\varepsilon}{2}=\varepsilon.$$

Thus

$$\lim_{x \to x_0} (\alpha+\beta)=0.$$

Note: The sum of an infinite number of infinitesimals is not necessarily infinitesimal. For example, when $n \to \infty$, $\frac{1}{n}$ is infinitesimal, but the sum of an infinite number of infinitesimals is

$$\lim_{n \to \infty} \left(\overbrace{\frac{1}{n}+\frac{1}{n}+\cdots+\frac{1}{n}}^{n \text{ terms}} \right)=\lim_{n \to \infty} 1=1.$$

1.7.2 Infinite Limits

Definition 1.7.2 Let the function $f(x)$ be defined in a deleted neighborhood of the point x_0. If $\forall M>0$, $\exists \delta>0$, when $0<|x-x_0|<\delta$, then

$$|f(x)|>M.$$

The function $f(x)$ is called an infinite large (or infinity) as $x \to x_0$, or $f(x)$ tends to infinity as $x \to x_0$, denoted as

$$\lim_{x \to x_0} f(x)=\infty \text{ or } f(x) \to \infty \quad (x \to x_0).$$

If $|f(x)|>M$ in the above definition is replaced by

$$f(x)>M \text{ or } f(x)<-M,$$

then the function $f(x)$ is respectively called positive infinity and negative infinity as $x \to x_0$, and is respectively denoted as

$$\lim_{x \to x_0} f(x)=+\infty \text{ or } f(x) \to +\infty \quad (x \to x_0)$$

and

$$\lim_{x \to x_0} f(x)=-\infty \text{ or } f(x) \to -\infty \quad (x \to x_0).$$

Note:

(1) Different forms of "infinity" can be defined by replacing $x \to x_0$ with $x \to x_0^+$, $x \to x_0^-$, $x \to +\infty$, $x \to -\infty$, $x \to \infty$ and $n \to \infty$.

(2) Infinity must be unbounded, but unbounded is not necessarily infinity. For example, the sequence $\{n^{(-1)^n}\}$ is unbounded, but it is not infinite.

(3) There is a fundamental difference between finite limits and infinite limits. Infinity (∞) is not a number. For infinite limits, one can only say that it tends to infinity or diverges to infinity, not that it converges to infinity.

The proof methods for function limits being infinite are essentially the same as those for function limits being finite.

Example 1.7.1 Prove: $\lim_{x \to 1} \frac{1}{x-1}=\infty$.

Proof $\forall M>0$, solve the inequality

$$\left|\frac{1}{x-1}\right|=\frac{1}{|x-1|}>M,$$

to get $|x-1|<\frac{1}{M}$. Take $\delta=\frac{1}{M}$, then when $0<|x-1|<\delta$, there is

$$\left|\frac{1}{x-1}\right|>M,$$

namely

$$\lim_{x\to1}\frac{1}{x-1}=\infty.$$

Theorem 1.7.3 In the same limiting process, the reciprocal of an infinite is infinitesimal, and the reciprocal of a non-zero infinitesimal is infinite.

Proof We only prove this for $x\to x_0$ (finite points). Suppose $\lim\limits_{x\to x_0}f(x)=\infty$, then $\forall\varepsilon>0$, for $M=\frac{1}{\varepsilon}$, $\exists\delta>0$, when $0<|x-x_0|<\delta$, there is

$$|f(x)|>M \text{ or } |f(x)|>\frac{1}{\varepsilon},$$

thereby

$$\left|\frac{1}{f(x)}\right|<\varepsilon,$$

namely

$$\lim_{x\to x_0}\frac{1}{f(x)}=0.$$

Additionally, suppose $\lim\limits_{x\to x_0}f(x)=0$, then $\forall M>0$, for $\varepsilon=\frac{1}{M}$, $\exists\delta>0$, when $0<|x-x_0|<\delta$, there is

$$|f(x)|<\varepsilon \text{ or } |f(x)|<\frac{1}{M},$$

thereby

$$\left|\frac{1}{f(x)}\right|>\frac{1}{M},$$

namely

$$\lim_{x\to x_0}\frac{1}{f(x)}=\infty.$$

Example 1.7.2 Calculate $\lim\limits_{x\to\infty}\frac{a_0x^n+a_1x^{n-1}+\cdots+a_n}{b_0x^m+b_1x^{m-1}+\cdots+b_m}$, where $a_0,\cdots,a_n$ and $b_0,\cdots,b_m$ are constants, and $a_0\neq0, b_0\neq0, m, n\in\mathbf{N}$.

Solution $$\lim_{x\to\infty}\frac{a_0x^n+a_1x^{n-1}+\cdots+a_n}{b_0x^m+b_1x^{m-1}+\cdots+b_m}=\lim_{x\to\infty}x^{n-m}\frac{a_0+\frac{a_1}{x}+\cdots+\frac{a_n}{x^n}}{b_0+\frac{b_1}{x}+\cdots+\frac{b_m}{x^m}}.$$

When $n=m$, $\lim\limits_{x\to\infty}x^{n-m}=1$, when $n<m$, $\lim\limits_{x\to\infty}x^{n-m}=0$, when $n>m$, $\lim\limits_{x\to\infty}x^{n-m}=\infty$.

Thus

$$\lim_{x\to\infty}\frac{a_0x^n+a_1x^{n-1}+\cdots+a_n}{b_0x^m+b_1x^{m-1}+\cdots+b_m}=\begin{cases}\dfrac{a_0}{b_0}, & n=m\\ 0, & n<m\\ \infty, & n>m\end{cases}.$$

1.7.3 Comparison of Infinitesimals

The sum and product of two infinitesimals are still infinitesimals, but the quotient of two infinitesimals is not necessarily an infinitesimal. For example, when $x\to 0$, the functions x, x^2, $\sin x$ are infinitesimals, however,

$$\lim_{x\to 0}\frac{x^2}{x}=0,\ \lim_{x\to 0}\frac{\sin x}{x}=1,\ \lim_{x\to 0}\frac{x}{x^2}=\infty.$$

This indicates that the rate at which the function x, x^2, $\sin x$ approaches 0 varies when $x\to 0$.

Definition 1.7.3 Let α, β be infinitesimals in the same limiting process.

(1) If $\lim\dfrac{\beta}{\alpha}=0$, then β is called a higher-order infinitesimal of α, denoted as $\beta=o(\alpha)$.

(2) If $\lim\dfrac{\beta}{\alpha}=\infty$, then β is called a lower-order infinitesimal of α.

(3) If $\lim\dfrac{\beta}{\alpha}=c\neq 0$, then β and α are called infinitesimals of the same order. Specifically, if $c=1$, then β and α are called equivalent infinitesimals, denoted as $\beta\sim\alpha$.

(4) If $\lim\dfrac{\beta}{\alpha^k}=c\neq 0, k\in\mathbf{R}^+$, then β is called a k-order infinitesimal with respect to α.

When $x\to 0$, x^n, $\sin^m x$ are all infinitesimals. Since

$$\lim_{x\to 0}\frac{\sin^m x}{x^n}=\begin{cases}1, & m=n\\ 0, & m>n\\ \infty, & m<n\end{cases}.$$

Thus, when $m=n$, $\sin^m x\sim x^n$, when $m>n$, $\sin^m x=o(x^n)$, when $m<n$, $\sin^m x$ is a lower-order infinitesimal of x^n, or $x^n=o(\sin^n x)$.

Since

$$\lim_{x\to 0}\frac{1-\cos x}{x^2}=\frac{1}{2},$$

thus, when $x\to 0$, $1-\cos x$ and x^2 are infinitesimals of the same order.

The notation $\beta=o(\alpha)$ represents $\lim\dfrac{\beta}{\alpha}=0$, and $\beta=o(\alpha)$ should not be treated as an equation. For example, one cannot deduce $\beta-o(\alpha)=0$ or $o(\alpha)=\beta$ from $\beta=o(\alpha)$, nor can one deduce $\beta=\gamma$ from $\beta=o(\alpha)$ and $\gamma=o(\alpha)$. One often needs to perform operations on expressions containing the o notation, and the common rules for such operations are as follows:

(1) $o(\alpha)\pm o(\alpha)=o(\alpha)$;

(2) $o(\alpha)o(\beta)=o(\alpha\beta)$.

If a function is continuous, then its inverse function $y=f^{-1}(x)$ is also strictly monotonically increasing (or strictly monotonically decreasing) over the corresponding domain.

(3) $o(o(\alpha))=o(\alpha)$.

Take $\alpha\cdot o(\beta)=o(\alpha\beta)$ as an example to demonstrate, the rest of the proofs are omitted.

Because

$$\lim\frac{\alpha\cdot o(\beta)}{\alpha\beta}=\lim\frac{o(\beta)}{\beta}=0,$$

this means

$$\alpha\cdot o(\beta)=o(\alpha\beta).$$

Theorem 1.7.4 If $\alpha\sim\beta,\beta\sim\gamma$, then $\alpha\sim\gamma$.

Proof From $\alpha\sim\beta,\beta\sim\gamma$, it is known that

$$\lim\frac{\alpha}{\beta}=1,\lim\frac{\beta}{\gamma}=1.$$

Therefore

$$\lim\frac{\alpha}{\gamma}=\lim\frac{\alpha\beta}{\beta\gamma}=\lim\frac{\alpha}{\beta}\lim\frac{\beta}{\gamma}=1.$$

That is

$$\alpha\sim\gamma.$$

Equivalent infinitesimals play a significant role in limit calculations.

Theorem 1.7.5 If $\alpha\sim\alpha',\beta\sim\beta'$, and $\lim\frac{\beta'}{\alpha}$ exists, then

$$\lim\frac{\beta}{\alpha}=\lim\frac{\beta'}{\alpha'}.$$

Proof $\lim\frac{\beta}{\alpha}=\lim\frac{\beta}{\beta'}\frac{\beta'}{\alpha'}\frac{\alpha'}{\alpha}=\lim\frac{\beta}{\beta'}\lim\frac{\beta'}{\alpha'}\lim\frac{\alpha'}{\alpha}=\lim\frac{\beta'}{\alpha'}$.

Some commonly used equivalent infinitesimals: When $x\to 0$,

$$\sin x\sim x,\tan x\sim x,\arcsin x\sim x,\arctan x\sim x,$$

$$a^x-1\sim x\ln a,1-\cos x\sim\frac{x^2}{2},(1+x)^a-1\sim ax,\ln(1+x)\sim x.$$

Example 1.7.3 Find the following limits:

(1) $\lim\limits_{x\to 0}\frac{\tan x-\sin x}{\sin^3 x}$; (2) $\lim\limits_{x\to 0}\frac{\sin(nx)}{\tan(mx)}$.

Solution

(1) $\lim\limits_{x\to 0}\frac{\tan x-\sin x}{\sin^3 x}=\lim\limits_{x\to 0}\frac{\tan x(1-\cos x)}{x^3}$ $(\sin^3 x\sim x^3)$

$$=\lim_{x\to 0}\frac{x\cdot\frac{x^2}{2}}{x^3}\left(\tan x\sim x, 1-\cos x\sim\frac{x^2}{2}\right)=\lim_{x\to 0}\frac{1}{2}=\frac{1}{2}.$$

(2) $\lim\limits_{x\to 0}\frac{\sin(nx)}{\tan(mx)}=\lim\limits_{x\to 0}\frac{nx}{mx}=\frac{n}{m}$ $(\sin(nx)\sim nx, \tan(mx)\sim mx)$.

When the limit is of the form "$\frac{0}{0}$", one can follow the example above to replace infinitesimal factors in the numerator or denominator with their equivalent infinitesimals. This method greatly simplifies the process of finding limits.

Exercise 1.7

1. Under what conditions is the following function infinitesimal, and under what conditions is infinite?

(1) $f(x)=\frac{1}{x^2}$; (2) $f(x)=\frac{1}{x-1}$; (3) $f(x)=\ln x$.

2. Compare the orders of the following infinitesimal pairs:

(1) When $x\to 0$, $x(\sin x+1)$ and x;

(2) When $x\to 0$, $\tan x-\sin x$ and x;

(3) When $x\to 0$, $1-\cos x$ and x;

(4) When $x\to+\infty$, 2^{-x} and a^{-x} $(a>0, a\neq 1)$.

3. Calculate the following limits:

(1) $\lim\limits_{x\to+\infty}\frac{2\arctan\frac{1}{x}}{\sin\frac{1}{x}}$;

(2) $\lim\limits_{x\to 0}\frac{\tan x-\sin x}{\sin^3 x}$;

(3) $\lim\limits_{x\to 0}\frac{1-\cos(3x)}{x^2}$;

(4) $\lim\limits_{x\to 0}\frac{\sqrt{1+x^2}-1}{1-\cos x}$;

(5) $\lim\limits_{x\to 0}\frac{x\ln(1+x)}{\sin^2 x}$;

(6) $\lim\limits_{x\to 0}\frac{e^{x^2}-1}{x\sin x}$.

4. Prove:

(1) $\lim\limits_{x\to 4}\frac{2}{(x-4)^3}=\infty$;

(2) $\lim\limits_{x\to 1}\frac{1}{(x-1)^2}=+\infty$;

(3) $\lim\limits_{x\to 0^+}\ln x=-\infty$;

(4) $\lim\limits_{x\to-\infty}a^x=+\infty$ $(0<a<1)$.

5. Prove: If $\lim\limits_{x\to a}f(x)=+\infty$, $\lim\limits_{x\to a}g(x)=A$, then $\lim\limits_{x\to a}[f(x)+g(x)]=+\infty$.

6. Prove: $f(x)\sim g(x)$ $(x\to a)\Leftrightarrow f(x)-g(x)=o(g(x))$ $(x\to a)$.

7. Prove:

(1) $x^2\sin x=o(x)$ $(x\to 0)$;

(2) $\frac{x+2}{x^4+3}=o\left(\frac{1}{x^2}\right)$;

(3) $\tan x\sim\sin x$ $(x\to 0)$;

(4) $x^n-1\sim n(x-1)$ $(x\to 1, n\in\mathbf{N}^+)$.

1.8 Continuity of Functions

Many phenomena in the physical world undergo continuous changes, such as the continuous flow of fluids, the gradual rise and fall of temperature, and the continuous variation of pressure. These phenomena are modeled mathematically by continuous functions. The essence of continuity lies in the stability of changes, where a small change in the independent variable leads to only a small change in the dependent variable. In contrast, discontinuity signifies the disruption of stability, where a small change in the independent variable results in a drastic change in the dependent variable. The concept of limits is used below to describe this characteristic of function changes.

1.8.1 Concept of Continuity

Definition 1.8.1 Let the function $y=f(x)$ be defined in a neighborhood of point x_0, with the change in the independent variable $\Delta x=x-x_0$, and the change in the function $\Delta y=f(x)-f(x_0)=f(x_0+\Delta x)-f(x_0)$. If

$$\lim_{\Delta x\to 0}\Delta y=0,$$

then the function $f(x)$ is said to be continuous at point x_0. If

$$\lim_{\Delta x\to 0^-}\Delta y=0,$$

then the function $f(x)$ is said to be left-continuous at point x_0. If

$$\lim_{\Delta x\to 0^+}\Delta y=0,$$

then the function $f(x)$ is said to be right-continuous at point x_0.

An equivalent definition for the continuity of a function at a point is:

Definition 1.8.2 Let the function $f(x)$ be defined in a neighborhood of point x_0. If

$$\lim_{x\to x_0}f(x)=f(x_0),$$

the function $f(x)$ is said to be continuous at the point x_0. If

$$\lim_{x\to x_0^-}f(x)=f(x_0-0)=f(x_0),$$

the function $f(x)$ is said to be left-continuous at the point x_0. If

$$\lim_{x\to x_0^+}f(x)=f(x_0+0)=f(x_0),$$

the function $f(x)$ is said to be right-continuous at the point x_0.

If the function $f(x)$ is continuous at every point in the interval I, then $f(x)$ is said to be continuous on the interval I, or $f(x)$ is referred to as a continuous function on I.

Note that the continuity of $f(x)$ at point x_0 must satisfy the following three conditions simultaneously:

(1) $f(x)$ is defined at point x_0;

(2) $f(x)$ has a limit at point x_0;

(3) The limit of $f(x)$ at point x_0 equals the function value of $f(x)$ at point x_0 $f(x_0)$.

From the properties of function limits, the following conclusions about the continuity of function $f(x)$ at point x_0 can be drawn:

Theorem 1.8.1 The function $f(x)$ is continuous at point x_0, if and only if it is both left-continuous and right-continuous at point x_0.

From the definition of continuous functions, it follows that

$$\lim_{x \to x_0} f(x) = f(x_0) = f(\lim_{x \to x_0} x).$$

That is, when taking the limit of a continuous function, the limit symbol can be exchanged with the function symbol.

Example 1.8.1 Prove: The function $f(x)=\sin x$ is continuous on $\mathbf{R}$.

Proof $\forall x_0 \in \mathbf{R}$, $\forall \varepsilon > 0$, solve the inequality

$$|\sin x - \sin x_0| = 2\left|\cos\frac{x+x_0}{2}\sin\frac{x-x_0}{2}\right|$$
$$= 2\left|\cos\frac{x+x_0}{2}\right|\left|\sin\frac{x-x_0}{2}\right| \leqslant 2 \cdot \frac{|x-x_0|}{2} = |x-x_0| < \varepsilon.$$

Obtain $|x-x_0|<\varepsilon$, take $\delta=\varepsilon$, then when $|x-x_0|<\delta$, there is

$$|\sin x - \sin x_0| < \varepsilon.$$

That is

$$\lim_{x \to x_0} \sin x = \sin x_0.$$

Thus, the function $f(x)=\sin x$ is continuous at point x_0, and by the arbitrariness of point x_0, it is known that $f(x)=\sin x$ is continuous on $\mathbf{R}$.

Example 1.8.2 Let $f(x)=\begin{cases} 3x^2+2, & x \geqslant 1 \\ e^x, & x<1 \end{cases}$, discuss the continuity of $f(x)$.

Solution When $x>1$, $f(x)=3x^2+2$, $\forall x_0>1$, there is

$$\lim_{x \to x_0} f(x) = \lim_{x \to x_0} (3x^2+2) = 3x_0^2+2 = f(x_0),$$
$$\lim_{x \to 1^+} f(x) = \lim_{x \to 1^+} (3x^2+2) = 5 = f(1).$$

Hence, $f(x)$ is continuous within $(1,+\infty)$, and right-continuous at point $x_0=1$.

When $x<1$, $f(x)=e^x$, $\forall x_0<1$, there is

$$\lim_{x \to x_0} f(x) = \lim_{x \to x_0} e^x = e^{x_0} = f(x_0),$$
$$\lim_{x \to 1^-} f(x) = \lim_{x \to 1^-} e^x = e \neq f(1).$$

Hence, $f(x)$ is continuous within $(-\infty,1)$, and not left-continuous at point $x_0=1$.

In summary, the function $f(x)$ is continuous on $(-\infty,1)\cup(1,+\infty)$, and discontinuous at point $x_0=1$.

1.8.2 Properties of Continuous Functions

Theorem 1.8.2 Let $f(x)$, $g(x)$ be continuous at point x_0, then

$$f(x) \pm g(x),\ f(x)g(x),\ \frac{f(x)}{g(x)} \quad (g(x) \neq 0)$$

continuous at point x_0.

Proof From $\lim\limits_{x\to x_0} f(x)=f(x_0)$, $\lim\limits_{x\to x_0} g(x)=g(x_0)$, there is

$$\lim_{x\to x_0}[f(x)\pm g(x)]=\lim_{x\to x_0}f(x)\pm\lim_{x\to x_0}g(x)=f(x_0)\pm g(x_0),$$

$$\lim_{x\to x_0}[f(x)g(x)]=\lim_{x\to x_0}f(x)\lim_{x\to x_0}g(x)=f(x_0)g(x_0),$$

$$\lim_{x\to x_0}\frac{f(x)}{g(x)}=\frac{\lim\limits_{x\to x_0}f(x)}{\lim\limits_{x\to x_0}g(x)}=\frac{f(x_0)}{g(x_0)}.$$

This theorem can be extended to the case of a finite number of functions, i.e., the sum and product of a finite number of continuous functions at a point are also continuous at that point.

Regarding the continuity of composite functions, the following conclusions hold:

Theorem 1.8.3 Let $g(x)$ be continuous at point x_0, and $f(u)$ be continuous at $u_0=g(x_0)$, then the composite function $f[g(x)]$ is continuous at point x_0.

Proof Given $\lim\limits_{x\to x_0} g(x)=g(x_0)$, $\lim\limits_{u\to u_0} f(u)=f(u_0)$, we have

$$\lim_{x\to x_0} f[g(x)]=\lim_{u\to u_0} f(u)=f(u_0)=f[g(x_0)],$$

that is, $f[g(x)]$ is continuous at point x_0.

Note: Under the condition of the continuity of function $f(x)$, the limit of the composite function can be rewritten as

$$\lim f[g(x)]=f[\lim g(x)].$$

Regarding the continuity of the inverse function, the following conclusion holds:

Theorem 1.8.4 Let function $y=f(x)$ be a strictly monotonically increasing (or decreasing) continuous function on interval I, then its inverse function $y=f^{-1}(x)$ is also a strictly monotonically increasing (or decreasing) continuous function on its corresponding domain.

Proof Omitted.

Utilizing the above conclusions to study the continuity of elementary functions.

In summary, basic elementary functions are continuous on their respective domains.

Elementary functions are formed by a finite number of arithmetic operations and compositions from basic elementary functions, hence elementary functions are continuous on their defined intervals.

Note: The defined interval refers to an interval within the domain, and elementary functions are not necessarily continuous within their entire domain. For example, the domain of $f(x)=\sqrt{x^2(x-1)^3}$ is $\{0\}\cup[1,+\infty)$, $f(x)$ which is discontinuous at point $x=0$.

Example 1.8.3 Find the following limits:

(1) $\lim\limits_{x\to 0}\dfrac{a^x-1}{x}$ $(a>0, a\neq 1)$; (2) $\lim\limits_{x\to 0}\dfrac{\ln(1+x)}{x}$; (3) $\lim\limits_{x\to 0}\dfrac{(1+x)^\alpha-1}{x}$ $(\alpha\neq 0)$.

Solution

(1) Let $a^x-1=t$, then $x=\log_a(1+t)$, and as $x\to 0$ approaches, $t\to 0$, thus

$$\lim_{x\to 0}\frac{a^x-1}{x}=\lim_{t\to 0}\frac{t}{\log_a(1+t)}=\frac{1}{\log_a[\lim\limits_{t\to 0}(1+t)^{\frac{1}{t}}]}=\frac{1}{\log_a e}=\ln a.$$

(2) $\lim\limits_{x\to 0}\frac{\ln(1+x)}{x}=\lim\limits_{x\to 0}\ln(1+x)^{\frac{1}{x}}=\ln[\lim\limits_{x\to 0}(1+x)^{\frac{1}{x}}]=\ln e=1.$

(3) $\lim\limits_{x\to 0}\frac{(1+x)^{\alpha}-1}{x}=\lim\limits_{x\to 0}\frac{e^{\alpha\ln(1+x)}-1}{x}=\lim\limits_{x\to 0}\frac{\alpha\ln(1+x)}{x}=\alpha\lim\limits_{x\to 0}\frac{\ln(1+x)}{x}=\alpha.$

1.8.3 Discontinuities of Functions and Their Classification

A point of discontinuity x_0 of a function $f(x)$ is referred to as a discontinuity point. According to the definition of function continuity, x_0 is a discontinuity point of $f(x)$ if at least one of the following conditions holds:

(1) $f(x)$ is undefined at point x_0;

(2) The limit $\lim\limits_{x\to x_0}f(x)$ of $f(x)$ at point x_0 does not exist;

(3) The limit $\lim\limits_{x\to x_0}f(x)$ of $f(x)$ at point x_0 exists, but is not equal to $f(x_0)$.

Based on the possible scenarios of discontinuities, one can classify the discontinuity points of a function using limits.

Definition 1.8.3 Let x_0 be a discontinuity point of $f(x)$.

(1) If $\lim\limits_{x\to x_0}f(x)$ exists, i.e., $f(x_0-0)=f(x_0+0)$, then x_0 is called a removable discontinuity of $f(x)$;

(2) If both $f(x_0-0)$, $f(x_0+0)$ exist, but $f(x_0-0)\neq f(x_0+0)$, then x_0 is called a jump discontinuity of $f(x)$, and $f(x_0+0)-f(x_0-0)$ is the jump at point x_0 of $f(x)$.

Removable and jump discontinuities are collectively referred to as first-kind discontinuities, and all other discontinuities are referred to as second-kind discontinuities.

Example 1.8.4 Let $f(x)=\frac{\sin x}{x}$, discuss the type of discontinuity point $x=0$.

Solution Since $\lim\limits_{x\to 0}\frac{\sin x}{x}=1$, therefore $x=0$ is a removable discontinuity of $f(x)$.

If supplemented $f(0)=\lim\limits_{x\to 0}f(x)=1$, then

$$f(x)=\begin{cases}\frac{\sin x}{x}, & x\neq 0\\ 1, & x=0\end{cases}.$$

Continuous at point $x=0$. Therefore, for the removable discontinuity x_0 of function $f(x)$, by supplementing or modifying the definition of function $f(x)$ at point x_0, letting $f(x_0)=\lim\limits_{x\to x_0}f(x)$, thereby making function $f(x)$ continuous at point x_0, which is precisely the essence of "removable".

Example 1.8.5 Determine the type of discontinuity for the following functions:

(1) $f(x)=\cos\frac{1}{x}$; (2) $f(x)=\begin{cases} x, & x\geqslant 1 \\ -x, & x<1 \end{cases}$; (3) $f(x)=\sqrt{\cos x-1}$; (4) $f(x)=\frac{1}{x}$.

Solution

(1) $f(x)=\cos\frac{1}{x}$ is undefined at point $x=0$, and as $x\to 0$, the function value $\cos\frac{1}{x}$ oscillates between -1 and 1, hence $x=0$ is called an oscillatory discontinuity of $f(x)$, belonging to the second kind of discontinuity.

(2) Since $\lim\limits_{x\to 1^-} f(x)=-1$, $\lim\limits_{x\to 1^+} f(x)=1$, i.e., $f(1-0)\neq f(1+0)$, therefore $x=1$ is a jump discontinuity of $f(x)$, belonging to the first kind of discontinuity.

(3) Since the domain of $f(x)=\sqrt{\cos x-1}$ is $D_f=\{2k\pi, k\in \mathbf{Z}\}$, and thus for all points within $\mathbf{R}$, neither the left nor right limit exists, hence it belongs to the second kind of discontinuity.

(4) Since $\lim\limits_{x\to 0} f(x)=\lim\limits_{x\to 0}\frac{1}{x}=\infty$, in this case $x=0$ is called an infinite discontinuity, belonging to the second kind of discontinuity.

Exercise 1.8

1. Which of the following propositions are correct?

(1) If $f(x)$ is continuous at point x_0, then $f(x)$ is defined at point x_0;

(2) If $f(x)$ is defined at point x_0 and the limit of $f(x)$ at point x_0 exists, then $f(x)$ is continuous at point x_0;

(3) If the limit of $f(x)$ at point x_0 exists, then $f(x)$ is continuous at point x_0.

2. Prove that the following functions are continuous within their domains:

(1) $f(x)=|x|$; (2) $f(x)=\frac{1}{x}$; (3) $f(x)=\sin\frac{1}{x}$.

3. Prove: If function $f(x)$ is continuous at point a, then function $|f(x)|$ is also continuous at point a. Does the converse hold?

4. Identify the intervals on which the following functions are continuous and the points at which they are discontinuous.

(1) $f(x)=\frac{2}{x+1}+2$; (2) $f(x)=\frac{1}{|x|+1}-\frac{x^2}{3}$;

(3) $f(x)=\frac{\sin x}{x}$; (4) $f(x)=\tan\frac{\pi x}{2}$;

(5) $f(x)=\ln(x^2-4)$; (6) $f(x)=\sqrt{x^2-a^2}$ $(a>0)$.

5. Given $f(x)=\begin{cases} e^x, & x<0 \\ a+x, & x\geqslant 0 \end{cases}$, for what value of a is $f(x)$ continuous at $(-\infty,+\infty)$?

6. Given $f(x)=\begin{cases}\dfrac{\sqrt{x+1}-1}{x}, & x\neq 0\\ 0, & x=0\end{cases}$, discuss the continuity of $f(x)$.

7. Identify the points of discontinuity and their types for the following functions:

(1) $f(x)=\sqrt{|x|}\sin\dfrac{1}{x}$;

(2) $f(x)=\dfrac{x^2-x}{|x|(x^2-1)}$;

(3) $f(x)=\begin{cases}2x+3, & x<-1\\ 0, & x=-1\\ \dfrac{1}{x}, & -1<x<0\\ \sin x, & x>0\end{cases}$;

(4) $f(x)=\lim\limits_{n\to\infty}\dfrac{nx}{nx^2+1}$.

8. Find the following limits:

(1) $\lim\limits_{x\to\pi}\sin(x-\sin x)$;

(2) $\lim\limits_{x\to 0}\sin\left[\dfrac{\pi}{2}\cos(\tan x)\right]$;

(3) $\lim\limits_{x\to 1}\sec(x\sec^2 x-\tan^2 x-1)$;

(4) $\lim\limits_{x\to\frac{\pi}{2}}\ln(\sin x)$;

(5) $\lim\ln(2\cos 2x)$;

(6) $\lim\limits_{x\to a}\dfrac{\sin x-\sin a}{x-a}$ $\left(x\to\dfrac{\pi}{6}\right)$;

(7) $\lim\limits_{x\to 1}\dfrac{\sqrt{5x-4}-\sqrt{x}}{x-1}$;

(8) $\lim(\sin 2x)^3$ $\left(x\to\dfrac{\pi}{4}\right)$.

1.9 Properties of Continuous Functions on Closed Intervals

In practical application problems, a function is often established first, and then the issues of its maximum and minimum values, as well as boundedness, are discussed. Another important issue is to study the existence of roots of an equation and the interval in which the roots exist, so that finding the roots or approximate solutions of the equation becomes meaningful. The following focuses on several important properties of continuous functions on closed intervals.

A function $f(x)$ is said to be continuous on a closed interval $[a,b]$, if the function $f(x)$ is continuous on the open interval (a,b), right-continuous at the left endpoint a, and left-continuous at the right endpoint b.

1.9.1 Extreme Value Theorem

Let $f(x)$ be a function defined on an interval I. If $\exists x_0\in I$, $\forall x\in I$ holds, such that $f(x)\leqslant f(x_0)$ $(f(x)\geqslant f(x_0))$, then x_0 is called the maximum (minimum) point of $f(x)$ on the interval I, and $f(x_0)$ is the maximum (minimum) value of $f(x)$ on the interval I.

For example, the function $f(x)=\sin x$ has a maximum value of 1 and a minimum value of -1 on the interval $[0,2\pi]$. The function $f(x)=\frac{1}{x}$ has a minimum value of 1 but no maximum value on the interval $(0,1]$. The function $f(x)=\tan x$ has neither a minimum nor a maximum value on the interval $\left(-\frac{\pi}{2},\frac{\pi}{2}\right)$. Therefore, the extreme values of a function on an interval may or may not exist. However, the extreme values of a continuous function on a closed interval always exist.

Theorem 1.9.1 If the function $f(x)$ is continuous on the closed interval $[a,b]$, then $f(x)$ is guaranteed to attain both a maximum and a minimum value on $[a,b]$.

This theorem indicates that if the function $f(x)$ is continuous on the closed interval $[a,b]$, then there exists at least one point $\xi_1\in[a,b]$, such that $\forall x\in[a,b]$, and $f(x)\leqslant f(\xi_1)=M$. There also exists at least one point $\xi_2\in[a,b]$, such that $\forall x\in[a,b]$ and $f(x)\geqslant f(\xi_2)=m$. Consequently, $\forall x\in[a,b]$ and $m\leqslant f(x)\leqslant M$, thus proving that a continuous function on a closed interval is bounded. This is illustrated in Figure 1-11.

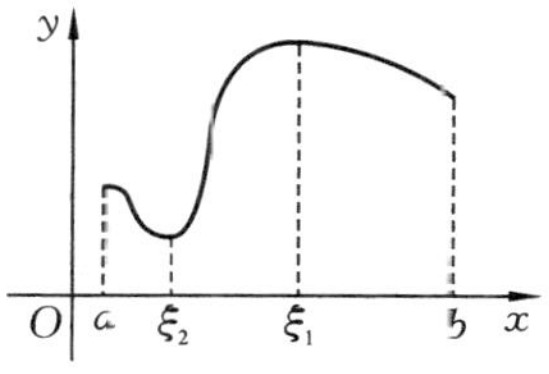

Figure 1-11

Note:

(1) The continuity condition in the theorem cannot be omitted. For instance, the function

$$f(x)=\begin{cases}-x, & -1\leqslant x<0\\ 1, & x=0\\ x, & 0<x\leqslant 1\end{cases}$$

is defined on the closed interval $[-1,1]$, is discontinuous at point $x=0$, has a maximum value of 1, but no minimum value. This is shown in Figure 1-12.

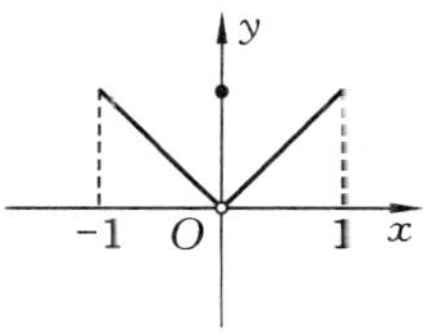

Figure 1-12

(2) The closed interval condition in the theorem cannot be omitted either. For example, the function

$$f(x)=x \quad [x\in(0,1)]$$

is continuous in the open interval $(0,1)$, but without a maximum or minimum value. As shown in Figure 1-13.

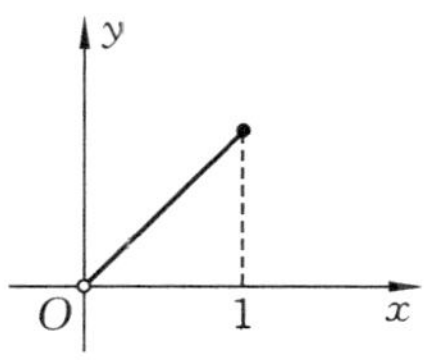

Figure 1-13

1.9.2 Intermediate Value Theorem

Theorem 1.9.2 (Zero Theorem) If the function $f(x)$ is continuous on the closed interval $[a,b]$ and $f(a)$, $f(b)$ are of opposite signs, i.e., $f(a)f(b)<0$, then there exists at least one point $\xi\in(a,b)$, such that $f(\xi)=0$. ξ is called the zero of the function $f(x)$, i.e., ξ is the root of the equation $f(x)=0$.

Note:

(1) The continuity condition cannot be removed. As shown in Figure 1-14.

(2) The closed interval condition cannot be removed. As shown in Figure 1-15.

The geometric meaning of the Zero Theorem is: For the continuous curve $y=f(x)$ on the closed interval $[a,b]$, if the endpoint values are of opposite signs $f(a)f(b)<0$, then this continuous curve $y=f(x)$ intersects the x axis at least once. As shown in Figure 1-16.

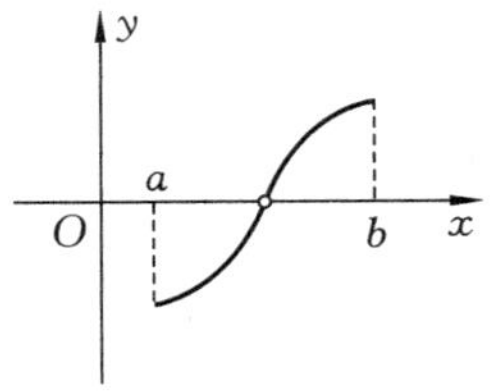

Figure 1-14

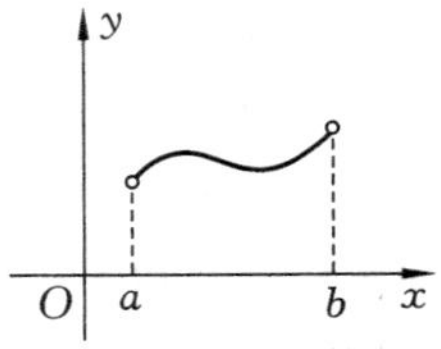

Figure 1-15

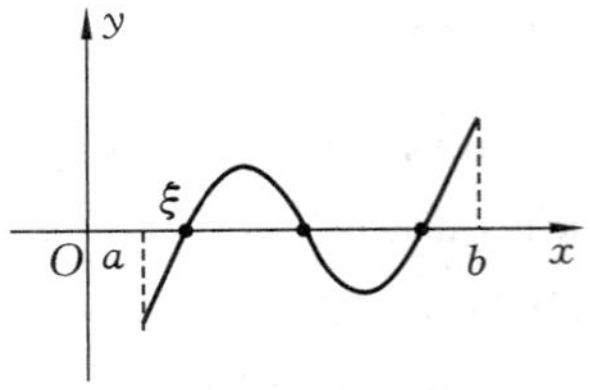

Figure 1-16

The Zero Theorem is also known as the Existence Theorem for Equation Roots, and it can be used to determine the existence of roots of an equation.

Theorem 1.9.3 (Intermediate Value Theorem) If the function $f(x)$ is continuous on the closed interval $[a,b]$, and $f(a)\neq f(b)$, then for any number μ between $f(a)$ and $f(b)$, there exists at least one point ξ in (a,b), such that $f(\xi)=\mu$.

Proof Assume $f(a)<f(b)$, and $f(a)<\mu<f(b)$, let

$$F(x)=f(x)-\mu,$$

then $F(x)$ is continuous on the closed interval $[a,b]$, and

$$F(a)=f(a)-\mu<0, F(b)=f(b)-\mu>0.$$

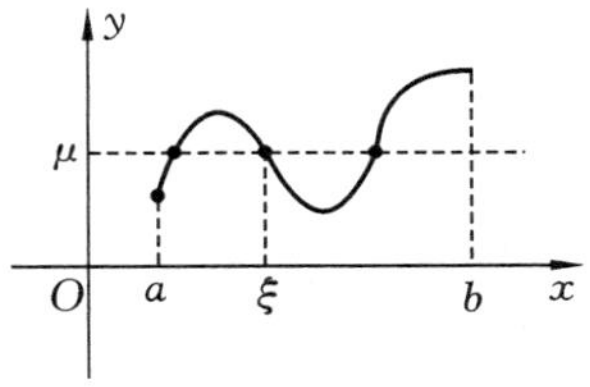

Figure 1-17

By the Zero Theorem, there exists $\exists\xi\in(a,b)$, such that $F(\xi)=0$, i.e., $f(\xi)=\mu$.

The geometric interpretation of the Intermediate Value Theorem is: On the closed interval $[a,b]$, the continuous curve $y=f(x)$, for any line $y=\mu$ parallel to the x axis between $f(a)$ and $f(b)$, must intersect the continuous curve at least once. As shown in Figure 1-17.

Corollary 1.9.1 If the function $f(x)$ is continuous on $[a,b]$, and m,M are the minimum and maximum values of $f(x)$ on $[a,b]$, respectively, then there exists $\forall\mu\ (m\leqslant\mu\leqslant M)$, $\exists\xi\in[a,b]$, such that $f(\xi)=\mu$.

Proof When $m=M$, $\forall x\in[a,b]$, $f(x)=\mu$, the conclusion is correct.

When $m\neq M$, $\exists x_1, x_2\in[a,b]$, and there exists $x_1\neq x_2$, such that $f(x_1)=m$, $f(x_2)=M$. Applying the Intermediate Value Theorem on the closed interval $[x_1,x_2]$ (or $[x_2,x_1]$), there exists $\forall\mu\ (m\leqslant\mu\leqslant M)$, $\exists\xi\in[x_1,x_2]$ (or $[x_2,x_1]$) $\subseteq[a,b]$), such that $f(\xi)=\mu$.

Example 1.9.1 Prove: A polynomial equation of odd degree must have at least one real root.

Proof Let the polynomial be of odd degree,

$$f(x)=x^{2k+1}+a_1x^{2k}+\cdots+a_{2k+1},$$

then

$$\lim_{x\to-\infty} f(x)=\lim_{x\to-\infty} x^{2k+1}\left(1+\frac{a_1}{x}+\cdots+\frac{a_{2k+1}}{x^{2k+1}}\right)=-\infty,$$

$$\lim_{x\to+\infty} f(x)=\lim_{x\to+\infty} x^{2k+1}\left(1+\frac{a_1}{x}+\cdots+\frac{a_{2k+1}}{x^{2k+1}}\right)=+\infty.$$

Consequently, $\exists a$ and $b\,(a<b)$, such that $f(a)<0, f(b)>0, f(x)$ is continuous on $[a,b]$. By the Zero Theorem, there exists $\xi\in(a,b)$, such that $f(\xi)=0$, i. e., the polynomial equation of odd degree must have at least one real root.

Example 1.9.2 Prove: The equation $x^5-3x-1=0$ has a real root within $(1,2)$.

Proof Let $f(x)=x^5-3x-1$, then $f(x)$ is continuous on $[1,2]$ and $f(1)=-3<0$, $f(2)=25>0$. By the Zero Theorem, there exists $\exists\xi\in(1,2)$, such that $f(\xi)=0$, i.e., ξ is a root of the original equation.

Example 1.9.3 Suppose $f(x)$ is continuous on $[0,2a]$ and $f(0)=f(2a)$. Prove: There exists at least one point $[0,a]$ on ξ, such that $f(\xi)=f(\xi+a)$.

Proof Let $F(x)=f(x)-f(x+a)$, then $F(x)$ is continuous on $[0,a]$ and has

$$F(0)=f(0)-f(a), F(a)=f(a)-f(2a)=f(a)-f(0).$$

When $f(0)-f(a)=0$, it is obvious that there exists $\xi=0\in[0,a]$ such that $f(\xi)=f(\xi+a)$, when $f(0)-f(a)\neq0$, $F(0)F(a)<0$, i.e., $F(0)$ and $F(a)$ have opposite signs. By the Zero Theorem, there exists $\exists\xi\in(0,a)$, such that $F(\xi)=0$, i.e., $f(\xi)=f(\xi+a)$.

Exercise 1.9

1. Prove: The equation $x=a\sin x+b\,(a>0,b>0)$ has at least one positive root, and it does not exceed $a+b$.

2. Proof: If $f(x)$ is continuous on $[a,b]$, $x_1,x_2,\cdots,x_n\in(a,b)$, then $\exists\xi\in(a,b)$, such that

$$f(\xi)=\frac{f(x_1)+f(x_2)+\cdots+f(x_n)}{n}.$$

3. Proof: If the function $f(x)$ is continuous on $[a,+\infty)$ and $\lim_{x\to+\infty} f(x)=A$, then the function $f(x)$ is bounded on $[a,+\infty)$.

4. Let $f(x)$ be continuous on $[0,1]$, and $\forall x\in[0,1]$ have $0\leqslant f(x)\leqslant1$, prove that $\exists\xi\in[0,1]$, such that $f(\xi)=\xi$, the point ξ is called a fixed point of $f(x)$.

5. Proof:

(1) $x^5-2x^2+x+1=0$ has at least one real root within $(-1,1)$;

(2) $\frac{1}{x-1}+\frac{2}{x-2}+\frac{3}{x-3}=0$ has a real root in both $(1,2)$ and $(2,3)$;

(3) $x^2\cos x-\sin x=0$ has at least one real root within $\left(\pi,\frac{3}{2}\pi\right)$.

6. Proof: If the functions $f(x)$ and $g(x)$ are continuous on $[a,b]$ and $f(a)<g(a)$, $f(b)>g(b)$, then $\exists \xi \in (a,b)$, such that $f(\xi)=g(\xi)$.

Comprehensive Review 1

1. Do the f, g in the following questions represent the same function? Explain the reason and indicate in which interval they are the same.

(1) $f(x)=\dfrac{x^2-1}{x-1}$, $g(x)=x+1$; (2) $f(x)=\sqrt{x^2}$, $g(x)=(\sqrt{x})^2$;

(3) $f(x)=\sin^2 x+\cos^2 x$, $g(x)=1$; (4) $f(x)=\ln x^3$, $g(x)=3\ln x$.

2. Find the domain of the following functions:

(1) $f(x)=\dfrac{1}{1-x^2}+\sqrt{x+2}$; (2) $f(x)=\dfrac{1}{x^2-4}+\sqrt{2\log_{\frac{1}{2}}(x-1)}$.

3. Identify which elementary functions the following functions are composed of:

(1) $y=\sin\dfrac{1}{\sqrt{x^2+1}}$; (2) $y=\sin\left(\log_a \arctan\dfrac{1}{\sqrt{x}}\right)$; (3) $y=\tan\dfrac{1}{\sqrt{\cos^2\left(\sin\dfrac{1}{\sqrt[3]{x}}\right)}}$.

4. Choose the correct term among "sufficient" "necessary" "sufficient and necessary" to fill in the blanks below.

(1) The boundedness of sequence $\{a_n\}$ is a ________ condition for the convergence of sequence $\{a_n\}$. The convergence of sequence $\{a_n\}$ is a ________ condition for the boundedness of sequence $\{a_n\}$.

(2) The boundedness of $f(x)$ in a deleted neighborhood of point x_0 is a ________ condition for the existence of $\lim\limits_{x\to x_0} f(x)$. The existence of $\lim\limits_{x\to x_0} f(x)$ is a ________ condition for the boundedness of $f(x)$ in a deleted neighborhood of point x_0.

(3) The unboundedness of $f(x)$ in a deleted neighborhood of point x_0 is a ________ condition for $\lim\limits_{x\to x_0} f(x)=\infty$. $\lim\limits_{x\to x_0} f(x)=\infty$ is a ________ condition for the unboundedness of $f(x)$ in a deleted neighborhood of point x_0.

5. Thought-provoking questions:

(1) Do functions f and g exist such that $f(g)=g(f)$?

(2) Are the sum and product of two unbounded functions necessarily unbounded?

(3) Is the sum of two strictly monotonic functions also strictly monotonic? Is their composition monotonic?

(4) If the sum of functions f, g and $f+g$ has a limit at point x_0, does f, g necessarily have a limit at point x_0? If one function has a limit at point x_0, does the other function necessarily have a limit at point x_0?

(5) If the left and right limits of f both exist at point x_0, does the limit of f at point x_0 necessarily exist or necessarily not exist?

6. Find the following limits:

(1) $\lim\limits_{n\to\infty}\dfrac{\sqrt[3]{n^2}\sin n!}{n+1}$; (2) $\lim\limits_{n\to\infty}\sin^2(\pi\sqrt{n^2+n})$;

(3) $\lim\limits_{n\to\infty}\left(1+\dfrac{1}{1+2}+\dfrac{1}{1+2+3}+\cdots+\dfrac{1}{1+2+\cdots+n}\right)$;

(4) $\lim\limits_{n\to\infty}(1+x)(1+x^2)(1+x^4)\cdots(1+x^{2n})$;

(5) $\lim\limits_{x\to 0}\ln\dfrac{\sin x}{x}$; (6) $\lim\limits_{x\to\infty}\left(\dfrac{x^2+1}{x^2-1}\right)^{x^2}$;

(7) $\lim\limits_{x\to 1}(1-x)\tan\dfrac{\pi x}{2}$; (8) $\lim\limits_{x\to 1}\dfrac{x^m-1}{x^n-1}$;

(9) $\lim\limits_{x\to 0+}\left(\sqrt{\dfrac{1}{x^2}+\dfrac{1}{x}}-\sqrt{\dfrac{1}{x^2}-\dfrac{2}{x}}\right)$; (10) $\lim\limits_{x\to 0+}\dfrac{\tan x}{\sqrt{x+\sqrt{x+\sqrt{x}}}}$;

(11) $\lim\limits_{x\to\infty}\left(\dfrac{3x+1}{3x-2}\right)^x$; (12) $\lim\limits_{x\to 0}(1+3\tan^2 x)^{\cot x}$.

7. For what value of a is the function $f(x)=\begin{cases}\dfrac{\cos 2x-\cos 3x}{x^2}, & x\neq 0\\ a, & x=0\end{cases}$ continuous at point $x=0$?

8. Find the constant a,b, such that the function $f(x)=\begin{cases}1+x^2, & x<0\\ ax+b, & 0\leqslant x\leqslant 1,\\ x^3-1, & x>1\end{cases}$ is continuous on $(-\infty,+\infty)$.

9. Identify the points of discontinuity and their types for the following functions:

(1) $f(x)=e^{x+\frac{1}{x}}$; (2) $f(x)=\dfrac{\frac{1}{x}-\frac{1}{x+1}}{\frac{1}{x-1}\cdot\frac{1}{x}}$; (3) $f(x)=\dfrac{1}{1-e^{x-\frac{1}{x}}}$; (4) $f(x)=\lim\limits_{t\to+\infty}\dfrac{x+e^{tx}}{1+e^{tx}}$.

10. Given that for any x,y, there exists $f(x+y)=f(x)+f(y)$, and $f(x)$ is continuous at point $x=0$, prove that $f(x)$ is continuous within $(-\infty,+\infty)$.

11. Given $f(x)=\sqrt{16-x}$, $g(x)=x^4$, find the expressions of the following functions and their domains: (1) $f\circ g$; (2) $g\circ f$; (3) $f+g$; (4) fg.

12. Given $f(x)=\begin{cases}x\sin x, & x\leqslant 0\\ \dfrac{\sin x}{x}, & x>0\end{cases}$, find:

(1) $\lim\limits_{x\to 0^-}f(x)$; (2) $\lim\limits_{x\to 0^+}f(x)$; (3) $\lim\limits_{x\to+\infty}f(x)$; (4) $\lim\limits_{x\to -1}f(x)$.

13. Suppose $f(x)$ is continuous on $[a,b]$ and $a<f(x)<b$, prove that there exists at least one point ξ in (a,b), such that $f(\xi)=\xi$.

14. If $f(x)$ is an odd function and continuous at point $x=0$, prove $f(0)=0$.

15. Prove that the equation $x^3\cos x+6\sin^5 x-3=0$ has a solution between 0 and 2π.

Chapter 2 Derivatives and Differentiation

The Dujiang Weir designed by Li Bing and his son regulates the flow of the Minjiang River through the mouth of the diverting fish. Its core is to control the rate of change (derivative) of the flow velocity:

If the water separation ratio of fish mouth is $f(t)$, then the change rate of water separation velocity $f'(t)$ determines the efficiency of drought and flood regulation.

At that time $f'(t)=0$, the water flow reached a dynamic balance ("Tao law nature"), to achieve "floods and droughts from people".

Chapter 1 discussed the trend of functions changing with the independent variable, i.e., the limit problem. However, in real life, it is also necessary to discuss the rate at which functions change with respect to the independent variable, i.e., the rate of change problem. For example, the speed of an object in motion, labor productivity, the slope of a tangent line, etc. To better and more systematically solve these problems, it is essential to study differential calculus. Differential calculus is a crucial part of calculus, with its fundamental concepts being derivatives and differentials. This chapter primarily introduces the concepts of derivatives and differentials and their calculation methods.

2.1 Concept of Derivative

2.1.1 Examples and Definitions

One example is about Velocity of Variable Rectilinear Motion. When an object undergoes uniform rectilinear motion, its velocity at any moment is the same, and it can be expressed using the ratio

$$\frac{\text{distance}}{\text{time}}$$

to calculate. If an object is undergoing uniformly accelerated rectilinear motion, its equation of motion is $s=s(t)$, the distance traveled by the object over the time interval $[t_0, t_0+\Delta t]$ is $\Delta s=s(t_0+\Delta t)-s(t_0)$, and the time taken is $(t_0+\Delta t)-t_0=\Delta t$. The ratio

$$\frac{\text{distance}}{\text{time}}=\frac{s(t_0+\Delta t)-s(t_0)}{\Delta t}$$

is a function of Δt, related to the magnitude of Δt, and it represents only the average speed of the object over the time period $[t_0, t_0+\Delta t]$, not the speed at any instant within that period. How can one determine the instantaneous speed of the object at a particular moment

t_0? The basic idea is: For uniformly accelerated rectilinear motion, although the speed appears to be variable overall, if the change in time Δt is very small, i.e., within a very short time period $[t_0, t_0+\Delta t]$ (assuming $\Delta t>0$), the change in speed is minimal and can be approximated as constant motion. Therefore, the average speed during this period can be considered an approximation of the speed of the object at time t_0. That is,

$$v(t_0)\approx\overline{v}=\frac{s(t_0+\Delta t)-s(t_0)}{\Delta t}=\frac{\Delta s}{\Delta t}.$$

Clearly, the smaller Δt is, the better the approximation. It is therefore natural to define the limit of $\overline{v}=\dfrac{\Delta s}{\Delta t}$ as $\Delta t\to 0$ approaches zero (if it exists) as the speed of the object at time t_0. That is,

$$v(t_0)=\lim_{\Delta t\to 0}\frac{\Delta s}{\Delta t}=\lim_{\Delta t\to 0}\frac{s(t_0+\Delta t)-s(t_0)}{\Delta t}.$$

Another example is about Slope of the Tangent Line. The tangent to a circle can be defined as "a line that intersects the circle at only one point", but this definition is clearly inappropriate for a general curve. As shown in Figure 2-1, the line l intersects the curve C at only point P, yet l is not the tangent to C. Whereas the line l_1 intersects the curve C at two points Q, R, but it is the tangent to the curve C at point R. For any curve, the tangent can be defined as the limiting position of a secant line.

As shown in Figure 2-2, let $M_0(x_0, y_0)$ be a fixed point on the curve $C: y=f(x)$, and $M(x_0+\Delta x, f(x_0+\Delta x))$ be another point on C. Draw the secant line M_0M. As point M along the curve C approaches point M_0, the secant line M_0M rotates around point M_0 and approaches a limiting position M_0T, which is called the tangent to the curve C at point M_0.

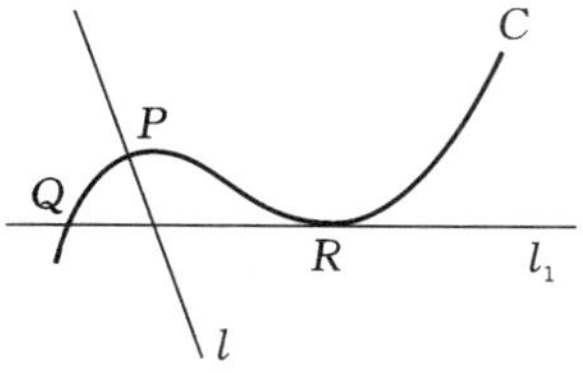

Figure 2-1

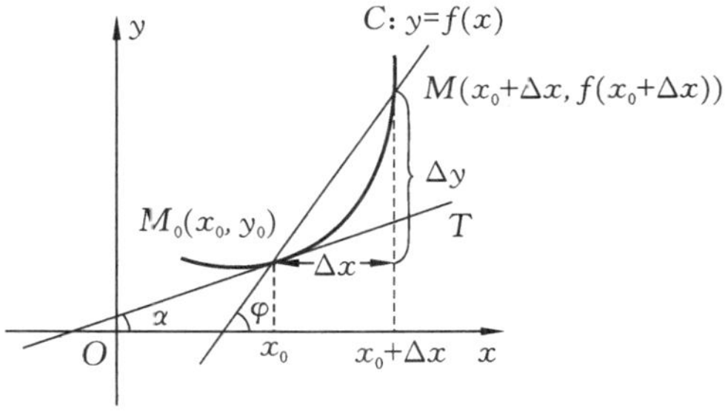

Figure 2-2

Obviously, the slope of the secant line M_0M is

$$\tan\varphi=\frac{\Delta y}{\Delta x}=\frac{f(x_0+\Delta x)-f(x_0)}{\Delta x}.$$

When the point M approaches the fixed point M_0 along the curve C (i.e., $\Delta x\to O$), the inclination of the secant line φ approaches the inclination of the tangent line M_0T (i.e., α). If $\lim\limits_{\varphi\to\alpha}\tan\varphi$ exists, the slope of the tangent line to the curve C at the point M_0 is defined.

$$k=\tan\alpha=\lim_{\varphi\to\alpha}\tan\varphi=\lim_{\Delta x\to 0}\frac{\Delta y}{\Delta x}=\lim_{\Delta x\to 0}\frac{f(x_0+\Delta x)-f(x_0)}{\Delta x}.$$

Although the specific meanings of the two examples differ, they share the same mathematical structure, which is the limit of the ratio of the function increment to the independent variable increment as the independent variable increment approaches zero.

Definition 2.1.1 Let the function $y=f(x)$ be defined in a neighborhood $U(x_0)$ of the point x_0. When x has an increment Δx $(x_0+\Delta x\in U(x_0))$ at the point x_0, the function y has a corresponding increment $\Delta y=f(x_0+\Delta x)-f(x_0)$. If

$$\lim_{\Delta x\to 0}\frac{\Delta y}{\Delta x}=\lim_{\Delta x\to 0}\frac{f(x_0+\Delta x)-f(x_0)}{\Delta x},$$

the limit exists, the function $y=f(x)$ is said to be differentiable at the point x_0, and this limit value is called the derivative of the function $y=f(x)$ at the point x_0, denoted as $f'(x_0)$ or $y'|_{x=x_0}$, or alternatively as $\left.\frac{dy}{dx}\right|_{x=x_0}$ or $\left.\frac{df(x)}{dx}\right|_{x=x_0}$. That is,

$$f'(x_0)=\lim_{\Delta x\to 0}\frac{\Delta y}{\Delta x}=\lim_{\Delta x\to 0}\frac{f(x_0+\Delta x)-f(x_0)}{\Delta x}.$$

If the limit of the ratio of the function increment to the independent variable increment $\frac{\Delta y}{\Delta x}$ does not exist when $\Delta x\to 0$, the function $f(x)$ is said to be non-differentiable at point x_0, or the derivative of $f(x)$ at point x_0 does not exist.

The increment Δx can be replaced with other symbols or analytic expressions. Thus, the definition of the derivative of function $f(x)$ at point x_0 can also take different forms. For example,

$$f'(x_0)=\lim_{h\to 0}\frac{f(x_0+h)-f(x_0)}{h}=\lim_{t\to 0}\frac{f(x_0-2t)-f(x_0)}{-2t}.$$

If $x=x_0+\Delta x$ is set, then $\Delta x\to 0$ is equivalent to $x\to x_0$, and another common form of the derivative is obtained:

$$f'(x_0)=\lim_{x\to x_0}\frac{f(x)-f(x_0)}{x-x_0}.$$

If function $y=f(x)$ is differentiable at every point within the open interval (a,b), then function $y=f(x)$ is said to be differentiable on the interval (a,b). In this case, for each point x within the interval, there is a definite derivative value $f'(x)$. $f'(x)$ varies with x and is a function of x, referred to as the derivative function of function $y=f(x)$, simply called the derivative. It is denoted as

$$y',f'(x),\frac{dy}{dx}\text{ or }\frac{df(x)}{dx}.$$

By replacing x_0 with x, the definition of the derivative function can be obtained:

$$f'(x)=\lim_{\Delta x\to 0}\frac{f(x+\Delta x)-f(x)}{\Delta x}.$$

Obviously, the derivative $f'(x_0)$ of function $f(x)$ at point x_0 is the function value of the derivative function $f'(x)$ at point x_0.

The ratio of the function increment to the independent variable increment $\frac{\Delta y}{\Delta x}=\frac{f(x+\Delta x)-f(x)}{\Delta x}$, representing the average rate at which the function changes over the interval ending at $x.x+\Delta x$, is called the average rate of change of the function. The derivative $f'(x)$ is the limit of the average rate of change $\frac{\Delta y}{\Delta x}$ of the function $f(x)$ as $\Delta x\to 0$, hence it is also referred to as the rate of change of the function at point x. It reflects the rate at which the function changes with respect to the independent variable. When the function has different practical meanings, its rate of change takes on various practical significance.

For example, when a particle undergoes variable motion, its instantaneous velocity is the derivative of the distance function $s=s(t)$ with respect to time t, i.e., $v(t)=\frac{\mathrm{d}s}{\mathrm{d}t}$. The slope of the tangent to the curve $y=f(x)$ at point $(x_0, f(x_0))$ is the derivative of the function $f(x)$ at point x_0, i.e., $k=f'(x_0)$.

2.1.2 Elementary Functions' Derivative

Calculating derivatives based on the definition.

Example 2.1.1 Find the derivative of the constant function $y=C$.

Solution $y'=\lim\limits_{\Delta x\to 0}\frac{f(x+\Delta x)-f(x)}{\Delta x}=\lim\limits_{\Delta x\to 0}\frac{C-C}{\Delta x}=0$, i.e., $C'=0$.

This means that the derivative of a constant is zero.

Example 2.1.2 Find the derivative of the function $y=x^3$.

Solution $y'=\lim\limits_{\Delta x\to 0}\frac{f(x+\Delta x)-f(x)}{\Delta x}=\lim\limits_{\Delta x\to 0}\frac{(x+\Delta x)^3-x^3}{\Delta x}$

$$=\lim_{\Delta x\to 0}\frac{3x^2\Delta x+3x\Delta x^2+\Delta x^3}{\Delta x}=\lim_{\Delta x\to 0}(3x^2+3x\Delta x+\Delta x^2)=3x^2.$$

Therefore,

$$(x^3)'=3x^2.$$

The above result can be generalized to $(x^n)'=nx^{n-1}$.

Indeed, for any power function $y=x^\alpha$ ($\alpha\in\mathbf{R}$), there exists $(x^\alpha)'=\alpha x^{\alpha-1}$. This formula will be proven in the next section.

Specifically, by taking $\alpha=1,\frac{1}{2},-1$, we obtain

$$x'=1,\ (\sqrt{x})'=\frac{1}{2\sqrt{x}},\ \left(\frac{1}{x}\right)'=-\frac{1}{x^2}.$$

Example 2.1.3 Find the derivative of the function $f(x)=\sin x$.

Solution　$f'(x)=\lim\limits_{h\to 0}\dfrac{f(x+h)-f(x)}{h}=\lim\limits_{h\to 0}\dfrac{\sin(x+h)-\sin x}{h}$

$$=\lim_{h\to 0}\frac{2\cos\left(x+\frac{h}{2}\right)\sin\frac{h}{2}}{h}=\lim_{h\to 0}\frac{\sin\frac{h}{2}}{\frac{h}{2}}\cos\left(x+\frac{h}{2}\right)=\cos x.$$

That is

$$(\sin x)'=\cos x.$$

Using the same method, it can be deduced that

$$(\cos x)'=-\sin x.$$

Example 2.1.4　Find the derivative of the function $y=a^x\ (a>0, a\neq 1)$.

Solution　$y'=\lim\limits_{h\to 0}\dfrac{f(x+h)-f(x)}{h}=\lim\limits_{h\to 0}\dfrac{a^{x+h}-a^x}{h}=a^x\lim\limits_{h\to 0}\dfrac{a^h-1}{h}$.

And $a^h=e^{h\ln a}$, when $h\to 0, h\ln a\to 0, e^{h\ln a}-1\sim h\ln a$, therefore

$$y'=a^x\lim_{h\to 0}\frac{e^{h\ln a}-1}{h}=a^x\lim_{h\to 0}\frac{h\ln a}{h}=a^x\ln a.$$

That is

$$(a^x)'=a^x\ln a.$$

Specifically, when $a=e$, there is

$$(e^x)'=e^x.$$

Example 2.1.5　Find the derivative of the function $y=\ln x$.

$$y'=\lim_{h\to 0}\frac{f(x+h)-f(x)}{h}=\lim_{h\to 0}\frac{\ln(x+h)-\ln x}{h}=\lim_{h\to 0}\frac{\ln\left(1+\frac{h}{x}\right)}{h}=\lim_{h\to 0}\frac{\frac{h}{x}}{h}=\frac{1}{x}.$$

That is

$$(\ln x)'=\frac{1}{x}.$$

Generally, when $a>0$ and $a\neq 1$, there is

$$(\log_a x)'=\frac{1}{x\ln a}.$$

By definition, the derivative

$$f'(x_0)=\lim_{\Delta x\to 0}\frac{\Delta y}{\Delta x}=\lim_{\Delta x\to 0}\frac{f(x_0+\Delta x)-f(x_0)}{\Delta x}$$

is the limit of the ratio $\dfrac{\Delta y}{\Delta x}$ as $\Delta x\to 0$. From the relationship between the limit of a function at a point and its left and right limits, it is known that $f'(x_0)$ exists if and only if both $\lim\limits_{\Delta x\to 0^-}\dfrac{\Delta y}{\Delta x}$ and $\lim\limits_{\Delta x\to 0^+}\dfrac{\Delta y}{\Delta x}$ exist and are equal. These two limits are respectively called the left derivative and the right derivative of the function $f(x)$ at point x_0, denoted as $f'_-(x_0)$ and $f'_+(x_0)$, that is

$$f'_-(x_0)=\lim_{\Delta x\to 0^-}\frac{\Delta y}{\Delta x}=\lim_{\Delta x\to 0^-}\frac{f(x_0+\Delta x)-f(x_0)}{\Delta x}=\lim_{x\to x_0}\frac{f(x)-f(x_0)}{x-x_0},$$

$$f'_+(x_0)=\lim_{\Delta x\to 0^+}\frac{\Delta y}{\Delta x}=\lim_{\Delta x\to 0^+}\frac{f(x_0+\Delta x)-f(x_0)}{\Delta x}=\lim_{x\to x_0^+}\frac{f(x)-f(x_0)}{x-x_0}.$$

The left derivative and the right derivative are collectively referred to as one-sided derivatives. Obviously, $f'(x_0)$ exists if and only if both $f'_-(x_0)$ and $f'_+(x_0)$ exist, and $f'_-(x_0)=f'_+(x_0)$. At this time, $f'(x_0)=f'_-(x_0)=f'_+(x_0)$.

If the function $f(x)$ is differentiable within the open interval (a,b), and both $f'_+(a)$ and $f'_-(b)$ exist, then the function $f(x)$ is said to be differentiable on the closed interval $[a,b]$.

Example 2.1.6 Discuss the differentiability of the piece-wise function

$$f(x)=\begin{cases}3x, & x<0\\ x^2, & 0\leqslant x<1\\ 2x-1, & x\geqslant 1\end{cases}\text{, at point } x=0, x=1.$$

Solution When $x=0$, $f(0)=0$,

$$f'_-(0)=\lim_{x\to 0^-}\frac{f(x)-f(0)}{x-0}=\lim_{x\to 0^-}\frac{3x-0}{x}=3,$$

$$f'_+(0)=\lim_{x\to 0^+}\frac{f(x)-f(0)}{x-0}=\lim_{x\to 0^+}\frac{x^2-0}{x}=0.$$

Since $f'_-(0)\neq f'_+(0)$, $f(x)$ is not differentiable at point $x=0$.

When $x=1$, $f(1)=2\cdot 1-1=1$,

$$f'_-(1)=\lim_{x\to 1^-}\frac{f(x)-f(1)}{x-1}=\lim_{x\to 1^-}\frac{x^2-1}{x-1}=\lim_{x\to 1}(x+1)=2,$$

$$f'_+(1)=\lim_{x\to 1^+}\frac{f(x)-f(1)}{x-1}=\lim_{x\to 1^+}\frac{(2x-1)-1}{x-1}=2,$$

$f'_-(1)=f'_+(1)$, so $f(x)$ is differentiable at point $x=1$, and $f'(1)=2$.

Note: $f'(x_0)$ is the value of the derivative function $f'(x)$ at point $x=x_0$. For ordinary elementary functions, one typically first finds the derivative function $f'(x)$ and then substitutes $x=x_0$ to obtain $f'(x_0)$. However, for piece-wise functions at the break points, one must discuss according to the definition of the derivative.

2.1.3 Geometrical Meaning of the Derivative

From the introductory example and the definition of the derivative, it is known that the derivative $f'(x_0)$ of the function $y=f(x)$ at point $x=x_0$ geometrically represents the slope k of the tangent to the curve $y=f(x)$ at the corresponding point $(x_0, f(x_0))$. That is,

$$k=f'(x_0).$$

If $f'(x_0)$ exists, according to the point-slope form of the equation of a line, the tangent line equation to the curve $y=f(x)$ at point $(x_0, f(x_0))$ is

$$y-f(x_0)=f'(x_0)(x-x_0).$$

The line passing through the point of tangent $(x_0, f(x_0))$ and perpendicular to the tangent line is called the normal line to the curve at that point. If $f'(x_0)\neq 0$, then the slope of the normal line is $-\frac{1}{f'(x_0)}$, and thus the equation of the normal line is

$$y-f(x_0)=-\frac{1}{f'(x_0)}(x-x_0).$$

Example 2.1.7 Find the equation of the tangent line and the normal line to the curve $y=\ln x$ at the point $(e,1)$.

Solution Since $y'=\frac{1}{x}$, therefore, at the point $(e,1)$, the slope of the tangent line is

$$k=y'|_{x=e}=\frac{1}{e},$$

the equation of the tangent line is

$$y-1=\frac{1}{e}(x-e),$$

that is

$$x-ey=0.$$

The slope of the normal line sought is $k'=-e$, and the equation of the normal line is

$$y-1=-e(x-e),$$

that is

$$ex+y-e^2-1=0.$$

2.1.4 Relationship Between Differentiability and Continuity

Theorem 2.1.1 If the function $y=f(x)$ is differentiable at the point x_0, then the function $y=f(x)$ is necessarily continuous at the point x_0.

Proof Since $f'(x_0)=\lim\limits_{\Delta x\to 0}\frac{\Delta y}{\Delta x}$ exists, as $\Delta x\neq 0$ approaches zero, $\Delta y=\frac{\Delta y}{\Delta x}\cdot\Delta x$.

Therefore,

$$\lim_{\Delta x\to 0}\Delta y=\lim_{\Delta x\to 0}\left(\frac{\Delta y}{\Delta x}\cdot\Delta x\right)=\lim_{\Delta x\to 0}\frac{\Delta y}{\Delta x}\cdot\lim_{\Delta x\to 0}\Delta x=f'(x_0)\cdot 0=0,$$

hence, the function $y=f(x)$ is continuous at the point x_0.

Note: Continuity of a function at a point is a necessary condition for its differentiability at that point, but not a sufficient condition. That is, the converse of the above theorem does not hold.

Example 2.1.8 Discuss the continuity and differentiability of the function $f(x)=|x|$ at the point $x=0$.

Solution

(1) Continuity. Since

$$\lim_{x\to 0} f(x)=\lim_{x\to 0}|x|=0=f(0),$$

therefore, $f(x)$ is continuous at the point $x=0$.

(2) Differentiability. Since

$$f'_-(0)=\lim_{x\to 0^-}\frac{f(x)-f(0)}{x-0}=\lim_{x\to 0^-}\frac{|x|-0}{x-0}=\lim_{x\to 0^-}\frac{-x}{x}=-1,$$

$$f'_+(0)=\lim_{x\to 0^+}\frac{f(x)-f(0)}{x-0}=\lim_{x\to 0^+}\frac{|x|-0}{x}=\lim_{x\to 0^+}\frac{x}{x}=1.$$

Therefore, $f'_-(0)\neq f'_+(0)$, hence $f(x)=|x|$ is not differentiable at point $x=0$.

As shown in Figure 2-3, the curve $y=|x|$ has no tangent line at point $x=0$.

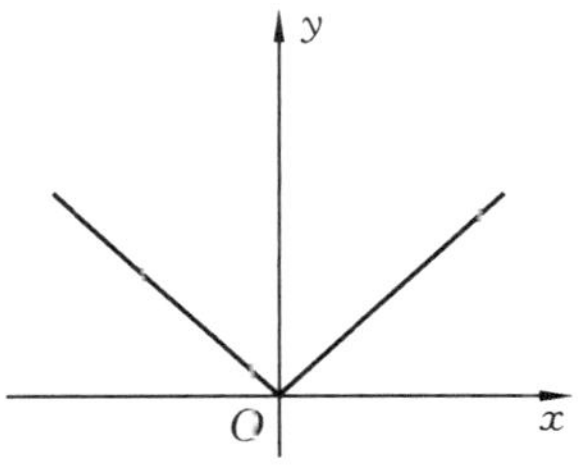

Figure 2-3

Example 2.1.9 Discusses the continuity and differentiability of function $f(x)=\sqrt[3]{x}$ at point $x=0$.

Solution Since $f(x)=\sqrt[3]{x}$ is an elementary function, it is continuous within its defined interval, hence it is continuous at point $x=0$. However,

$$\lim_{x\to 0}\frac{f(x)-f(0)}{x-0}=\lim_{x\to 0}\frac{\sqrt[3]{x}-0}{x}=\lim_{x\to 0}\frac{1}{\sqrt[3]{x^2}}=\infty.$$

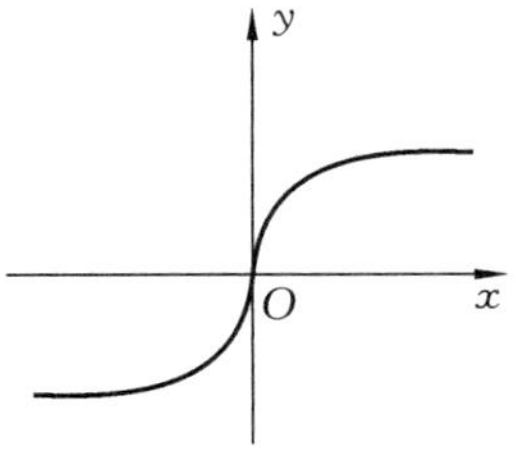

Figure 2-4

Therefore, $f(x)=\sqrt[3]{x}$ is not differentiable at point $x=0$, as shown in Figure 2-4.

If function $y=f(x)$ is continuous but not differentiable at point x_0, but $\lim\limits_{x\to x_0}\frac{\Delta y}{\Delta x}=\infty$, this can be written as $f'(x_0)=\infty$.

At this point, the curve $y=f(x)$ has a tangent line perpendicular to the x axis at the corresponding point $(x_0, f(x_0))$. As shown in Figure 2-4.

Exercise 2.1

1. A ball is thrown from a bridge into the air, and after t seconds, its height relative to the ground y (in units: m) is

$$y=f(t)=-5t^2+15t+12.$$

(1) What is the average velocity of the ball during the 1 s interval?

(2) Calculate the velocity of the ball at $t=1$ s.

(3) Draw the graph of $y=f(t)$, find the maximum height the ball can reach. What is the velocity of the ball at its maximum height?

2. When the temperature of an object is higher than that of the surrounding medium,

the object continues to cool. If the relationship between the object's temperature T and time t is given by $T=T(t)$, how should the cooling rate of the object at time t be determined?

3. Given $f(x)=6x^2-2x+3$, find $f'(2)$, $f'(-1)$ and $[f(-1)]'$.

4. Use the definition to find the derivative of the following functions.

(1) $y=mx+b$; (2) $y=\frac{1}{x^2}$; (3) $y=x^2+3x-1$.

5. Find the equation of the tangent line and the normal line to the curve $y=\ln x$ at point $(e,1)$.

6. Find the points on the parabola $y=x^2$ such that the tangent line at these points is parallel to the line $y=4x-2$.

7. Let $f(x)$ be an odd function, and $f'(x_0)=3$, find $f'(-x_0)$.

8. Prove:

(1) The derivative of a differentiable even function is an odd function;

(2) The derivative of a differentiable odd function is an even function.

9. Let $f(x)$ be differentiable at points x_0 and $2x_0$. Find the following limits:

(1) $\lim\limits_{x\to x_0}\frac{f(2x)-f(2x_0)}{x-x_0}$;

(2) $\lim\limits_{\Delta x\to 0}\frac{f(x_0-\Delta x)-f(x_0)}{\Delta x}$;

(3) $\lim\limits_{h\to 0}\frac{f(x_0+mh)-f(x_0-nh)}{h}$ (m, n are constants);

(4) $\lim\limits_{\Delta x\to 0}\frac{f^2(x_0+\Delta x)-f^2(x_0)}{\Delta x}$.

10. Discuss the continuity and differentiability of the following functions at the specified points:

(1) $y=|\sin x|$ at point $x=0$;

(2) $y=\begin{cases} x^2\sin\frac{1}{x}, & x\neq 0 \\ 0, & x=0 \end{cases}$, at point $x=0$.

(3) $y=\begin{cases} x^2, & x\leqslant 0 \\ xe^x, & x>0 \end{cases}$, at the point $x=0$;

(4) $y=\begin{cases} 2x+1, & x\leqslant 1 \\ x^2+2, & x>1 \end{cases}$, at the point $x=1$.

11. Let $f(x)$ be a function defined within $(-\infty,+\infty)$, and for any $x_1, x_2\in\mathbf{R}$, there exists

$$f(x_1+x_2)=f(x_1)f(x_2).$$

If $f'(0)=1$, prove that for any $x\in(-\infty,+\infty)$, there exists $f'(x)=f(x)$.

12. Prove that the area of the triangle formed by the tangent to the curve $xy=1$ at any

point and the two coordinate axes is equal to 2.

13. If the function $f(x)$ is continuous at the point $x=0$, and $\lim\limits_{x\to 0}\frac{f(x)}{x}$ exists, prove that $f(x)$ is differentiable at the point $x=0$.

2.2 Rules of Differentiation

In the previous section, the derivatives of several basic elementary functions were derived using the definition of the derivative. However, for more complex functions, finding their derivatives based on the definition can often be very cumbersome. Since elementary functions are common, and elementary functions are constructed from basic elementary functions through finite arithmetic operations and composition operations, this section will provide the rules for differentiating functions through arithmetic and composition operations.

2.2.1 Rules for Differentiating Arithmetic Operations

Theorem 2.2.1 If the functions $u=u(x)$, $v=v(x)$ are differentiable at the point x, then their sum, difference, product and quotient (excluding points where the denominator is zero) are also differentiable at point x, and

(1) $(u\pm v)'=u'\pm v'$;

(2) $(uv)'=u'v+uv'$;

(3) $\left(\frac{u}{v}\right)'=\frac{u'v-uv'}{v^2}$ $(v\neq 0)$.

Proof Here, we only prove the differentiation rule for products.

Because $uv=u(x)v(x)$,

$$\frac{\Delta(uv)}{\Delta x}=\frac{u(x+\Delta x)v(x+\Delta x)-u(x)v(x)}{\Delta x}$$
$$=\frac{[u(x+\Delta x)v(x+\Delta x)-u(x)v(x+\Delta x)]+[u(x)v(x+\Delta x)-u(x)v(x)]}{\Delta x}$$
$$=\frac{u(x+\Delta x)-u(x)}{\Delta x}v(x+\Delta x)+u(x)\frac{v(x+\Delta x)-v(x)}{\Delta x}.$$

Since $v(x)$ is differentiable at point x, it follows that $v(x)$ is continuous at point x, thus

$$\lim_{\Delta x\to 0}v(x+\Delta x)=v(x).$$

Given that both $u(x)$, $v(x)$ are differentiable at point x and by the definition of the derivative, we know

$$\lim_{\Delta x\to 0}\frac{u(x+\Delta x)-u(x)}{\Delta x}=u'(x),\ \lim_{\Delta x\to 0}\frac{v(x+\Delta x)-v(x)}{\Delta x}=v'(x).$$

Taking the limit on both sides of this equation, we obtain

$$(uv)'=\lim_{\Delta x\to 0}\frac{\Delta(uv)}{\Delta x}=u'(x)v(x)+u(x)v'(x)=u'v+uv'.$$

Similarly, rules (1) and (3) can be proven.

It is easy to see that the differentiation rules for sums and products in the theorem can be extended to finite sums and products of multiple functions.

Corollary 2.2.1 If $u_1, u_2, \cdots, u_n$ are all differentiable functions, then

(1) $(u_1 \pm u_2 \pm \cdots \pm u_n)'=u_1' \pm u_2' \pm \cdots \pm u_n'$;

(2) $(u_1u_2\cdots u_n)'=u_1'u_2\cdots u_n+u_1u_2'\cdots u_n+\cdots+u_1u_2\cdots u_n'$.

In the rule (2) of Theorem 2.2.1, setting $u(x)=k$ yields Corollary 2.2.1.

Corollary 2.2.2 If $u=u(x)$ is differentiable and k is any constant, then ku is differentiable, and

$$(ku)'=ku'.$$

Example 2.2.1 Find the derivative of the function $y=3x^2-x+\ln 2$.

Solution

$$y'=(3x^2-x+\ln 2)'=(3x^2)'-(x)'+(\ln 2)'=3(x^2)'-1-0=6x-1.$$

Example 2.2.2 Given $y=2e^x-\frac{1}{x}+3\sqrt[3]{x}+\cos\frac{\pi}{5}$, find y' and $y'|_{x=1}$.

Solution

$$y'=\left(2e^x-\frac{1}{x}+3\sqrt[3]{x}+\cos\frac{\pi}{5}\right)'=(2e^x)'-\left(\frac{1}{x}\right)'+(3\sqrt[3]{x})'+\left(\cos\frac{\pi}{5}\right)'$$

$$=2e^x+\frac{1}{x^2}+3\cdot\frac{1}{3}x^{-\frac{2}{3}}+0=2e^x+\frac{1}{x^2}+\frac{1}{\sqrt[3]{x^2}},$$

$$y'|_{x=1}=2e^1+\frac{1}{1^2}+\frac{1}{\sqrt[3]{1^2}}=2e+2.$$

Example 2.2.3 Find the derivative of the function $y=2^x\sin x$.

Solution

$$y'=(2^x\sin x)'=(2^x)'\sin x+2^x(\sin x)'=2^x\ln 2\cdot\sin x+2^x\cos x.$$

Example 2.2.4 Find the derivative of the function $y=\frac{\cos x}{1+\sin x}$.

Solution

$$y'=\left(\frac{\cos x}{1+\sin x}\right)'=\frac{(\cos x)'(1+\sin x)-\cos x(1+\sin x)'}{(1+\sin x)^2}$$

$$=\frac{-\sin x(1+\sin x)-\cos x\cos x}{(1+\sin x)^2}=-\frac{1}{1+\sin x}.$$

Example 2.2.5 Find the derivative of $y=\tan x$.

Solution

$$y'=(\tan x)'=\left(\frac{\sin x}{\cos x}\right)'=\frac{(\sin x)'\cos x-\sin x(\cos x)'}{\cos^2 x}=\frac{\cos^2 x+\sin^2 x}{\cos^2 x}=\frac{1}{\cos^2 x}=\sec^2 x.$$

That is

$$(\tan x)'=\sec^2 x.$$

Example 2.2.6 Find the derivative of $y=\sec x$.

Solution

$$y'=(\sec x)'=\left(\frac{1}{\cos x}\right)'=\frac{1'\cdot\cos x-1\cdot(\cos x)'}{\cos^2 x}=\frac{\sin x}{\cos^2 x}=\sec x\tan x.$$

That is

$$(\sec x)'=\sec x\tan x.$$

Similarly, one can obtain

$$(\cot x)'=-\csc^2 x,\ (\csc x)'=-\csc x\cot x.$$

2.2.2 Derivative Rules for Inverse Functions

Theorem 2.2.2 Let the function $y=f(x)$ be continuous and strictly monotonic in a neighborhood of point x. If the function $y=f(x)$ is differentiable at point x and $f'(x)\neq 0$, then its inverse function $x=\varphi(y)$ is differentiable at the corresponding point $y=f(x)$, and

$$[\varphi(y)]'=\frac{1}{f'(x)}\ \text{or}\ \frac{dx}{dy}=\frac{1}{\frac{dy}{dx}}.$$

Proof Since $y=f(x)$ is continuous at point x, let x take an increment $\Delta x\ (\Delta x\neq 0)$, so when $\Delta x\to 0$, there is $\Delta y=f(x+\Delta x)-f(x)\to 0$.

Because $y=f(x)$ is continuous and strictly monotonic, its inverse function $x=\varphi(y)$ is continuous and strictly monotonic at the corresponding point y. When $\Delta y\neq 0$, there is $\Delta x\neq 0$, when $\Delta y\to 0$, there is $\Delta x\to 0$. Thus,

$$\frac{\Delta x}{\Delta y}=\frac{1}{\frac{\Delta y}{\Delta x}}.$$

Also, since $y=f(x)$ is differentiable at point x and $f'(x)\neq 0$, we have

$$\lim_{\Delta y\to 0}\frac{\Delta x}{\Delta y}=\lim_{\Delta y\to 0}\frac{1}{\frac{\Delta y}{\Delta x}}=\frac{1}{\lim\limits_{\Delta x\to 0}\frac{\Delta y}{\Delta x}}=\frac{1}{f'(x)}.$$

That is

$$\varphi'(y)=\frac{1}{f'(x)}.$$

This indicates that the derivative of the inverse function is the reciprocal of the derivative of the direct function.

Example 2.2.7 Find the derivative of the function $y=\arcsin x\ (-1<x<1)$.

Solution Since the inverse function of $y=\arcsin x\ (-1<x<1)$ is

$$x=\sin y\quad\left(-\frac{\pi}{2}<y<\frac{\pi}{2}\right).$$

Clearly

$$\frac{dx}{dy}=(\sin y)'=\cos y>0.$$

By Theorem 2.2.2,

$$\frac{dy}{dx}=\frac{1}{\frac{dx}{dy}}=\frac{1}{\cos y}=\frac{1}{\sqrt{1-x^2}},$$

that is,

$$(\arcsin x)'=\frac{1}{\sqrt{1-x^2}},$$

similarly,

$$(\arccos x)'=-\frac{1}{\sqrt{1-x^2}}.$$

Example 2.2.8 Find the derivative of the function $y=\arctan x$.

Solution Since $y=\arctan x$ is the inverse function of $x=\tan y$, and

$$\frac{dx}{dy}=(\tan y)'=\sec^2 y=1+\tan^2 y=1+x^2,$$

by Theorem 2.2.2,

$$\frac{dy}{dx}=\frac{1}{\frac{dx}{dy}}=\frac{1}{1+x^2},$$

that is,

$$(\arctan x)'=\frac{1}{1+x^2},$$

similarly,

$$(\text{arccot}\, x)'=-\frac{1}{1+x^2}.$$

2.2.3 Chain Rules for Composite Functions

How to find the derivative of a composite function, the chain rule states:

Theorem 2.2.3 If the function $u=\varphi(x)$ is differentiable at point x, and $y=f(u)$ is differentiable at point $u=\varphi(x)$, then the composite function $y=f[\varphi(x)]$ is differentiable at point x, and

$$\{f[\varphi(x)]\}'=f'(u)\varphi'(x) \text{ or } \frac{dy}{dx}=\frac{dy}{du}\frac{du}{dx}.$$

Proof Let x take an increment Δx, then u takes a corresponding increment Δu, and thus y takes a corresponding increment Δy.

Since $\lim\limits_{\Delta x\to 0}\frac{\Delta u}{\Delta x}=\varphi'(x)$, when $\Delta x\to 0$,

$$\frac{\Delta u}{\Delta x}=\varphi'(x)+\alpha(\Delta x), \lim_{\Delta x\to 0}\alpha(\Delta x)=0,$$

$$\Delta u=\varphi'(x)\cdot\Delta x+o(\Delta x).$$

Similarly, when $\Delta x \to 0$,

$$\Delta y = f'(u) \cdot \Delta u + \beta(\Delta u) \cdot \Delta u, \lim_{\Delta x \to 0} \beta(\Delta u) = 0,$$

$$\lim_{\Delta x \to 0} \frac{\Delta y}{\Delta x} = \lim_{\Delta x \to 0} \frac{\Delta y}{\Delta x} \left[f'(u) \cdot \frac{\Delta u}{\Delta x} + \beta(\Delta u) \cdot \frac{\Delta u}{\Delta x} \right] = \lim_{\Delta x \to 0} f'(u) \cdot \frac{\Delta u}{\Delta x} + \lim_{\Delta x \to 0} \frac{\Delta u}{\Delta x} \cdot \lim_{\Delta x \to 0} \beta(\Delta u)$$

$$= f'(u) \cdot \varphi'(x) + f'(x) \cdot \lim_{\Delta x \to 0} \beta(\Delta u) = f'(u) \cdot \varphi'(x) + \varphi'(x) \cdot 0 = f'(u) \cdot \varphi'(x).$$

That is,

$$\{f[\varphi(x)]\}' = f'(u)\varphi'(x) \text{ or } \frac{dy}{dx} = \frac{dy}{du}\frac{du}{dx}.$$

When $\Delta u = 0$, it can be proven that the above equation still holds.

Theorem 2.2.3 indicates that the derivative of the composite function y is equal to the derivative of y with respect to the intermediate variable u multiplied by the derivative of the intermediate variable u with respect to the independent variable x.

When there are multiple layers of composition, a similar chain rule applies. For example, if $y = f(u)$, $u = \varphi(v)$, $v = \psi(x)$ are all differentiable, then the composite function $y = f\{\varphi[\psi(x)]\}$ is differentiable, and $\frac{dy}{dx} = \frac{dy}{du}\frac{du}{dv}\frac{dv}{dx}$.

Example 2.2.9 Find the derivative of the function $y = \ln\cos x$.

Solution $y = \ln\cos x$ can be considered as a composition of the function $y = \ln u$, $u = \cos x$. By the chain rule, we have

$$\frac{dy}{dx} = \frac{dy}{du}\frac{du}{dx} = (\ln u)'(\cos x)' = \frac{1}{u}(-\sin x) = \frac{1}{\cos x}(-\sin x) = -\tan x.$$

Example 2.2.10 Find the derivative of the function $y = e^{x^2 - 4x}$.

Solution $y = e^{x^2 - 4x}$ can be considered as a composition of $y = e^u$, $u = x^2 - 4x$. By the chain rule, we have

$$\frac{dy}{dx} = \frac{dy}{du}\frac{du}{dx} = (e^u)'(x^2 - 4x)' = e^u(2x - 4) = (2x - 4)e^{x^2 - 4x}.$$

Example 2.2.11 Find the derivative of the function $y = \sin^2 x$.

Solution $y = \sin^2 x$ can be considered as a composition of $y = u^2$, $u = \sin x$, so

$$\frac{dy}{dx} = \frac{dy}{du}\frac{du}{dx} = (u^2)'(\sin x)' = 2u\cos x = 2\sin x\cos x = \sin 2x.$$

After becoming proficient in decomposing composite functions, intermediate variables can be omitted, and the chain rule can be directly applied to express the derivative of the function as the product of the derivative of the function with respect to the intermediate variable and the derivative of the intermediate variable.

Example 2.2.12 Find the derivative of the function $y = \tan\frac{1}{x}$.

Solution $y' = \left(\tan\frac{1}{x}\right)' = \sec^2\frac{1}{x} \cdot \left(\frac{1}{x}\right)' = -\frac{1}{x^2}\sec^2\frac{1}{x}.$

Example 2.2.13 Find the derivative of the function $y=\sqrt[3]{x^2-5}$.

Solution

$$y'=[(x^2-5)^{\frac{1}{3}}]'=\frac{1}{3}(x^2-5)^{-\frac{2}{3}}(x^2-5)'=\frac{1}{3}(x^2-5)^{-\frac{2}{3}}\cdot 2x=\frac{2}{3}x\,(x^2-5)^{-\frac{2}{3}}.$$

Example 2.2.14 Given $y=\arctan\frac{1-x}{1+x}$, find y'.

Solution

$$y'=\left(\arctan\frac{1-x}{1+x}\right)'=\frac{1}{1+\left(\frac{1-x}{1+x}\right)^2}\left(\frac{1-x}{1+x}\right)'$$

$$=\frac{(1+x)^2}{(1+x)^2+(1-x)^2}\frac{(1-x)'(1+x)-(1-x)(1+x)'}{(1+x)^2}=\frac{-2}{2+2x^2}=-\frac{1}{1+x^2}.$$

If the composite function involves more than one intermediate variable, i.e., multiple layers of composition, the chain rule can still be applied without explicitly writing out the intermediate variables. Derivatives should be taken layer by layer from the outermost to the innermost, ensuring no omissions or repetitions, until the derivative with respect to the independent variable is obtained.

Example 2.2.15 Given $y=\ln\cos x^2$, find y'.

Solution

$$y'=(\ln\cos x^2)'=\frac{1}{\cos x^2}(\cos x^2)'$$

$$=\frac{1}{\cos x^2}(-\sin x^2)(x^2)'=-\tan x^2\cdot 2x=-2x\tan x^2.$$

Example 2.2.16 Given $y=e^{\arcsin\sqrt{x}}$, find y'.

Solution

$$y'=(e^{\arcsin\sqrt{x}})'=e^{\arcsin\sqrt{x}}(\arcsin\sqrt{x})'=e^{\arcsin\sqrt{x}}\frac{1}{\sqrt{1-x}}(\sqrt{x})'$$

$$=e^{\arcsin\sqrt{x}}\frac{1}{\sqrt{1-x}}\frac{1}{2\sqrt{x}}=\frac{1}{2\sqrt{x}\sqrt{1-x}}e^{\arcsin\sqrt{x}}.$$

Example 2.2.17 Given $y=\ln\left(x+\sqrt{1+x^2}\right)$, find y'.

Solution

$$y'=\frac{1}{x+\sqrt{1+x^2}}(x+\sqrt{1+x^2})'=\frac{1}{x+\sqrt{1+x^2}}[x'+(\sqrt{1+x^2})']$$

$$=\frac{1}{x+\sqrt{1+x^2}}\left[1+\frac{1}{2\sqrt{1+x^2}}(1+x^2)'\right]$$

$$=\frac{1}{x+\sqrt{1+x^2}}\left(1+\frac{x}{\sqrt{1+x^2}}\right)=\frac{1}{\sqrt{1+x^2}}.$$

Example 2.2.18 Given $f(u)$ is differentiable, and $y=[xf(x^2)]^2$, find $\frac{dy}{dx}$.

Solution

$$\frac{dy}{dx}=2[xf(x^2)]\cdot[xf(x^2)]'=2xf(x^2)[f(x^2)+xf'(x^2)\cdot 2x]$$
$$=2xf^2(x^2)+4x^3f(x^2)f'(x^2).$$

2.2.4 Basic Derivative Formulas

For ease of memorization, the derivative formulas for basic elementary functions are summarized as follows:

(1) $(C')=0$ (C is a constant);
(2) $(x^\alpha)'=\alpha x^{\alpha-1}$ (α is any real number);
(3) $(a^x)'=a^x\ln a$ $(a>0, a\neq 1)$;
(4) $(e^x)'=e^x$;
(5) $(\log_a x)'=\frac{1}{x\ln a}$ $(a>0, a\neq 1)$;
(6) $(\ln x)'=\frac{1}{x}$;
(7) $(\sin x)'=\cos x$;
(8) $(\cos x)'=-\sin x$;
(9) $(\tan x)'=\sec^2 x$;
(10) $(\cot x)'=-\csc^2 x$;
(11) $(\sec x)'=\sec x\tan x$;
(12) $(\csc x)'=-\csc x\cot x$;
(13) $(\arcsin x)'=\frac{1}{\sqrt{1-x^2}}(-1<x<1)$;
(14) $(\arccos x)'=-\frac{1}{\sqrt{1-x^2}}(-1<x<1)$;
(15) $(\arctan x)'=\frac{1}{1+x^2}$;
(16) $(\text{arccot}\, x)'=-\frac{1}{1+x^2}$.

Note:

(1) When $x>0$, since $x^\alpha=(e^{\ln x})^\alpha=e^{\alpha\ln x}$, therefore

$$(x^\alpha)'=(e^{\alpha\ln x})'=e^{\alpha\ln x}(\alpha\ln x)'=x^\alpha\cdot\alpha\cdot\frac{1}{x}=\alpha x^{\alpha-1}.$$

(2) Formula (6) can be generalized to $(\ln|x|)'=\frac{1}{x}$. Please prove this yourself.

Exercise 2.2

1. Find the derivatives of the following functions:

(1) $y=2\sqrt{x}-\frac{1}{x}+4\sqrt{3}$;
(2) $y=\frac{1-x^2}{\sqrt{x}}$;
(3) $y=2^x(x^2-3x+1)$;
(4) $y=t^a-a^t+\ln t-\sin t$;
(5) $y=\sqrt{x\sqrt{x\sqrt{x}}}$;
(6) $y=\frac{1+x-x^2}{1-x+x^2}$;
(7) $y=\frac{1+\sin t}{1+\cos t}$;
(8) $y=\frac{1}{\csc x+\cot x}$;
(9) $y=\frac{\sin x}{x}+\frac{x}{\sin x}$;
(10) $y=xe^x\cos x$.

2. Find the derivative of the following functions at the given points:

(1) $f(x)=\dfrac{x}{\cos x}$, find $f'(0)$, $f'(\pi)$;

(2) $y=t^2-3\sin t+\tan\dfrac{\pi}{5}$, find $y'\Big|_{t=\frac{\pi}{3}}$.

3. At which point does the tangent to the curve $y=x^{\frac{3}{2}}$ parallel the line $y=3x-1$?

4. Let $x=\varphi(y)$ and $y=f(x)$ be inverse functions of each other, $\varphi(2)=1$, and $f'(1)=3$, find $\varphi'(2)$.

5. Find the derivative of the following functions:

(1) $y=\cos(\ln x)$; (2) $y=(1-x^2)^{100}$;

(3) $y=\sqrt[3]{4-3x}$; (4) $y=e^{-x2+2x}$;

(5) $y=x^2\sin\dfrac{1}{x}$; (6) $y=\ln\sqrt{x}+\sqrt{\ln x}$;

(7) $y=\dfrac{x}{\sqrt{1-x^2}}$; (8) $y=\sin(nx)\cos''x$;

(9) $y=\arctan e^x$; (10) $y=\sqrt{\arccos x}$.

6. Find the derivative of the following functions:

(1) $y=\ln\left(\tan\dfrac{x}{2}\right)$; (2) $y=\sqrt{1+\ln^2\dfrac{x}{2}}$;

(3) $y=2^{\frac{x}{\ln x}}$; (4) $y=\cos^2(x^3)$;

(5) $y=\sin\left(\cos\dfrac{1}{x}\right)$; (6) $y=\arcsin x^2-xe^{x^2}$;

(7) $y=x\arctan\dfrac{x}{2}$; (8) $y=x\sqrt{1-x^2}+\arcsin x$.

7. Given that $f(x)$ is differentiable, find the derivatives of the following functions:

(1) $y=f(e^x+x^e)$; (2) $y=f\left(\arcsin\dfrac{1}{x}\right)$;

(3) $y=f(\ln x)+\ln f(x)$; (4) $y=f(\sin^2x)+f(\cos^2x)$.

8. Given $y=f\left(\dfrac{3x-2}{3x+2}\right)$, $f'(x)=\arctan x^2$, find $\dfrac{dy}{dx}\Big|_{x=0}$.

9. Find the derivative of the function $f(x)=\begin{cases} e^{2x}-1, & x\geqslant 0 \\ \sin^2x, & x<0 \end{cases}$.

10. The lengths of the two sides of a rectangle are represented by x and y, respectively. If the side x is decreasing at a rate of 0.01 m/s, and the side y is increasing at a rate of 0.02 m/s, find the rate of change of the rectangle's area and the rate of change of its diagonal at time.

11. Given $g(x)=a^{f^2(x)}$ and $f'(x)=\dfrac{1}{f(x)\ln a}$, prove that: $g'(x)=2g(x)$.

2.3 Higher-Order Derivatives

The instantaneous velocity v of an object in motion is the rate of change of the distance function $s=s(t)$ with respect to time:

$$v=v(t)=s'(t)=\frac{ds}{dt}.$$

The acceleration $a(t)$ of the object is the rate of change of the velocity function $v(t)$ with respect to time t, that is,

$$a=a(t)=v'(t)=[s'(t)]'.$$

In other words, the acceleration is the derivative of the derivative of the distance function $s=s(t)$, referred to as the second derivative of s with respect to t.

Generally, if the derivative $f'(x)$ of the function $y=f(x)$ is differentiable at point x, then the derivative $[f'(x)]'$ at point x is called the second derivative of the function $y=f(x)$ at point x. It is denoted as

$$y'', f''(x), \frac{d^2y}{dx^2} \text{ or } \frac{d^2f(x)}{dx^2}.$$

To find the nth derivative, simply apply the differentiation process n times.

$$f''(x)=\lim_{\Delta x\to 0}\frac{f'(x+\Delta x)-f'(x)}{\Delta x}.$$

Similarly, the derivative of the second derivative of a function $f(x)$ is called the third derivative of $f(x)$. The derivative of the third derivative is called the fourth derivative. The derivative of the $n-1$th derivative is called the nth derivative, denoted respectively as

$$y''', y^{(4)}, \cdots, y^{(n)} \text{ or } f'''(x), f^{(4)}(x), \cdots, f^{(n)}(x).$$

It can also be denoted as

$$\frac{d^3y}{dx^3}, \frac{d^4y}{dx^4}, \cdots, \frac{d^ny}{dx^n}.$$

Derivatives of the second order and higher are collectively referred to as higher-order derivatives.

When a function $y=f(x)$ has a nth derivative, it can also be said that $f(x)$ is n times differentiable. If the function $f(x)$ is n times differentiable at point x, then $f(x)$ has all derivatives of orders lower than n in some neighborhood of point x.

From the definition of higher-order derivatives, to find the nth derivative of a function, one simply needs to apply the previous differentiation methods n times in succession.

Example 2.3.1 Given $y=e^{2x}$, find $\frac{d^2y}{dx^2}$.

Solution

$$y'=2e^{2x}, y''=(y')'=(2e^{2x})'=4e^{2x}.$$

Example 2.3.2 Given $y=\sin^2 x$, find $y''|_{x=\frac{\pi}{4}}$.

Solution

$$y' = 2\sin x \cdot (\sin x)' = 2\sin x\cos x = \sin 2x$$
$$y'' = (y')' = (\sin 2x)' = 2\cos 2x.$$

Therefore

$$y''|_{x=\frac{\pi}{4}} = 2\cos\frac{\pi}{2} = 0.$$

Example 2.3.3 Find the nth derivative of $y = e^x$.

Solution

$$y' = e^x, y'' = e^x, y''' = e^x,$$

by mathematical induction, it is known that

$$y^{(n)} = (e^x)^{(n)} = e^x.$$

Example 2.3.4 Find the nth derivative of $y = \sin x$.

Solution

$$y' = \cos x = \sin\left(x + \frac{\pi}{2}\right),$$
$$y'' = \cos\left(x + \frac{\pi}{2}\right) = \sin\left(x + 2\cdot\frac{\pi}{2}\right),$$
$$y''' = \cos\left(x + 2\cdot\frac{\pi}{2}\right) = \sin\left(x + 3\cdot\frac{\pi}{2}\right),$$
$$y^{(4)} = \cos\left(x + 3\cdot\frac{\pi}{2}\right) = \sin\left(x + 4\cdot\frac{\pi}{2}\right),$$
$$\cdots$$
$$y^{(n)} = \sin\left(x + n\cdot\frac{\pi}{2}\right).$$

That is

$$(\sin x)^{(n)} = \sin\left(x + n\cdot\frac{\pi}{2}\right).$$

Example 2.3.5 Find the nth derivative of $y = \ln(1+x)$.

Solution

$$y' = \frac{1}{1+x},$$
$$y'' = [(1+x)^{-1}]' = -\frac{1}{(1+x)^2},$$
$$y''' = -[(1+x)^{-2}]' = (-1)\cdot(-2)\cdot\frac{1}{(1+x)^3}.$$

By mathematical induction, it is known that

$$y^{(n)} = (-1)^{n-1}\frac{(n-1)!}{(1+x)^n}.$$

Example 2.3.6 Let $x = \varphi(y)$ be the inverse function of $y = x^4 + x - 1$, find $\frac{d^2x}{dy^2}$.

Solution Since $y'=4x^3+1$, therefore

$$\frac{dx}{dy}=\frac{1}{y'}=\frac{1}{4x^3+1},$$

$$\frac{d^2x}{dy^2}=\frac{d}{dy}\left(\frac{1}{4x^3+1}\right)=\left(\frac{1}{4x^3+1}\right)'_x\frac{dx}{dy}=\frac{-12x^2}{(4x^3+1)^2}\frac{1}{4x^3+1}=-\frac{12x^2}{(4x^3+1)^3}.$$

According to the definition of higher-order derivatives, it is easy to obtain the following differentiation rules for higher-order derivatives:

If the functions $u=u(x)$ and $v=v(x)$ are both nth-order differentiable at x, then,

(1) $(u\pm v)^{(n)}=u^{(n)}\pm v^{(n)}$;

(2) $(cu)^{(n)}=cu^{(n)}$ (c is a constant);

(3) $(u\cdot v)^{(n)}=\sum\limits_{k=0}^{n}C_n^k u^{(k)}v^{(n-k)}$.

Where equation (3) is known as Leibniz's formula.

Example 2.3.7 Let $y=x^2e^{2x}$, find $y^{(20)}$.

Solution Let $u=x^2$, $v=e^{2x}$, then

$$v^{(k)}=2^ke^{2x}\ (k=0,1,2,\cdots,20),$$
$$u'=2x,u''=2,u^{(k)}=0\ (k=3,4,\cdots,20),$$

thus

$$y^{(20)}=x^2(e^{2x})^{(20)}+20(x^2)'(e^{2x})^{(19)}+\frac{20\times 19}{2!}(x^2)''(e^{2x})^{(18)}=2^{20}e^{2x}(x^2+20x+95).$$

Exercise 2.3

1. Find the second derivative of the following functions:

(1) $y=\ln(1+5x^2)$; (2) $y=x\ln x$;

(3) $y=(1+x^2)\arctan x$; (4) $y=\ln\left(x+\sqrt{1+x^2}\right)$;

(5) $y=\cos x+\tan x$; (6) $y=xe^{x^2}$.

2. Find the second derivative value of the following functions:

(1) $f(x)=(x^3+10)^4$, find $f''(0)$; (2) $y=\dfrac{\arcsin x}{\sqrt{1-x^2}}$, find $y''\Big|_{x=0}$;

(3) $y=\dfrac{\ln x}{x}$, find $y''(e)$.

3. Given $y=e^{-x^2}$, find y'''.

4. Find the nth derivative of the following functions:

(1) $y=x\ln x$; (2) $y=xe^x$;

(3) $y=\dfrac{1-x}{1+x}$; (4) $y=\cos^2 x$.

5. Given that $f(x)$ is twice differentiable, find the second derivative of the following functions $\dfrac{d^2y}{dx^2}$.

(1) $y=f(x^n)$; (2) $y=f(\ln x)$;

(3) $y=f(e^{-x^2})$; (4) $y=f[f(x)]$.

6. Suppose the function $y=f(x)$ has a third derivative at point x, and $y'=f'(x)\neq 0$. If $f(x)$ has an inverse function $x=f^{-1}(y)$. Derive from $\frac{dx}{dy}=\frac{1}{y'}$:

(1) $\frac{d^2x}{dy^2}=\frac{-y''}{(y')^3}$; (2) $\frac{d^3x}{dy^3}=\frac{3(y'')^2-y'y'''}{(y')^5}$.

7. Verify that $y=C_1e^{2x}+C_2e^{-x}$ (C_1, C_2 is a constant) satisfies the relation

$$y''-y'-2y=0.$$

8. Verify that the function $y=e^{2x}(\sin x+\cos x)$ satisfies the following equation:

$$y''-4y'+5y=0.$$

2.4 Implicit Differentiation

2.4.1 Derivative of Implicit Functions

Let y be a function of x. If the dependent variable y is directly expressed in terms of the independent variable x through a mathematical analytical expression, then y is called an explicit function of x, such as $y=\sin x$, $y=x^3+\ln x-e^{2x}$, etc. If the functional relationship between the dependent variable y and the independent variable x is determined by the equation $F(x,y)=0$, that is, whenever x takes any value on a set I, there is always a unique y value that satisfies the equation $F(x,y)=0$ corresponding to it, then y is called an implicit function of x, such as $x^3-4y=0$, $e^{x+y}+xy-1=0$, etc.

Converting an implicit function into an explicit function is referred to as the explicitization of the implicit function. For example, solving for $y=\frac{1}{4}x^3$ from the equation $x^3-4y=0$ converts the implicit function into an explicit function. However, the explicitization of implicit functions is often very difficult, or even impossible. For instance, the implicit function $y=f(x)$ determined by the equation $e^{x+y}+xy-1=0$ cannot be explicitly expressed. How then can the derivative of an implicit function be found?

Indeed, for the implicit function $y=f(x)$ determined by the equation $F(x,y)=0$, to find $\frac{dy}{dx}$, one should note that y is a function of x. By applying the chain rule for derivatives, differentiate both sides of the equation with respect to x, and solve for $\frac{dy}{dx}$. Below, this method is illustrated through examples.

Example 2.4.1 Find the derivative y' and $y'|_{x=0}$ of the implicit function y determined by the equation $e^y+xy-e=0$.

Solution Noticing that y is a function of x, so e^y is a composite function of x. Substituting $y=y(x)$ into the equation yields an identity in terms of x, indicating that the two sides of the equation represent the same function, hence their derivatives are equal. Therefore, differentiate both sides of the equation with respect to x, yielding

$$(e^y+xy-e)'=0',$$

that is

$$e^y y'+1\cdot y+xy'-0=0.$$

Therefore

$$y'=-\frac{y}{e^y+x}.$$

When $x=0$, from the original equation we know $y=1$, thus

$$y'|_{x=0}=-\frac{1}{e^1+0}=-\frac{1}{e}.$$

Example 2.4.2 Given $\sin(x+y)+y^2-x^2=0$, find $\frac{dy}{dx}$.

Solution Since y is a function of x, both $\sin(x+y)$, y^2 are composite functions of x. Differentiate both sides of the original equation with respect to x, yielding

$$\cos(x+y)(x+y)'+2y\cdot y'-2x=0,$$

that is

$$\cos(x+y)(1+y')+2yy'-2x=0.$$

Therefore

$$y'=\frac{2x-\cos(x+y)}{2y+\cos(x+y)}.$$

Example 2.4.3 Find the tangent and normal equations to the curve $(5y+2)^3=(2x+1)^5$ at the point $\left(0,-\frac{1}{5}\right)$.

Solution Differentiate both sides of the equation with respect to x, yielding

$$3(5y+2)^2(5y+2)'=5(2x+1)^4(2x+1)',$$

that is

$$15(5y+2)^2y'=10(2x+1)^4.$$

Solving for

$$y'=\frac{2(2x-1)^4}{3(5y+2)^2}.$$

At the point $\left(0,-\frac{1}{5}\right)$, the slopes of the tangent and normal are respectively

$$k=y'|_{\left(0,-\frac{1}{5}\right)}=\frac{2}{3},\quad k_{1}=-\frac{1}{k}=-\frac{3}{2}.$$

Therefore, the equation of the tangent is $y+\frac{1}{5}=\frac{2}{3}x$, that is

$$10x-15y-3=0.$$

The equation of the normal is $y+\frac{1}{5}=-\frac{3}{2}x$, that is

$$15x+10y+2=0.$$

Example 2.4.4 Given $y-xe^y=1$, find $y''|_{x=0}$.

Solution Differentiate both sides of the equation with respect to x, yielding

$$y'-e^y-xe^y y'=0.$$

Solving for

$$y'=\frac{e^y}{1-xe^y}.$$

Differentiate both sides of equation with respect to x, yielding

$$y''-e^y y'-e^y y'-xe^y(y')^2-xe^y y''=0,$$

that is

$$y''=\frac{2e^y y'+xe^y(y')^2}{1-xe^y}.$$

Also, when $x=0, y=1$, substituting into equation yields

$$y'|_{x=0}=e.$$

Substituting $x=0, y=1, y'|_{x=0}=e$ into equation yields

$$y''|_{x=0}=2e^2.$$

Some functions are cumbersome to differentiate directly and often require transforming using the properties of logarithms before applying the method of implicit differentiation, thus leading to the method of logarithmic differentiation.

The method of logarithmic differentiation involves taking the logarithm of both sides of the original equation, simplifying the equation using the properties of logarithms, and then applying the method of implicit differentiation. This method is commonly used in two situations:

(1) Exponential function $y=u^v(u=u(x), v=v(x))$, find $\frac{dy}{dx}$;

(2) $y=f(x)$ is composed of several factors through multiplication, division, exponentiation, and root extraction, find $\frac{dy}{dx}$.

Example 2.4.5 Given $y=(1+x^2)^{\sin x}$, find y'.

Solution Take the logarithm on both sides of the original equation and simplify, yielding

$$\ln y=\ln(1+x^2)^{\sin x}=\sin x\ln(1+x^2).$$

Differentiate both sides of the equation with respect to x, obtaining

$$\frac{1}{y}\cdot y'=\cos x\ln(1+x^2)+\frac{2x\sin x}{1+x^2}.$$

Therefore

$$y'=y\left[\cos x\ln(1+x^2)+\frac{2x\sin x}{1+x^2}\right]=(1+x^2)^{\sin x}\left[\cos x\ln(1+x^2)+\frac{2x\sin x}{1+x^2}\right].$$

If the function is composed of an exponential function u^v and the sum of other functions, it is not possible to take the logarithm of both sides. For convenience, the logarithmic identity can be used to transform u^v into a composite function $e^{v\ln u}$ before differentiating.

Example 2.4.6 Find the derivative of the function $y=x^x+e^x+x^{e^x}$.

Solution The right-hand side of the function y consists of three terms added together. The first and third terms are both exponential functions, but it is not possible to directly take the logarithm of both sides. Instead, the logarithmic identity can be used to transform and then differentiate.

$$\begin{aligned}y'&=(x^x+e^x+x^{e^x})'=(x^x)'+(e^x)'+(x^{e^x})'=(e^{x\ln x})'+e^x+(e^{e^x\ln x})'\\&=e^{x\ln x}(x\ln x)'+e^x+e^{e^x\ln x}(e^x\ln x)'=x^x(\ln x+1)+e^x+x^{e^x}\left(e^x\ln x+\frac{e^x}{x}\right).\end{aligned}$$

Example 2.4.7 Find the derivative of $y=\dfrac{(2x-1)^2\sqrt[3]{x+3}}{(x^2+1)\sqrt{5-x}}$.

Solution Take the absolute value of both sides of the equation, then take the logarithm, and simplify to obtain

$$\ln|y|=2\ln|2x-1|+\frac{1}{3}\ln|x+3|-\ln(x^2+1)-\frac{1}{2}\ln|5-x|.$$

Differentiate both sides of the equation with respect to x, since $(\ln|u|)'=\dfrac{1}{u}$, we have

$$\frac{1}{y}\cdot y'=\frac{4}{2x-1}+\frac{1}{3(x+3)}-\frac{2x}{x^2+1}+\frac{1}{2(5-x)},$$

thus

$$\begin{aligned}y'&=y\left[\frac{4}{2x-1}+\frac{1}{3(x+3)}-\frac{2x}{x^2+1}+\frac{1}{2(5-x)}\right]\\&=\frac{(2x-1)^2\sqrt[3]{x+3}}{(x^2+1)\sqrt{5-x}}\left[\frac{4}{2x-1}+\frac{1}{3(x+3)}-\frac{2x}{x^2+1}+\frac{1}{2(5-x)}\right].\end{aligned}$$

Note: When encountering similar problems using logarithmic differentiation in the future, for the sake of brevity, the step of taking the absolute value can be omitted.

2.4.2 Related Rates of Change

Suppose $x=\varphi(t)$, $y=\psi(t)$ are all differentiable functions, and there exists a certain relationship between variables x and y. Then, their derivatives (rates of change) with respect to t $\dfrac{dx}{dt}$ and $\dfrac{dy}{dt}$ also exhibit a certain relationship. These interdependent rates of change are termed related rates. In practice, it is often necessary to determine one rate of change based on another.

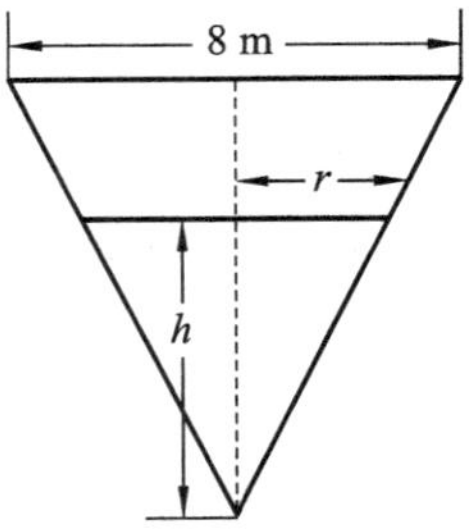

Figure 2-5

Example 2.4.8 Water is being poured into a right circular cone with a depth of 8 m and a top diameter of 8 m at a rate of 4 m^3/min. What is the rate at which the water level rises when the depth is 5 m (Figure 2-5)?

Solution Let the depth of the water be h m, when the volume of water poured is V m^3. It is evident that

$$V=\frac{1}{3}\pi r^2 h=\frac{1}{3}\pi\left(\frac{h}{2}\right)^2 h=\frac{1}{12}\pi h^3.$$

Here, V, h are all functions of t. Differentiating both sides of the equation with respect to t, we get

$$\frac{\mathrm{d}V}{\mathrm{d}t}=\frac{1}{4}\pi h^2\frac{\mathrm{d}h}{\mathrm{d}t}.$$

Given $\frac{\mathrm{d}V}{\mathrm{d}t}=4$ m^3/min, so when $h=5$ m, the rate at which the surface rises is

$$\frac{\mathrm{d}h}{\mathrm{d}t}=4\frac{\frac{\mathrm{d}V}{\mathrm{d}t}}{\pi h^2}=4\cdot\frac{4}{\pi 5^2}=\frac{16}{25\pi}\ (\mathrm{m/min}).$$

Exercise 2.4

1. Find the derivative of the implicit function y defined by the following equations.

(1) $x^2+xy=1$; (2) $\mathrm{e}^{x+y}+\cos(xy)=0$;

(3) $x^3+y^3-3axy=0$; (4) $\arctan\frac{x}{y}=\ln(x^2+y^2)$.

2. Find the derivative $\left.\frac{\mathrm{d}y}{\mathrm{d}x}\right|_{x=0}$ of the implicit function y determined by the equation $\sin(xy)+\ln(y-x)=x$ at the point $x=0$.

3. Given $xy-\sin(\pi y^2)=0$, find $\left.\frac{\mathrm{d}^2y}{\mathrm{d}x^2}\right|_{\substack{x=0\\y=1}}$.

4. Find the second derivative of the implicit function y determined by the following equation $\frac{\mathrm{d}^2y}{\mathrm{d}x^2}$.

(1) $x^2-y^2=1$; (2) $y=\tan(x+y)$.

5. Using logarithmic differentiation to find the derivative of the following functions.

(1) $y=(\sin x)^{\cos x}$; (2) $y=\left(\frac{x}{1+x}\right)^x$;

(3) $y=\sqrt[3]{\frac{x(x-1)}{(x-2)(x+3)}}$; (4) $y=(x-a_1)^{a_1}(x-a_2)^{a_2}\cdots(x-a_n)^{a_n}$.

6. Given $x^y+y^x=3$, find $\left.\frac{\mathrm{d}y}{\mathrm{d}x}\right|_{x=1}$.

7. Find the derivative of the function determined by the following parametric equations $\frac{dy}{dx}$.

(1) $\begin{cases} x=\theta(1-\sin\theta) \\ y=\theta\cos\theta \end{cases}$;

(2) $\begin{cases} x=2t-t^2 \\ y=3t-t^3 \end{cases}$;

(3) $\begin{cases} x=e^t\sin t \\ y=e^t(\sin t-\cos t) \end{cases}$.

8. Find the derivative of the function $y=f(x)$ determined by the parametric equation $\begin{cases} x=\cos^4 t \\ y=\sin^4 t \end{cases}$, at $t=0, t=\frac{\pi}{4}$.

9. Find the tangent and normal equations to the curve $\begin{cases} x=1-t^2 \\ y=t-t^2 \end{cases}$, at $t=1$.

10. Find the second derivative of the function $y=f(x)$ determined by the following parametric equations $\frac{d^2y}{dx^2}$.

(1) $\begin{cases} x=2t-t^2 \\ y=3t-t^3 \end{cases}$;

(2) $\begin{cases} x=a\cos t \\ y=b\sin t \end{cases}$.

11. Verify that the function $y=f(x)$ determined by the parametric equation $\begin{cases} x=e^t\sin t \\ y=e^t\cos t \end{cases}$, satisfies the relation

$$(x+y)^2\frac{d^2y}{dx^2}=2\left(x\frac{dy}{dx}-y\right).$$

12. Let $y=y(x)$ be determined by $\begin{cases} x=\arctan t \\ 2y-ty^2+e^t=5 \end{cases}$, find $\frac{dy}{dx}$.

13. Suppose $f''(t)$ exists, and $f''(t)\neq 0$. The function $y=y(x)$ is determined by the system of equations $\begin{cases} x=f'(t) \\ y=tf'(t)-f(t) \end{cases}$, find $\frac{d^2y}{dx^2}$.

14. An object falling on calm water produces concentric ripples, and if the rate of increase of the radius of the outermost circle is always 6 m/s, what is the rate of increase of the disturbed water area at the end of 2 s?

2.5 Differential of a Function

2.5.1 Concept of Differentiation

To study the rate of change of a function at point x, the concept of the derivative was introduced earlier. Sometimes, it is necessary to calculate the value of the corresponding increment Δy of the function y when the independent variable x obtains a small increment Δx. Generally, calculating the exact value of Δy is very difficult, and often, knowing an approximate value of Δy is sufficient. So, is it possible to calculate an approximate value of Δy in a simple yet relatively accurate manner? In other words, can Δy be approximated by a simple function of Δx (such as a linear function of Δx)? This leads to the concept of differentiation.

Let's first consider a specific problem.

The area S of a square is a function of its side length x:

$$S=x^2 \quad (x>0).$$

When the side length has an increment Δx, the area S has a corresponding increment:

$$\Delta S=(x+\Delta x)^2-x^2=2x\Delta x+\Delta x^2.$$

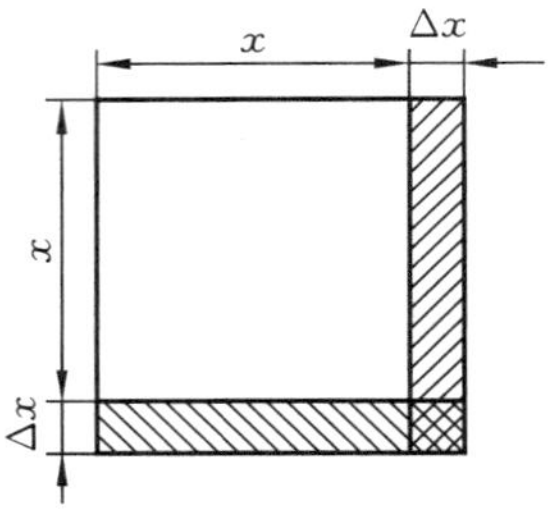

Figure 2-6

The above equation consists of two parts: The first part $2x\Delta x$ is a linear function of Δx, namely the sum of the areas of the two rectangles with slanted lines in Figure 2-6. The second part $(\Delta x)^2$ is a higher-order infinitesimal compared to Δx. Therefore, when Δx is small, $(\Delta x)^2$ can be neglected, and at this time, the first part $2x\Delta x$ can be used to approximate ΔS, with the resulting error being a higher-order infinitesimal of Δx. The term $2x\Delta x$ is referred to as the differential of the square area S, denoted as $\mathrm{d}S$. It can be seen that when $\Delta x\to 0$, there is

$$\Delta S\approx \mathrm{d}S=2x\Delta x.$$

Definition 2.5.1 Let the function $y=f(x)$ be defined in a neighborhood $U(x)$ of the point x, and $x+\Delta x\in U(x)$. If the increment of the function $\Delta y=f(x+\Delta x)-f(x)$ can be expressed as

$$\Delta y=A\Delta x+o(\Delta x),$$

where: A is independent of Δx, and $o(\Delta x)$ is a higher-order infinitesimal of Δx. Then the function $y=f(x)$ is said to be differentiable at the point x, and $A\Delta x$ is called the differential of the function $y=f(x)$ at the point x, denoted as $\mathrm{d}y$ or $\mathrm{d}f(x)$, that is

$$\mathrm{d}y=A\Delta x.$$

It can be seen that the differential of the function $\mathrm{d}y$ is a linear function of Δx. When

$A \neq 0$ and $\Delta x \to 0$, dy can be used as an approximation for Δy, i.e., dy is the principal part of Δy, thus, the differential dy is also referred to as the linear principal part of Δy.

The following needs to be discussed: What kind of functions are differentiable? How to determine the value of A in the differential form?

In the previous specific problem, the differential of the area of a square S is already known.

$$dS = 2x\Delta x.$$

Here, $A = 2x = (x^2)' = S'$, i.e., the A in the differential form equals the derivative of the function at point x.

Theorem 2.5.1 The necessary and sufficient condition for the function $y = f(x)$ to be differentiable at point x is that $f(x)$ is derivable at point x, and

$$dy = f'(x)\Delta x.$$

Proof Suppose $y = f(x)$ is differentiable at point x, by Definition 2.5.1,

$$\Delta y = A\Delta x + o(\Delta x).$$

Dividing both sides of the equation by Δx, we get

$$\frac{\Delta y}{\Delta x} = A + \frac{o(\Delta x)}{\Delta x}.$$

Thus

$$f'(x) = \lim_{\Delta x \to 0} \frac{\Delta y}{\Delta x} = A + 0 = A.$$

That is, $f(x)$ is derivable at point x, and $A = f'(x)$.

Conversely, if $y = f(x)$ is derivable at point x, i.e.,

$$\lim_{\Delta x \to 0} \frac{\Delta y}{\Delta x} = f'(x)$$

exists, according to the relationship between limits and infinitesimals,

$$\frac{\Delta y}{\Delta x} = f'(x) + \alpha \ (\lim_{\Delta x \to 0} \alpha = 0),$$

therefore,

$$\Delta y = f'(x)\Delta x + \alpha \Delta x.$$

Here, $f'(x)$ is independent of Δx, $f'(x)\Delta x$ is a linear function of Δx, $\alpha\Delta x = o(\Delta x)$, thus $f(x)$ is differentiable at point x, and

$$dy = f'(x)\Delta x.$$

If $y = x$, then $dy = dx = (x)'\Delta x = \Delta x$, that is, the differential of the independent variable is the increment of the independent variable. Thus, the differential of the function can also be expressed as

$$dy = f'(x)dx,$$

hence

$$\frac{dy}{dx} = f'(x).$$

That is, the quotient of the differential of the function and the differential of the independent variable equals the derivative of the function, hence the derivative is also called the "differential quotient". The methods of finding the derivative and the differential are collectively referred to as differential calculus.

Example 2.5.1 Find the increment and differential of the function $y=x^3$ at $x=1$, $\Delta x=0.01$.

Solution

$$\Delta y=f(x+\Delta x)-f(x)=1.01^3-1^3=0.030301$$

$$y'=3x^2, \mathrm{d}y=y'\mathrm{d}x=3x^2\mathrm{d}x.$$

Since $x=1, \Delta x=0.01$, therefore,

$$\mathrm{d}y=3\times 1^2\times 0.01=0.03.$$

It can be seen that: $|\Delta y-\mathrm{d}y|=0.000301$ is small, i.e., $\Delta y\approx \mathrm{d}y$.

Example 2.5.2 Find the differential of the function $y=\ln(1+x)$ at $x=1, x=2$.

Solution Because $\mathrm{d}y=y'\mathrm{d}x=\dfrac{1}{1+x}\mathrm{d}x$, therefore,

$$\mathrm{d}y\big|_{x=1}=\frac{1}{1+1}\mathrm{d}x=\frac{1}{2}\mathrm{d}x,$$

$$\mathrm{d}y\big|_{x=2}=\frac{1}{1+2}\mathrm{d}x=\frac{1}{3}\mathrm{d}x.$$

To gain a more intuitive understanding of the concept of differential, let us explore its geometric meaning.

Draw the graph of a differentiable function $y=f(x)$ (Figure 2-7), select a point $M(x, y)$ on the curve, and through point M draw the tangent MT to the curve, with an inclination angle of α, as shown in the Figure 2-7.

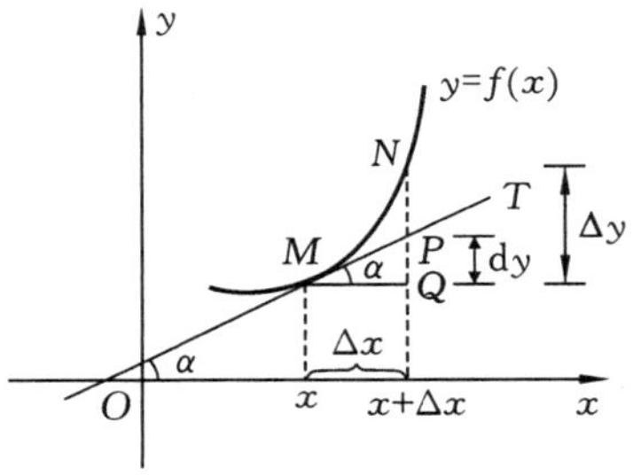

Figure 2-7

$$\mathrm{d}y=f'(x)\mathrm{d}x=\tan\alpha\cdot\Delta x=\frac{PQ}{MQ}\cdot MQ=PQ,$$

the differential of a function at point x, $\mathrm{d}y$ is exactly the increment in the ordinate of the point on the tangent line to the curve at the corresponding point. It can also be seen from the diagram that when $|\Delta x|$ is small, $|\Delta y-\mathrm{d}y|=NQ-PQ=NP$ is small and negligible compared to x, thus near point M, the tangent segment can be used to approximate the curve segment.

2.5.2 Differential Formulas and Rules

The formula $\mathrm{d}y=f'(x)\mathrm{d}x$ indicates that to find the differential of a function, one only needs to find the derivative of the function and then multiply it by the differential of the independent variable. Therefore, each derivative formula corresponds to a differential formula.

1. Differential Formulas for Basic Elementary Functions

(1) $dc=0$;

(2) $dx^{\alpha}=\alpha x^{\alpha-1}dx$ (α is a real number);

(3) $da^{x}=a^{x}\ln a\ dx$ $(a>0,a\neq 1)$;

(4) $de^{x}=e^{x}dx$;

(5) $d(\log_{a}x)=\dfrac{1}{x\ln a}dx$ $(a>0,a\neq 1)$;

(6) $d(\ln x)=\dfrac{1}{x}dx$;

(7) $d(\sin x)=\cos x\,dx$;

(8) $d(\cos x)=-\sin x\,dx$;

(9) $d(\tan x)=\sec^{2}x\,dx$;

(10) $d(\cot x)=-\csc^{2}x\,dx$;

(11) $d(\sec x)=\sec x\cdot\tan x\,dx$;

(12) $d(\csc x)=-\csc x\cot x\,dx$;

(13) $d(\arcsin x)=\dfrac{1}{\sqrt{1-x^{2}}}dx$;

(14) $d(\arccos x)=-\dfrac{1}{\sqrt{1-x^{2}}}dx$;

(15) $d(\arctan x)=\dfrac{1}{1+x^{2}}dx$;

(16) $d(\text{arccot}\,x)=-\dfrac{1}{1+x^{2}}dx$.

2. Differential Rules for Sums, Differences, Products, and Quotients of Functions

Let $u=u(x),v=v(x)$ be differentiable, then we have

(1) $d(u\pm v)=du\pm dv$;

(2) $d(uv)=v\,du+u\,dv$;

(3) $d\left(\dfrac{u}{v}\right)=\dfrac{v\,du-u\,dv}{v^{2}}$ $(v\neq 0)$.

3. Differentiation Rule for Composite Functions

Suppose $y=f(u)$ is differentiable, then

(1) When u is an independent variable, the differential of the function

$$dy=f'(u)du.$$

(2) When u is not an independent variable but a differentiable function $u=\varphi(x)$ of x, where y is a composite function of x with $y=f[\varphi(x)]$, u as an intermediate variable, by the chain rule, we have

$$dy=f'(u)\varphi'(x)dx=f'(u)du.$$

It can be seen that for the function $y=f(u)$, whether u is an independent variable or an intermediate variable in a composite function, the differential

$$dy=f'(u)du$$

remains unchanged. This property is called the invariance of the first-order differential form.

Example 2.5.3 Given $y=\ln(x^{2}+2x-3)$, find dy.

Solution Let $u=x^{2}+2x-3$, by the differentiation rule, we get

$$dy=d\ln u=\frac{1}{u}du=\frac{1}{x^{2}+2x-3}d(x^{2}+2x-3)=\frac{1}{x^{2}+2x-3}[d(x^{2})+d(2x)-d3]$$

$$=\frac{1}{x^{2}+2x-3}(2x\,dx+2dx)=\frac{2x+2}{x^{2}+2x-3}dx.$$

Example 2.5.4 Given $y=e^{\sin^2 x}$, find dy.

Solution First treat $\sin^2 x$ as an intermediate variable u, applying the invariance of the differential form, we get

$$\begin{aligned} dy &= de^u = e^u du = e^{\sin^2 x} d(\sin^2 x) \xrightarrow{\sin x = v} e^{\sin^2 x} dv^2 = e^{\sin^2 x} \cdot 2v dv \\ &= e^{\sin^2 x} 2\sin x d(\sin x) = e^{\sin^2 x} 2\sin x \cos x dx = \sin 2x \cdot e^{\sin^2 x} dx. \end{aligned}$$

Example 2.5.5 Find the differential of the implicit function y defined by the equation $xe^y - \ln y + 5 = 0$.

Solution Differentiate both sides of the equation, we get

$$d(xe^y) - d(\ln y) + d5 = 0,$$

$$e^y dx + x de^y - \frac{1}{y} dy = 0,$$

which is

$$e^y dx + xe^y dy - \frac{1}{y} dy = 0,$$

therefore

$$dy = \frac{e^y}{\frac{1}{y} - xe^y} dx = \frac{ye^y}{1 - xye^y} dx.$$

Note: In all three examples above, one could directly find y' first, and then use the formula $dy = y'dx$ to derive.

Example 2.5.6 Fill in the appropriate function in the following brackets to make the equation hold.

(1) $d(\quad) = e^{-x} dx$; (2) $d(\quad) = \frac{1}{x\ln x} dx$.

Solution

(1) Because

$$(e^{-x})' = -e^{-x}, d(e^{-x}) = -e^{-x} dx,$$

therefore

$$e^{-x} dx = -d(e^{-x}) = d(-e^{-x}),$$

namely

$$d(-e^{-x}) = e^{-x} dx.$$

(2) Because

$$\frac{1}{x\ln x} dx = \frac{1}{\ln x}\frac{1}{x} dx = \frac{1}{\ln x}(\ln x)' dx = \frac{1}{\ln x} d(\ln x),$$

let $u = \ln x$, then

$$\frac{1}{\ln x} d(\ln x) = \frac{1}{u} du = (\ln u)' du = d\ln u,$$

therefore

$$\frac{1}{x\ln x} dx = d(\ln u) = d[\ln(\ln x)],$$

namely

$$d[\ln(\ln x)]=\frac{1}{x\ln x}dx.$$

Clearly, the result still holds after adding C.

2.5.3 Application of Differentials in Approximation

When $f'(x_0)\neq 0$, the differential $dy=f'(x_0)dx$ of the function $y=f(x)$ at the point x_0 is the linear principal part of the increment $\Delta y=f(x_0+\Delta x)-f(x_0)$. When $|\Delta x|$ is small, dy can be used as an approximation for Δy, namely

$$\Delta y\approx dy=f'(x_0)\Delta x.$$

Example 2.5.7 A metal disc with a radius of 10 cm is heated, resulting in an increase in radius of 0.05 cm. Estimate the increase in the disc's area.

Solution Let the area of the circle be S and the radius be r, then

$$S=\pi r^2,\quad dS=2\pi r\,dr.$$

When $r=10$, $\Delta r=dr=0.05$, from the definition of dy we get

$$\Delta S\approx dS=2\pi\times 10\times 0.05=\pi(\text{cm}^2).$$

Equation can also be expressed as

$$f(x_0+\Delta x)-f(x_0)\approx f'(x_0)\Delta x,$$

namely

$$f(x_0+\Delta x)\approx f(x_0)+f'(x_0)\Delta x.$$

Equation indicates that to approximate the value of $f(x_0+\Delta x)$, one can find a point x_0 close to $x_0+\Delta x$, such that both $f(x_0)$ and $f'(x_0)$ are easy to compute, and then use the approximation formula.

Example 2.5.8 Find the approximate value of $\sqrt[3]{8.02}$.

Solution Let $f(x)=\sqrt[3]{x}$, then $f'(x)=\frac{1}{3}x^{-\frac{2}{3}}$. Taking $x_0=8$, $\Delta x=0.02$, from the approximation equation we get

$$\sqrt[3]{8.02}=f(x_0+0.02)\approx f(x_0)+f'(x_0)\Delta x=\sqrt[3]{8}-f'(8)\cdot 0.02=2+\frac{1}{3\times 4}\times 0.02\approx 2.00167.$$

In approximation formula, let $x_0+\Delta x=x$, then $\Delta x=x-x_0$, $\Delta x\to 0$, i.e., $x\to x_0$, thus obtaining an equivalent form of approximation formula:

$$f(x)\approx f(x_0)+f'(x_0)(x-x_0).$$

This equation indicates that when x is sufficiently close to x_0, $f(x)$ can be approximated by a linear function of $x-x_0$.

Specifically, taking $x_0=0$, when $x\to 0$, there is an approximate formula

$$f(x)\approx f(0)+f'(0)x.$$

When $|x|$ is small, applying approximation formula is easy to derive the following commonly used approximate formulas:

(1) $\sin x\approx x$;

(2) $\tan x \approx x$;

(3) $e^x \approx 1+x$;

(4) $\ln(1+x) \approx x$;

(5) $\sqrt[n]{1+x} \approx 1+\frac{1}{n}x$.

Exercise 2.5

1. Given the function $f(x)=x^2-3x+5$, when $x=0$, find the increment and differential of the function when $\Delta x=0.1$, $\Delta x=0.01$, $\Delta x=0.001$, and compare them. Can we conclude that the smaller Δx, the closer they are?

2. For the function $y=f(x)$, when its graph is as shown in Figure 2-8 (a), (b), (c), and (d), mark the point x_0 on the graph, indicate dy, Δy and $\Delta y - dy$, and explain the sign of $\Delta y - dy$.

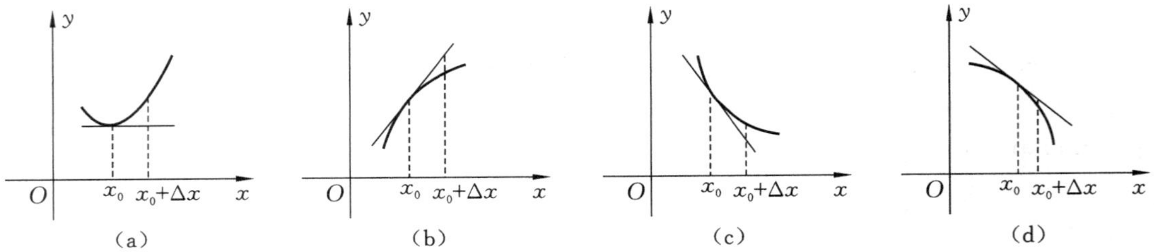

Figure 2-8

3. Differentiate the following functions:

(1) $y=x\ln x - x$;　　(2) $y=\dfrac{x}{\sqrt{x^2+1}}$;

(3) $y=\ln\tan\dfrac{x}{2}$;　　(4) $y=e^{-5x}\cos(2x)$;

(5) $y=\arcsin\sqrt{1-x^2}$;　　(6) $y=e^{\sin^2\frac{1}{x}}$.

4. Find the derivative of the function y determined by the following equations.

(1) $y=e^{-\frac{x}{y}}$;　　(2) $x^3+y^2\sin x - y=6$.

5. Given the edge length of a cube $x=10$ m, if the edge length increases by 0.1 m, find the exact and approximate increase in the volume of the cube.

6. There is a batch of spheres with radius 1 cm. To improve the smoothness of the surface, a layer of copper with thickness 0.01 cm is to be applied. Estimate the amount of copper (density of copper is 8.9 g/cm^3) required for each sphere.

7. Calculate the approximate values of the following expressions.

(1) $\cos 60°20'$;　　(2) $\arctan 1.02$;

(3) $e^{0.2}$;　　(4) $\sqrt[5]{31.9}$.

8. Let $\varphi(t)$ be differentiable everywhere, and $y=y(x)$ be determined by the equation $\varphi(\sin x)+\sin\varphi(y)=\varphi(x+y)$, Find dy.

9. Fill in the appropriate function in the brackets to make the following expressions hold true.

(1) $x\ dx = d(\quad)$;
(2) $\sin x\ dx = d(\quad)$;

(3) $\dfrac{1}{\sqrt{x}}\ dx = d(\quad)$;
(4) $e^{-3x}\ dx = d(\quad)$;

(5) $\dfrac{1}{1+x}\ dx = d(\quad)$;
(6) $\dfrac{1}{1+4x^2}\ dx = d(\quad)$;

(7) $\sec^2 \dfrac{1}{2}x\ dx = d(\quad)$;
(8) $\dfrac{1}{\sqrt{1-2x^2}}\ dx = d(\quad)$.

Comprehensive Review 2

1. Choose the correct term among "sufficient""necessary" "sufficient and necessary" to fill in the blank.

The function $f(x)$ is differentiable at point x_0 is a ________ condition of $f(x)$ continuous at point x_0 condition.

The function $f(x)$ continuous at point x_0 is a ________ condition of $f(x)$ differentiable at point x_0 condition.

The function $f(x)$ differentiable at point x_0 is a ________ condition of $f(x)$ differentiable at point x_0 condition.

2. Multiple choice question.

(1) Let $a>0$, a be a constant, $x>0$, then which of the following operations is correct? ( ).

A. $(x^x)'=xx^{x-1}$ B. $(x^x)'=x^x\ln x$ C. $(a^x)'=a^x\ln a$ D. $(a^x)'=a^x\ln x$

(2) Let $y=f(x)$, $f'(x_0)=\dfrac{1}{2}$, at $\Delta x \to 0$, at point $x=x_0$, the differential dy is ( ).

A. infinitesimal equivalent to Δy

B. infinitesimal of higher order than Δy

C. infinitesimal of lower order than Δy

D. infinitesimal of the same order but not equivalent to Δy

3. Discuss the continuity and differentiability of the following functions at point $x=0$.

(1) $f(x)=|x|^k\ (k>0)$;
(2) $f(x)=\begin{cases} \dfrac{x}{1+e^{\frac{1}{x}}}, & x\neq 0 \\ 0, & x=0 \end{cases}$.

4. Find the derivatives of the following functions:

(1) $y=\arcsin(\sin^2 x)$;
(2) $y=\dfrac{1}{4}\ln\dfrac{1+x}{1-x}-\dfrac{1}{2}\arctan x$;

(3) $y=\sqrt{x^2-a^2}-\arccos\dfrac{a}{x}\ (a>0)$;
(4) $y=\ln\left(e^x+\sqrt{1+e^{2x}}\right)$.

5. Find the differentials of the following functions:

(1) $y=\operatorname{arccot}\dfrac{1-x^2}{1+x^2}$;　　(2) $y=x\,[\sin(\ln x)-\cos(\ln x)]$.

6. Given the $n-2$th derivative $y^{(n-2)}=\dfrac{x}{\ln x}$ of y, find the nth derivative $y^{(n)}$ of y.

7. Given a function $y=y(x)$ determined by the equation $x^2y^2-\cos(xy)=0$, find $\dfrac{dy}{dx}$ and dy.

8. Given a function $y=y(x)$ determined by the equation $e^y+xy=e$, find $y''(0)$.

9. Find the first and second derivatives $\dfrac{dy}{dx}$ and $\dfrac{d^2y}{dx^2}$ of the function y determined by the following parametric equations.

(1) $\begin{cases}x=\theta-\sin\theta\\ y=1-\cos\theta\end{cases}$;　　(2) $\begin{cases}x=\ln\sqrt{1+t^2}\\ y=\arctan t\end{cases}$.

10. Find the tangent and normal equations to the curve $x^2+2xy^2+3y^4=6$ at the point $M(1,-1)$.

11. Given a function $y=y(x)$ determined by the parametric equations $\begin{cases}x=f(2\sin t)\\ y=f(\cos t+1)+f(t-\pi)\end{cases}$, where $f(u)$ is differentiable, and $f'(0)\neq 0$, find $\left.\dfrac{dy}{dx}\right|_{t=\pi}$.

12. If $\lim\limits_{x\to 0}\dfrac{x\,[f(x)-f(0)]}{1-\cos x}=1$, find $f'(0)$.

13. Suppose there is a snowball with a radius of r, and the rate of change of its volume V with respect to time t is proportional to the surface area of the snowball, with a proportionality constant $k>0$, i.e., $\dfrac{dV}{dt}=-k\cdot 4\pi r^2$. It is known that within the first 3 hours, the snowball melted $\dfrac{7}{8}$ of its volume. How long will it take for the snowball to completely melt?

14. Suppose $f(x)$ is differentiable and satisfies $af(x)+bf\left(\dfrac{1}{x}\right)=\dfrac{c}{x}$, where a,b,c is a constant, $|a|\neq|b|$, find $f'(x)$.

15. Given that $f(x)$ is a continuous function with a period of 5, it satisfies in a neighborhood of point $x=0$,

$$f(1+\sin x)-3f(1-\sin x)=8x+\alpha(x),$$

where $\alpha(x)$ is an infinitesimal of higher order than x as $x\to 0$, and $f(x)$ is differentiable at point $x=1$, find the tangent equation of the curve $y=f(x)$ at point $(6,f(6))$.

16. Suppose $u=f\,[\varphi(x)+y^2]$, where x,y satisfies the equation $y+e^y=x$, and both $f(x)$ and $\varphi(x)$ are twice differentiable, find $\dfrac{du}{dx},\dfrac{d^2u}{dx^2}$.

Chapter 3 Applications of Derivatives

The *I Ching* emphasizes the mutual transformation and balance between Yin and Yang, and believes that everything in the world is in constant change, and this change always tends to reach a state of balance between Yin and Yang. The differential mean value theorem describes that in the process of a function changing in a certain interval, there must be at least one intermediate point, so that the derivative of the function at that point (that is, the instantaneous rate of change) is equal to the ratio of the difference of the end function values of the interval to the length of the interval (that is, the average rate of change). This can be seen as the process of function change in the interval, the two forces of Yin and Yang (that is, the increase and decrease of the function value) reach a momentary equilibrium state.

Zen emphasizes the realization of truth through meditation and insight, believing that at some point in one's life, one will suddenly realize the true meaning of life. The "realization of the intermediate state" of the differential mean value theorem is reflected in that no matter how complex the change of the function in the interval, as long as certain conditions (such as continuity, derivability, etc.) are met, there must be at least one intermediate point, so that the derivative of the function at that point can reflect the average effect of the change of the function in the whole interval. This is like a mathematical "Epiphany" that allows us to suddenly realize that there is such a special intermediate point in the process of changing the function. Imagine a winding mountain road, and the differential mean value theorem tells us that no matter how winding the road is, there must be at least one point at which the slope (i.e., the derivative) represents the average slope of the entire mountain path. This "realization of the intermediate state" gives us a deeper understanding of how the function varies in the interval.

The previous chapter introduced the concept of derivatives and their calculation methods, understanding that derivatives represent the rate of change of a function at a point. This chapter will use derivatives to study the overall behavior of functions over an interval and to solve some practical problems. To do this, a bridge needs to be built between derivatives and functions, i.e., a link that connects the local and the global, which is the Mean Value Theorem.

3.1 Mean Value Theorems

We will start with simple and special cases, and then generalize.

3.1.1 Rolle's Mean Value Theorem

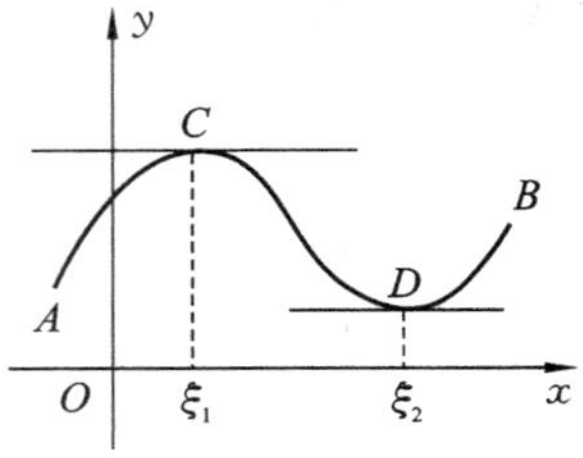

Figure 3-1

Analyze geometrically first. Draw a smooth curve arc $\overset{\frown}{AB}$ (Figure 3-1), on which there are peak point C and valley point D. It can be seen that the tangent lines at the peak and valley points are horizontal. Suppose the equation of $\overset{\frown}{AB}$ is $y=f(x)$, C, D with corresponding abscissas ξ_1, ξ_2, then

$$f'(\xi_1)=f'(\xi_2)=0.$$

That is, the derivative of a differentiable function at its peak and valley points is zero. This is known as Fermat's theorem.

Fermat's Theorem Let the function $f(x)$ be defined in a neighborhood $U(x_0)$ of point x_0, and let $f'(x_0)$ exist. If for any $x \in U(x_0)$, there is $f(x) \leqslant f(x_0)$ (or $f(x) \geqslant f(x_0)$), then

$$f'(x_0)=0.$$

Proof Suppose $x \in U(x_0)$ holds, $f(x) \leqslant f(x_0)$ (if $f(x) \geqslant f(x_0)$, the proof can be similarly constructed). That is,

$$f(x)-f(x_0) \leqslant 0.$$

When $x<x_0$,

$$\frac{f(x)-f(x_0)}{x-x_0} \geqslant 0.$$

When $x>x_0$,

$$\frac{f(x)-f(x_0)}{x-x_0} \leqslant 0.$$

From the definitions of left and right derivatives and the preservation of the limit's sign, we get

$$f'_-(x_0)=\lim_{x \to x_0}\frac{f(x)-f(x_0)}{x-x_0} \geqslant 0,$$

$$f'_+(x_0)=\lim_{x \to x_0^+}\frac{f(x)-f(x_0)}{x-x_0} \leqslant 0.$$

Also, since $f'(x_0)$ exists, it follows that $f'_-(x_0)=f'_+(x_0)=0$, hence $f'(x_0)=0$.

Points where the derivative is zero are commonly referred to as stationary points or critical points of the function.

Rolle's Theorem If the function $f(x)$ satisfies the following conditions:

(1) It is continuous on the closed interval $[a, b]$;

(2) It is differentiable on the open interval (a, b);

(3) The function values are equal at the endpoints of the interval, i.e., $f(a)=f(b)$.

Then there exists at least one point $\xi \in (a, b)$, such that $f'(\xi)=0$.

Proof Since $f(x)$ is continuous on $[a,b]$, by the properties of continuous functions on closed intervals, $f(x)$ must attain its maximum M and minimum m on $[a,b]$.

If $M=m$, then on $[a,b]$, $f(x)$ is a constant, i.e., $f(x)\equiv M$. In this case, for $\forall x\in(a,b)$, there is $f'(x)=0$.

If $M>m$, because $f(a)=f(b)$, at least one of M and m is not equal to the function value at the endpoints, i.e., at least one of M and m (assume M) is attained at an interior point ξ of (a,b), i.e., $f(\xi)=M$. Therefore, for $\forall x\in[a,b]$, there is

$$f(x)\leqslant f(\xi)=M.$$

By Fermat's Theorem, $f'(\xi)=0$.

Note: If any of the three conditions in the theorem is omitted, the conclusion of the theorem may not hold. As shown in Figure 3-2.

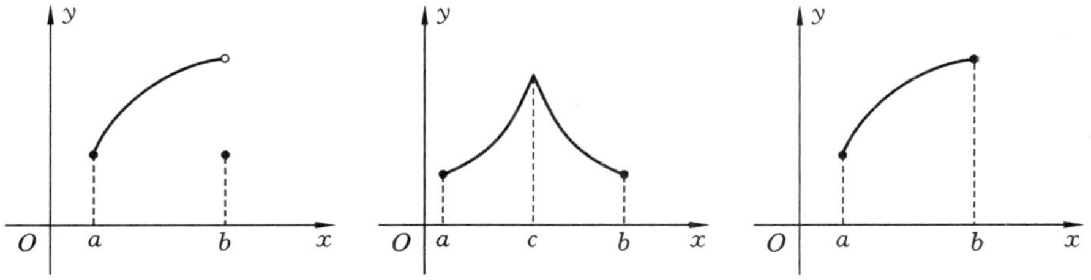

(a) (b) (c)

Figure 3-2

In Figure 3-2(a), $f(x)$ is discontinuous at the endpoint b. In Figure 3-2(b), $f(x)$ is not differentiable at some point c within (a,b). In Figure 3-2(c), the function values of $f(x)$ at the endpoints are not equal. The derivatives $f'(x)$ of these functions do not have a point within (a,b) where they equal 0, meaning Rolle's Theorem does not hold.

Example 3.1.1 Verify that the function $f(x)=x^3-3x^2+2$ satisfies the conditions of Rolle's Theorem on the interval $[0,3]$, and find the value of ξ that meets the theorem's conclusion.

Solution Since $f(x)$ is an elementary function, it is continuous within its defined interval **R** and continuous on $[0,3]$, also,

$$f'(x)=3x^2-6x.$$

$\forall x\in\mathbf{R}$, $f'(x)$ exists, so $f(x)$ is differentiable within $(0,3)$. It is easy to calculate

$$f(0)=f(3)=2.$$

Therefore, $f(x)$ satisfies the conditions of Rolle's Theorem on $[0,3]$.

Let $f'(x)=3x^2-6x=0$, yielding

$$x_1=0, x_2=2.$$

Clearly $x_1\in(0,3)$, which should be discarded, whereas $x_2\in(0,3)$, hence we take $\xi=x_2=2$, resulting in

$$f'(\xi)=0.$$

Example 3.1.2 Given the function $f(x)=(x-1)(x-2)(x-3)$, determine how many real roots $f'(x)=0$ has and specify the intervals in which they lie.

Solution Clearly, $f(x)$ is continuous and differentiable within **R**, and

$$f(1)=f(2)=f(3).$$

Therefore, $f(x)$ satisfies the conditions of Rolle's Theorem on $[1,2]$ and $[2,3]$, thus within $(1,2)$, there exists at least one point ξ_1, such that $f'(\xi_1)=0$, within $(2,3)$, there exists at least one point ξ_2, such that $f'(\xi_2)=0$.

On the other hand, $f'(x)=0$ is a quadratic equation and can have at most two real roots.

Combining the above two aspects, $f'(x)=0$ has two real roots, located within $(1,2)$ and $(2,3)$, respectively.

3.1.2 Lagrange's Mean Value Theorem

The geometric meaning of Rolle's Theorem: For a continuous, smooth curve arc $\overset{\frown}{AB}$, if the heights at both endpoints are the same, then there is at least one point C on the arc where the tangent is parallel to the x axis, which is also parallel to the chord AB [Figure 3-3(a)].

After rotating the curve in Figure 3-3(a) by an angle, the height at endpoint A, B changes, but the curve remains continuous and smooth. At this point, the tangent at point C is still parallel to the chord AB [Figure 3-3(b)]. In analytical terms, this leads to Lagrange's Mean Value Theorem.

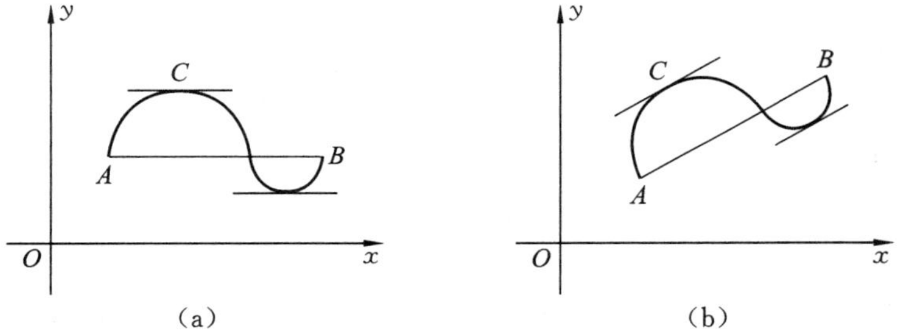

Figure 3-3

Lagrange's Mean Value Theorem If the function $y=f(x)$ satisfies the following conditions:

(1) Continuous on the closed interval $[a,b]$;

(2) Differentiable within the open interval (a,b).

Then there exists at least one point ξ within (a,b), such that

$$f'(\xi)=\frac{f(b)-f(a)}{b-a},$$

or

$$f(b)-f(a)=f'(\xi)(b-a).$$

To prove this theorem, first analyze its conclusion, transforming equation into

$$f'(\xi)-\frac{f(b)-f(a)}{b-a}=0.$$

The left-hand side of this equation represents the derivative of the function,

$$\varphi(x)=f(x)-\frac{f(b)-f(a)}{b-a}x.$$

At point ξ, $\varphi'(\xi)=0$. This aligns exactly with the conclusion of Rolle's Theorem, thus allowing us to use Rolle's Theorem to prove the Lagrange Mean Value Theorem.

Proof Construct an auxiliary function (Figure 3-4):

$$\varphi(x)=f(x)-\frac{f(b)-f(a)}{b-a}x.$$

Since $f(x)$ is continuous on $[a,b]$ and differentiable within (a,b), $\varphi(x)$ is also continuous on $[a,b]$ and differentiable within (a,b). It is easy to calculate that

$$\varphi(a)=\varphi(b)=\frac{bf(a)-af(b)}{b-a}.$$

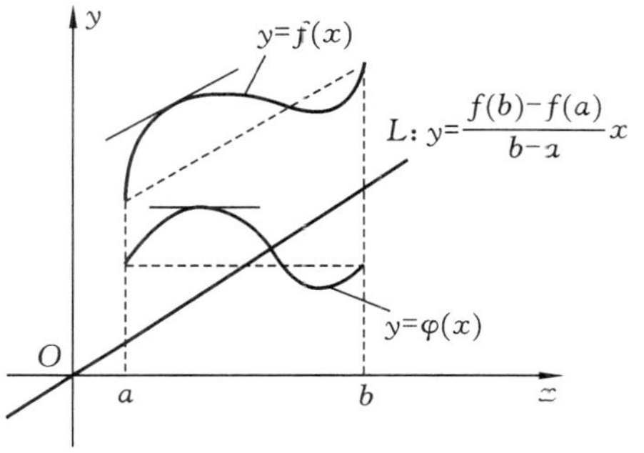

Figure 3-4

Therefore, $\varphi(x)$ satisfies the conditions of Rolle's Theorem on $[a,b]$. According to Rolle's Theorem, there exists at least one point ξ within (a,b), such that $\varphi'(\xi)=0$, i.e.,

$$f'(\xi)-\frac{f(b)-f(a)}{b-a}=0,$$

which is

$$f(b)-f(a)=f'(\xi)(b-a).$$

This is the Lagrange Mean Value Formula, which also holds for $b<a$. It plays a crucial role in differential calculus.

When $f(x)$ is differentiable within (a,b), if $x, x+\Delta x\in(a,b)$, by the Lagrange Mean Value Theorem, there exists ξ between x, $x+\Delta x$, such that

$$f(x+\Delta x)-f(x)=f'(\xi)\Delta x.$$

Since ξ can be expressed as

$$\xi=x+\theta\Delta x\ (0<\theta<1),$$

therefore, equation can be expressed as

$$f(x+\Delta x)-f(x)=f'(x+\theta\Delta x)\Delta x,$$

that is

$$\Delta y=f'(x+\theta\Delta x)\Delta x\ (0<\theta<1).$$

This equation is the finite increment formula.

When $|\Delta x|$ is small, the differential dy can be used as an approximate expression for Δy, that is

$$\Delta y \approx f'(x)\mathrm{d}x.$$

Compared to equation, the former is simpler and easier to compute, requiring $|\Delta x|$ to be small, resulting in an approximate value of Δy. While the latter is an exact expression, with no restriction on the size of Δx, primarily used for theoretical research.

From the Mean Value Theorem of Lagrange, two useful conclusions can be derived in integral calculus.

Corollary 3.1.1 If the derivative of function $f(x)$ is zero at any point within interval (a,b), then $f(x)$ is a constant within (a,b).

Proof Let $\forall x_1, x_2 \in (a,b)$, and $x_1 < x_2$. Obviously, $f(x)$ satisfies the conditions of the Mean Value Theorem of Lagrange on $[x_1, x_2]$, applying equation yields

$$f(x_2) - f(x_1) = f'(\xi)(x_2 - x_1) \quad (x_1 < \xi < x_2).$$

Given that $f'(\xi) = 0$, therefore

$$f(x_2) - f(x_1) = 0.$$

That is

$$f(x_2) = f(x_1).$$

This means that the function values at any two points within (a,b) are equal, so $f(x)$ is a constant within (a,b).

Corollary 3.1.2 If the derivatives of functions $f(x)$ and $g(x)$ are equal at every point within (a,b), then there exists a constant C, such that $f(x) - g(x) = C$, i.e., $f(x) = g(x) + C$.

Proof Given the condition, $\forall x \in (a,b)$, $f'(x) = g'(x)$. Let

$$F(x) = f(x) - g(x).$$

Then $\forall x \in (a,b)$, $F'(x) = 0$.

By Corollary 3.1.1, $F(x)$ is a constant within (a,b), let this constant be C, i.e.,

$$F(x) = f(x) - g(x) = C.$$

Example 3.1.3 Prove that when $x > 0$, $\ln(1+x) < x$.

Proof Let $f(x) = x - \ln(1+x)$. Obviously, when $x > 0$, $f(x)$ satisfies the conditions of Lagrange's Mean Value Theorem on $[0, x]$, thus there exists a point $\xi \in (0, x)$, such that

$$f'(\xi) = \frac{f(x) - f(0)}{x - 0} = \frac{x - \ln(1+x)}{x},$$

and

$$f'(\xi) = 1 - \frac{1}{1+\xi} = \frac{\xi}{1+\xi} > 0.$$

Therefore

$$\frac{x - \ln(1+x)}{x} > 0.$$

Hence, when $x > 0$, $\ln(1+x) < x$.

Exercise 3.1

1. Do the following functions satisfy all the conditions of Rolle's Theorem on the given intervals? If so, find the ξ value in the theorem.

(1) $f(x)=\frac{1}{1+x^2}, x\in[-2,2]$; (2) $f(x)=\ln(\sin x), x\in\left[\frac{\pi}{6},\frac{5\pi}{6}\right]$;

(3) $f(x)=\begin{cases} x, & x\in\left[0,\frac{1}{2}\right) \\ 1-x, & x\in\left[\frac{1}{2},1\right] \end{cases}$; (4) $f(x)=\begin{cases} \left(x-\frac{1}{2}\right)^2, & x\in[0,1) \\ \frac{1}{4x}, & x=1 \end{cases}$.

2. Verify the correctness of Lagrange's Mean Value Theorem for function $f(x)=x^3-3x^2+x-2$ on interval $[-1,0]$.

3. Verify that functions $f(x)=\sin x$ and $g(x)=x+\cos x$ satisfy the conditions of Cauchy's Theorem on $\left[0,\frac{\pi}{2}\right]$.

4. How many zeros does the derivative of the function $f(x)=x(x-1)(x-2)(x+1)(x+2)$ have? And in which intervals are they located?

5. Prove: $\arcsin x+\arccos x=\frac{\pi}{2}$ holds for all $[-1,1]$.

6. Prove: The equation $x^5+x-1=0$ has only one positive real root.

7. Suppose $y=f(x)$ is twice differentiable, and $f(x_1)=f(x_2)=f(x_3)$, where $x_1<x_2<x_3$. Prove: There exists at least one point $\xi\in(x_1,x_3)$, such that $f''(\xi)=0$.

8. Prove the following inequalities.

(1) When $x>0$, $\frac{x}{1+x}<\ln(1+x)<x$;

(2) $|\arctan x-\arctan y|\leqslant|x-y|$;

(3) When $x>1$, $e^x>ex$.

9. If $\lim\limits_{x\to0^+}f(x)=f(0)=0$, and when $x>0$, $f'(x)>0$, prove that when $x>0$, $f(x)>0$.

10. If the function $f(x)$ is continuous on the interval $[x_0,x_0+\delta]$, differentiable within $(x_0,x_0+\delta)$, and

$$\lim_{x\to x_0^+}f'(x)=A.$$

Prove that the right derivative of $f(x)$ at point x_0 exists and equals A.

11. Suppose $f(x)$, $g(x)$ are all differentiable, prove that between any two zeros of $f(x)$, there must be a real root of the equation

$$f'(x)+g'(x)f(x)=0.$$

12. Suppose $f(x)$ is continuous on $[a,b]$, differentiable within (a,b), $a>0$. Prove: There exists a point $\xi\in(a,b)$, such that

$$f(b)-f(a)=\xi f'(\xi)\ln\frac{b}{a}.$$

3.2 L'Hôpital's Rule

It is already known in limits: When the independent variable x has a certain trend (such as $x \to x_0$), if $f(x)$, $g(x)$ are all infinitesimals (or infinite quantities), then the limit $\lim\limits_{x \to x_0} \frac{f(x)}{g(x)}$ may or may not exist. This is usually referred to as an indeterminate form.

The limit is called an indeterminate form, denoted as "$\frac{0}{0}$" (or "$\frac{\infty}{\infty}$" type). Some methods for determining indeterminate forms have been introduced in Chapter 1, and formula $\lim\limits_{x \to 0} \frac{\sin x}{x} = 1$ is an example of the "$\frac{0}{0}$" type determination method. This section will introduce a new method for solving such problems using derivatives—L'Hôpital's Rule.

Theorem 3.2.1 Let $f(x)$ and $g(x)$ satisfy the conditions:

(1) $\lim\limits_{x \to x_0} f(x) = \lim\limits_{x \to x_0} g(x) = 0$;

(2) Differentiable in a neighborhood $\mathring{U}(x_0)$ around the point x_0, and $g'(x) \neq 0$;

(3) $\lim\limits_{x \to x_0} \frac{f'(x)}{g'(x)} = A\ (\infty)$.

Then we have

$$\lim_{x \to x_0} \frac{f(x)}{g(x)} = \lim_{x \to x_0} \frac{f'(x)}{g'(x)} = A\ (\infty).$$

Proof Since the limit value at a point is independent of the function value, we can supplement the definition of $f(x_0) = g(x_0) = 0$. Thus, $f(x)$, $g(x)$ is continuous at the point x_0. Let any $x \in U(x_0)$ satisfy the conditions of the Cauchy Mean Value Theorem on the interval with x_0, x as the endpoint, therefore we have

$$\frac{f(x)}{g(x)} = \frac{f(x) - f(x_0)}{g(x) - g(x_0)} = \frac{f'(\xi)}{g'(\xi)} \quad (\xi \text{ is between } x_0, x).$$

Let $x \to x_0$, then $\xi \to x_0$, taking the limit of both sides of the equation, combined with condition (3), we obtain

$$\lim_{x \to x_0} \frac{f(x)}{g(x)} = \lim_{x \to x_0} \frac{f'(\xi)}{g'(\xi)} = \lim_{x \to x_0} \frac{f'(x)}{g'(x)} = A\ (\infty).$$

Theorem 3.2.1 states: If $\lim\limits_{x \to x_0} \frac{f(x)}{g(x)}$ is an indeterminate form of type "$\frac{0}{0}$", and if the value of $\lim\limits_{x \to x_0} \frac{f'(x)}{g'(x)}$ can be found or determined, then the limit problem of the indeterminate form $\lim\limits_{x \to x_0} \frac{f(x)}{g(x)}$ is resolved, at this point,

$$\lim_{x \to x_0} \frac{f(x)}{g(x)} = \lim_{x \to x_0} \frac{f'(x)}{g'(x)}.$$

This method, which determines the value of an indeterminate form by separately finding the derivatives of the numerator and denominator and then taking the limit, is known as L'Hôpital's Rule.

If $\lim_{x \to x_0} \frac{f'(x)}{g'(x)}$ remains an indeterminate form of type "$\frac{0}{0}$", and if $f'(x), g'(x)$ satisfies the conditions required by Theorem 3.2.1 for $f(x), g(x)$, then L'Hôpital's Rule can be applied again.

Example 3.2.1 Find $\lim_{x \to \pi} \frac{1+\cos x}{\tan^2 x}$.

Solution This limit is of the form "$\frac{0}{0}$", applying L'Hôpital's Rule, we get

$$\lim_{x \to \pi} \frac{1+\cos x}{\tan^2 x} = \lim_{x \to \pi} \frac{(1+\cos x)'}{(\tan^2 x)'} = \lim_{x \to \pi} \frac{-\sin x}{2\tan x \sec^2 x} = -\lim_{x \to \pi} \frac{\cos^3 x}{2} = -\frac{(-1)^3}{2} = \frac{1}{2}.$$

Example 3.2.2 Find $\lim_{x \to 0} \frac{\sin 2x + 3x}{\sin x - 2x}$.

Solution This limit is of the form "$\frac{0}{0}$", applying L'Hôpital's Rule, we get

$$\lim_{x \to 0} \frac{\sin 2x + 3x}{\sin x - 2x} = \lim_{x \to 0} \frac{(\sin 2x + 3x)'}{(\sin x - 2x)'} = \lim_{x \to 0} \frac{2\cos 2x + 3}{\cos x - 2} = \frac{2 \times 1 + 3}{1-2} = -5.$$

Note: In the above equation, $\lim_{x \to 0} \frac{2\cos 2x + 3}{\cos x - 2}$ is no longer an indeterminate form, and applying L'Hôpital's Rule to it would lead to an error.

Example 3.2.3 Find $\lim_{x \to 0} \frac{e^x - \cos x}{x \ln(1+x)}$.

Solution Since $\ln(1+x) \sim x$ as $x \to 0$, therefore,

$$\lim_{x \to 0} \frac{e^x - \cos x}{x \ln(1+x)} = \lim_{x \to 0} \frac{e^x - \cos x}{x^2} = \lim_{x \to 0} \frac{(e^x - \cos x)'}{(x^2)'} = \lim_{x \to 0} \frac{e^x + \sin x}{2x} = \infty.$$

Example 3.2.4 Find $\lim_{x \to 0} \frac{x - \sin x}{x^3}$.

Solution

$$\lim_{x \to 0} \frac{x - \sin x}{x^3} = \lim_{x \to 0} \frac{(x - \sin x)'}{(x^3)'} = \lim_{x \to 0} \frac{1 - \cos x}{3x^2}$$
$$= \lim_{x \to 0} \frac{(1 - \cos x)'}{(3x^2)'} = \lim_{x \to 0} \frac{\sin x}{6x} = \frac{1}{6}.$$

In this problem, when solving for $\lim_{x \to 0} \frac{1 - \cos x}{3x^2}$, one can avoid using L'Hôpital's Rule and directly replace the numerator $1 - \cos x$ with its equivalent infinitesimal $\frac{1}{2}x^2$, which simplifies the calculation.

Note: For the "$\frac{0}{0}$" type, L'Hôpital's Rule only holds when $\lim\limits_{x\to x_0}\frac{f'(x)}{g'(x)}$ exists or tends to infinity. If the limit status of $\lim\limits_{x\to x_0}\frac{f'(x)}{g'(x)}$ cannot be determined or can be asserted that it oscillates without a limit, then L'Hôpital's Rule fails. In such cases, other methods are required to determine $\lim\limits_{x\to x_0}\frac{f(x)}{g(x)}$.

Example 3.2.5 Solve for $\lim\limits_{x\to 0}\frac{x^2\sin\frac{1}{x}}{\sin x}$.

Solution This problem falls into the "$\frac{0}{0}$" type. If L'Hôpital's Rule is applied, we get

$$\lim_{x\to 0}\frac{x^2\sin\frac{1}{x}}{\sin x}=\lim_{x\to 0}\frac{2x\sin\frac{1}{x}-\cos\frac{1}{x}}{\cos x}.$$

The right-hand side oscillates without a limit, indicating that the rule is ineffective here, and the equality does not hold. In fact,

$$\lim_{x\to 0}\frac{x^2\sin\frac{1}{x}}{\sin x}=\lim_{x\to 0}\left(\frac{x}{\sin x}\cdot x\sin\frac{1}{x}\right)=1\times 0=0.$$

There is also a corresponding L'Hôpital's Rule for the "$\frac{\infty}{\infty}$" type of indeterminate form.

Theorem 3.2.2 Suppose the function $f(x)$, $g(x)$ satisfies:

(1) $\lim\limits_{x\to x_0}f(x)=\infty$, $\lim\limits_{x\to x_0}g(x)=\infty$;

(2) It is differentiable within some deleted neighborhood $\mathring{U}(x_0)$ of point x_0, and $g'(x)\neq 0$;

(3) $\lim\limits_{x\to x_0}\frac{f'(x)}{g'(x)}=A\,(\infty)$.

Then it must follow that

$$\lim_{x\to x_0}\frac{f(x)}{g(x)}=\lim_{x\to x_0}\frac{f'(x)}{g'(x)}=A\,(\infty).$$

Example 3.2.6 Solve for $\lim\limits_{x\to\frac{\pi}{2}}\frac{\tan x}{\tan 3x}$.

Solution $$\lim_{x\to\frac{\pi}{2}}\frac{\tan x}{\tan 3x}=\lim_{x\to\frac{\pi}{2}}\frac{(\tan x)'}{(\tan 3x)'}=\lim_{x\to\frac{\pi}{2}}\frac{\sec^2 x}{3\sec^2 3x}=\lim_{x\to\frac{\pi}{2}}\frac{\cos^2 3x}{3\cos^2 x}=\frac{1}{3}\left(\lim_{x\to\frac{\pi}{2}}\frac{\cos 3x}{\cos x}\right)^2$$
$$=\frac{1}{3}\left(\lim_{x\to\frac{\pi}{2}}\frac{-3\sin 3x}{-\sin x}\right)^2=\frac{1}{3}\left[\frac{3\cdot(-1)}{1}\right]^2=3.$$

When replacing $x\to x_0$ with $x\to\infty$ or any other trend in Theorem 3.2.1 and Theorem 3.2.2, L'Hôpital's Rule remains valid.

Example 3.2.7 Find $\lim\limits_{x\to+\infty}\dfrac{x^{\alpha}}{\ln x}$ $(\alpha>0)$.

Solution $\lim\limits_{x\to+\infty}\dfrac{x^{\alpha}}{\ln x}=\lim\limits_{x\to+\infty}\dfrac{(x^{\alpha})'}{(\ln x)'}=\lim\limits_{x\to+\infty}\dfrac{\alpha x^{\alpha-1}}{\dfrac{1}{x}}=\lim\limits_{x\to+\infty}\alpha x^{\alpha}=+\infty.$

This result indicates that any positive power function converges to infinity much faster than the natural logarithm when $x\to+\infty$.

Example 3.2.8 Find $\lim\limits_{x\to+\infty}\dfrac{x^{n}}{e^{\lambda x}}$ $(n\in\mathbf{N},\lambda>0)$.

Solution $\lim\limits_{x\to+\infty}\dfrac{x^{n}}{e^{\lambda x}}=\lim\limits_{x\to+\infty}\dfrac{(x^{n})'}{(e^{\lambda x})'}=\lim\limits_{x\to+\infty}\dfrac{nx^{n-1}}{\lambda e^{\lambda x}}=\lim\limits_{x\to+\infty}\dfrac{n(n-1)x^{n-2}}{\lambda^{2}e^{\lambda x}}=\cdots=\lim\limits_{x\to+\infty}\dfrac{n!}{\lambda^{n}e^{\lambda x}}=0.$

Replacing n in this example with any positive number α does not change the conclusion. This demonstrates that, when $x\to+\infty$, the exponential function $e^{\lambda x}$ $(\lambda>0)$ converges to infinity much faster than any positive power function x^{α} $(\alpha>0)$. When using L'Hôpital's Rule, the following points should be noted:

(1) The form of application must be of the "$\dfrac{0}{0}$" or "$\dfrac{\infty}{\infty}$" type, otherwise, this rule cannot be used.

(2) When applying the rule, the derivatives of the numerator and the denominator must be found separately, not the derivative of the entire fraction.

(3) It is important to simplify the expression under the limit symbol promptly, and after simplification, check whether it remains of the "$\dfrac{0}{0}$" or "$\dfrac{\infty}{\infty}$" type. If not, L'Hôpital's Rule should be discontinued immediately.

(4) When applying the rule, it should be combined effectively with other methods of finding limits to simplify the calculations as much as possible.

In addition to the "$\dfrac{0}{0}$" type and "$\dfrac{\infty}{\infty}$" type, there are also "$0\cdot\infty$" type, "$\infty-\infty$" type, "1" type, "0" type, and "∞" type indeterminate forms. These indeterminate forms can be transformed into "$\dfrac{0}{0}$" type or "$\dfrac{\infty}{\infty}$" type through appropriate transformations for calculation.

Example 3.2.9 Find $\lim\limits_{x\to+\infty}x\left(\dfrac{\pi}{2}-\arctan x\right)$.

Solution Solving this limit is of the "$\infty\cdot 0$" type, transforming it into the "$\dfrac{0}{0}$" type before applying L'Hôpital's Rule (some "$\infty\cdot 0$" types can be first transformed into the "$\dfrac{\infty}{\infty}$" type before applying the rule).

$$\lim_{x\to+\infty} x\left(\frac{\pi}{2}-\arctan x\right)=\lim_{x\to+\infty}\frac{\frac{\pi}{2}-\arctan x}{\frac{1}{x}}=\lim_{x\to+\infty}\frac{\left(\frac{\pi}{2}-\arctan x\right)'}{\left(\frac{1}{x}\right)'}$$

$$=\lim_{x\to+\infty}\frac{-\frac{1}{1+x^2}}{-\frac{1}{x^2}}=\lim_{x\to+\infty}\frac{x^2}{1+x^2}=1.$$

Example 3.2.10 Find $\lim_{x\to1}\left(\frac{1}{\ln x}-\frac{x}{x-1}\right)$.

Solution Solving this problem is of the "$\infty-\infty$" type, first simplifying it into the "$\frac{0}{0}$" type, then applying L'Hôpital's Rule.

$$\lim_{x\to1}\left(\frac{1}{\ln x}-\frac{x}{x-1}\right)=\lim_{x\to1}\frac{x-1-x\ln x}{(x-1)\ln x}=\lim_{x\to1}\frac{1-\ln x-1}{\ln x+\frac{x-1}{x}}$$

$$=\lim_{x\to1}\frac{-\ln x}{\ln x+1-\frac{1}{x}}=\lim_{x\to1}\frac{-\frac{1}{x}}{\frac{1}{x}+\frac{1}{x^2}}=\frac{-1}{1+1}=-\frac{1}{2}.$$

Example 3.2.11 Find $\lim_{x\to0}(\cos x)^{\frac{1}{x^2}}$.

Solution Solving this is of the "1^∞" type, since $f^g=e^{g\ln f}\ (f>0)$, therefore

$$\lim_{x\to0}(\cos x)^{\frac{1}{x^2}}=\lim_{x\to0}e^{\frac{1}{x^2}\ln\cos x}=e^{\lim_{x\to0}\frac{1}{x^2}\ln\cos x},$$

and

$$\lim_{x\to0}\frac{1}{x^2}\ln\cos x=\lim_{x\to0}\frac{\ln\cos x}{x^2}=\lim_{x\to0}\frac{-\frac{\sin x}{\cos x}}{2x}=\lim_{x\to0}\left(-\frac{\sin x}{x}\frac{1}{2\cos x}\right)=-\frac{1}{2}.$$

thus

$$\lim_{x\to0}(\cos x)^{\frac{1}{x^2}}=e^{-\frac{1}{2}}.$$

For the "∞" and "0" type indeterminate forms, their evaluation methods can be analogized from Example 3.2.11.

Exercise 3.2

1. In the following limit calculations, L'Hôpital's Rule has been applied, are there any errors in the solutions?

(1) $\lim_{x\to0}\frac{x^2+1}{x-1}=\lim_{x\to0}\frac{(x^2+1)'}{(x-1)'}=\lim_{x\to0}\frac{2x}{1}=0$;

(2) $\lim\limits_{x\to\infty}\dfrac{\sin x+x}{x}=\lim\limits_{x\to\infty}\dfrac{(\sin x+x)'}{x'}=\lim\limits_{x\to\infty}\dfrac{\cos x+1}{1}$, the limit does not exist.

2. Use L'Hôpital's Rule to find the following limits.

(1) $\lim\limits_{x\to 0}\dfrac{e^x-e^{-x}}{x}$;

(2) $\lim\limits_{x\to 1}\dfrac{x^3-3x^2+2}{x^3-x^2-x+1}$;

(3) $\lim\limits_{x\to\frac{\pi}{2}}\dfrac{\ln(\sin x)}{(\pi-2x)^2}$;

(4) $\lim\limits_{x\to+\infty}\dfrac{\ln\left(1+\dfrac{1}{x}\right)}{\operatorname{arccot} x}$;

(5) $\lim\limits_{x\to\infty}\dfrac{\ln(x^4-x^3+1)}{\ln(x^2+2x+1)}$;

(6) $\lim\limits_{x\to 0}\left(\cot x-\dfrac{1}{x}\right)$;

(7) $\lim\limits_{x\to 0}\left(\dfrac{1}{x}-\dfrac{1}{e^x-1}\right)$;

(8) $\lim\limits_{x\to -1}\left[\dfrac{1}{x+1}-\dfrac{1}{\ln(x+2)}\right]$;

(9) $\lim\limits_{x\to\infty}x(e^{\frac{1}{x}}-1)$;

(10) $\lim\limits_{x\to 1}(1-x)\tan\dfrac{\pi x}{2}$;

(11) $\lim\limits_{x\to 0^+}\sin x\ln x$;

(12) $\lim\limits_{x\to 0}(1+\sin x)^{\frac{1}{x}}$;

(13) $\lim\limits_{x\to 0^+}\left(\ln\dfrac{1}{x}\right)^x$;

(14) $\lim\limits_{x\to\frac{\pi}{2}^-}(\cos x)^{\frac{\pi}{2}-x}$.

3. Verify the existence of the limit $\lim\limits_{x\to+\infty}\dfrac{e^x+e^{-x}}{e^x-e^{-x}}$, but it cannot be solved using L'Hôpital's Rule.

4. Discuss the continuity of the function $f(x)=\begin{cases}(\cos x)^{\frac{\pi}{2}-x}, & x<\dfrac{\pi}{2}\\ 1, & x\geqslant\dfrac{\pi}{2}\end{cases}$, at $x=\dfrac{\pi}{2}$.

5. Suppose $f(x)$ is twice differentiable, prove:

$$\lim_{h\to 0}\frac{f(x+2h)-2f(x+h)+f(x)}{h^2}=f''(x).$$

6. Find the limit $\lim\limits_{x\to+\infty}\left(\dfrac{a_1^{\frac{1}{n}}+a_2^{\frac{1}{n}}+\cdots+a_n^{\frac{1}{n}}}{n}\right)^{nx}$, where $a_1,a_2,\cdots,a_n>0$.

7. Given $\lim\limits_{x\to 0}\dfrac{2\arctan x-\ln\dfrac{1+x}{1-x}}{x^P}=C\neq 0$, find the values of P and C.

8. Suppose $f(x)=\begin{cases}\dfrac{g(x)}{x}, & x\neq 0\\ 0, & x=0\end{cases}$, and $g(x)$ is twice differentiable, $g(0)=g'(0)=0$, $g''(0)=17$, find $f'(0)$.

3.3 Monotonicity and Extremum of Functions

3.3.1 Monotonicity of Functions

The concept of function monotonicity was introduced in Chapter 1, but determining the monotonicity of a function using its definition is often difficult. This subsection will introduce a simple and practical method for determining monotonicity.

Geometrically, if the function $y=f(x)$ is monotonically increasing (decreasing) on $[a,b]$, then when viewed from left to right, its graph should be an ascending (descending) curve. As shown in Figure 3-5, the inclination of the tangent at each point on the ascending curve is an acute angle $\left(0<\alpha<\frac{\pi}{2}\right)$, i.e., the slope of the tangent $k=f'(x)=\tan\alpha>0$. Conversely, the inclination of the tangent at each point on the descending curve is an obtuse angle $\left(\frac{\pi}{2}<\alpha<\pi\right)$, i.e., the slope of the tangent $k=f'(x)=\tan\alpha<0$. It can be seen that the monotonicity of the function is related to the sign of the derivative. There is generally the following theorem.

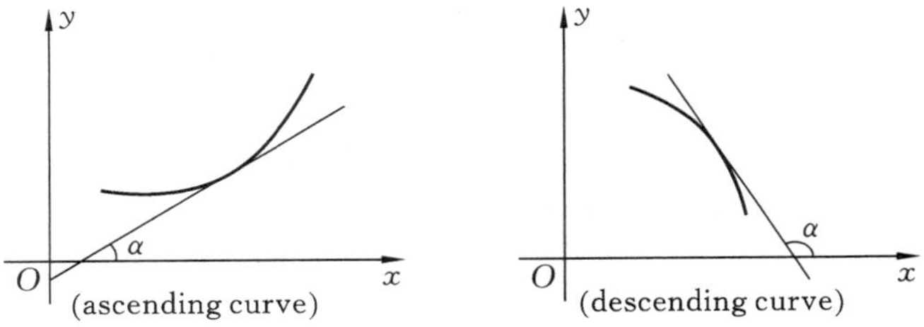

Figure 3-5

Theorem 3.3.1 Let the function $y=f(x)$ be continuous on $[a,b]$ and differentiable within (a,b).

(1) If for any $x\in(a,b)$ there is $f'(x)>0$, then $f(x)$ is monotonically increasing on $[a,b]$.

(2) If for any $x\in(a,b)$ there is $f'(x)<0$, then $f(x)$ is monotonically decreasing on $[a,b]$.

Proof We only prove the case of $f'(x)>0$.

Let x_1, x_2 be any two points on $[a,b]$ with $x_1<x_2$, and from the given conditions, it is known that: $f(x)$ satisfies the conditions of the Lagrange Mean Value Theorem on $[x_1,x_2]$. Therefore, there exists a point $\xi\in(x_1,x_2)$, such that

$$f(x_2)-f(x_1)=f'(\xi)(x_2-x_1).$$

Since $\forall x\in(a,b)$, $f'(x)>0$, it follows that $f'(\xi)>0$. Therefore,

$$f(x_2)-f(x_1)>0,$$

that is,

$$f(x_2)>f(x_1).$$

That is, the function $f(x)$ is monotonically increasing on $[a,b]$.

From the above proof process, it can be seen that replacing the interval $[a,b]$ in the theorem with other forms of intervals (including infinite intervals), the conclusion still holds.

Example 3.3.1 Determine the monotonicity of the function $y=x+\arctan x$.

Solution Since $y'=1+\dfrac{1}{1+x^2}>0$, $y=x+\arctan x$ is monotonically increasing within its domain $\mathbf{R}$. At this point, the function is also referred to as a monotonically increasing function.

Example 3.3.2 Discuss the monotonicity of the function $y=xe^x$.

Solution Clearly, the domain of $y=xe^x$ is $(-\infty,+\infty)$. Within this interval, $y=xe^x$ is continuous, differentiable, and

$$y'=e^x+xe^x=(1+x)e^x.$$

When $x\in(-\infty,-1)$, $y'<0$, so $y=xe^x$ is monotonically decreasing on $(-\infty,-1]$. When $x\in(-1,+\infty)$, $y'>0$, so $y=xe^x$ is monotonically increasing on $[-1,+\infty)$.

This example illustrates that the sign of the derivative of the same function $f(x)$ varies with changes in the independent variable. It may be positive in some small intervals and negative in others. Generally, to determine the sign of $f'(x)$, one can first find the points (stationary points) where $f'(x)=0$ and points where the derivative does not exist (if any), and use these points to divide the domain into several subintervals. Then, discuss the sign of $f'(x)$ within each subinterval to determine the monotonicity of the function over the corresponding intervals.

Example 3.3.3 Determine the monotonic intervals of the function $y=\dfrac{\ln x}{x}$.

Solution The domain of $y=\dfrac{\ln x}{x}$ is $(0,+\infty)$, and

$$y'=\frac{1-\ln x}{x^2}.$$

Let $y'=0$, yielding $x=e$. Using $x=e$ to divide the domain, $y=\dfrac{\ln x}{x}$ is monotonically decreasing within $(0,e]$ and monotonically decreasing within $[e,+\infty)$.

Example 3.3.4 Discuss the monotonicity of the function $f(x)=\dfrac{1}{3}x^3-x^2+\dfrac{1}{3}$.

Solution
$$f'(x)=x^2-2x=x(x-2).$$

Let $f'(x)=0$, yielding $x_1=0$, $x_2=2$. Clearly, $f(x)$ has no points of non-differentiability. Using the stationary points x_1, x_2 to divide the domain $\mathbf{R}$ of the function into three subintervals, the function $f(x)=\frac{1}{3}x^3-x^2+\frac{1}{3}$ is monotonically increasing on $(-\infty,0]$ and $[2,+\infty)$ and monotonically decreasing on $[0,2]$.

Note: In Theorem 3.3.1, any $x\in(a,b)$, $f'(x)>0\,(<0)$ is a sufficient condition for $f(x)$ to be monotonically increasing (decreasing) on $[a,b]$, but not a necessary condition.

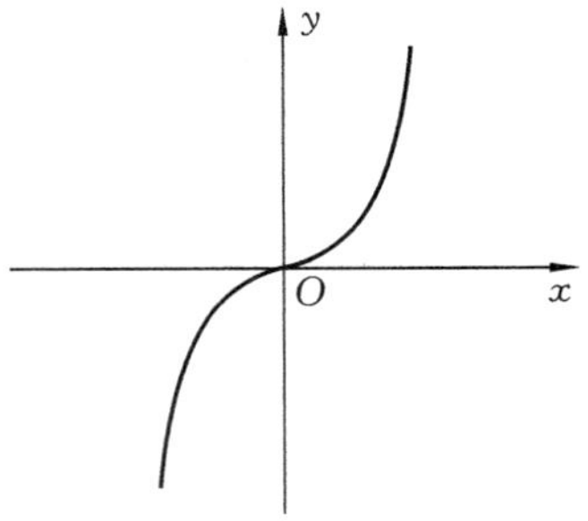

Figure 3-6

In fact, at individual points within the interval, $f'(x)=0$ does not affect the monotonicity of the function across the entire interval. That is, within (a,b), if $f'(x)\geqslant 0\,(\leqslant 0)$, the equality holds only at individual points, then $f(x)$ is still monotonically increasing (decreasing) within (a,b).

For example, the function $y=x^3$ (Figure 3-6), $y'=3x^2\geqslant 0$, when $x\neq 0$, $y'>0$, so $y=x^3$ is monotonically increasing on $(-\infty,0]$ and $[0,+\infty)$, and thus monotonically increasing on its domain **R**.

Example 3.3.5 Prove that when $x>0$, $x-\frac{x^2}{2}<\ln(1+x)$.

Proof Let $f(x)=x-\frac{x^2}{2}-\ln(1+x)$. When $x>0$,

$$f'(x)=1-x-\frac{1}{1+x}=\frac{-x^2}{1+x}<0.$$

Therefore, $f(x)$ is monotonically decreasing on $[0,+\infty)$. Since $f(0)=0$, thus

$$f(x)<f(0)=0,$$

that is,

$$x-\frac{x^2}{2}-\ln(1+x)<0.$$

Consequently, when $x>0$,

$$x-\frac{x^2}{2}<\ln(1+x).$$

3.3.2 Extreme Values of Functions

As shown in Figure 3-7, the continuous curve $y=f(x)$ has a vertical coordinate $f(x_0)$ at the peak point M that is larger than the vertical coordinates $f(x)$ of nearby points. That is, within a neighborhood $U(x_0)$ of point x_0, $f(x_0)$ is the maximum value of the function, and such a local maximum value is referred to as an extreme value of the function $f(x)$. Generally, the following definition applies:

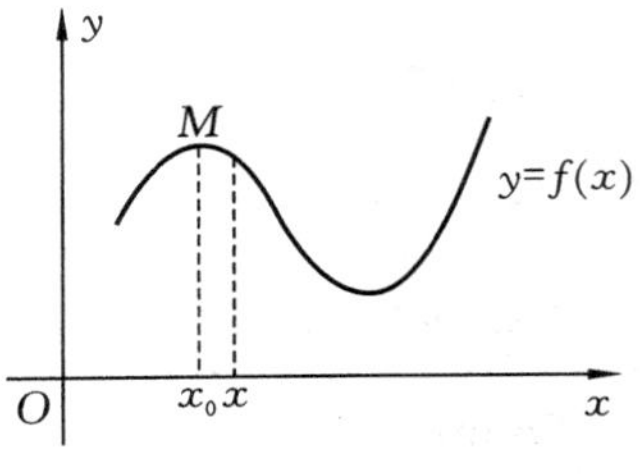

Figure 3-7

Definition 3.3.1 Let the function $f(x)$ be defined in a neighborhood $U(x_0)$ of point x_0. If for any $x \in U(x_0)$, there exists

(1) $f(x)<f(x_0)$, then $f(x_0)$ is called the maximum value of the function $f(x)$, and x_0 is the point of maximum value of $f(x)$.

(2) $f(x)>f(x_0)$, then $f(x_0)$ is called the minimum value of the function $f(x)$, and x_0 is the point of minimum value of $f(x)$.

The maximum and minimum values of a function are collectively referred to as extreme values. The points of maximum and minimum values are collectively referred to as extreme points.

Note: The concept of extreme values is local, differing from the meaning of the maximum or minimum values. As shown in Figure 3-8, a continuous function $y=f(x)$ on a closed interval $[a,b]$ has two maximum points $f(x_2)$, $f(x_4)$ and three minimum points $f(x_1)$, $f(x_3)$, $f(x_5)$. Among them, the minimum point $f(x_1)$ is larger than the maximum point $f(x_4)$. From this figure, it can also be seen that the minimum point $f(x_3)$ is also the smallest value of the function $f(x)$ on $[a,b]$, while the other extreme values are not the smallest or largest values, and the largest value of $f(x)$ is $f(a)$.

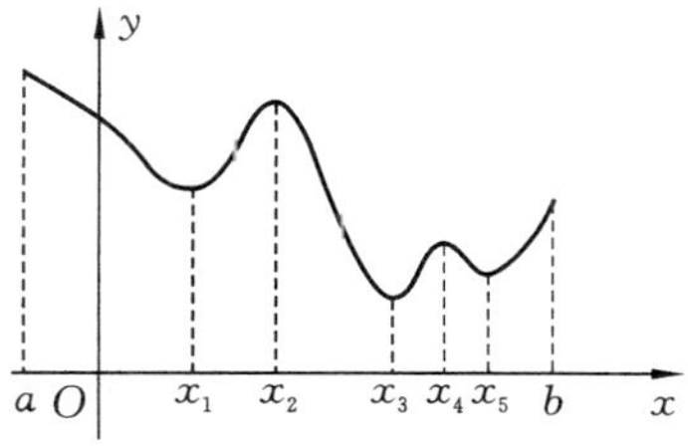

Figure 3-8

The following discusses how to find extreme values.

According to the definition of extreme values, if the function $y=f(x)$ is continuous in a neighborhood of the extreme point x_0, then the point $(x_0, f(x_0))$ must be a peak or a trough of the curve $y=f(x)$ (as shown in Figure 3-9 and Figure 3-10).

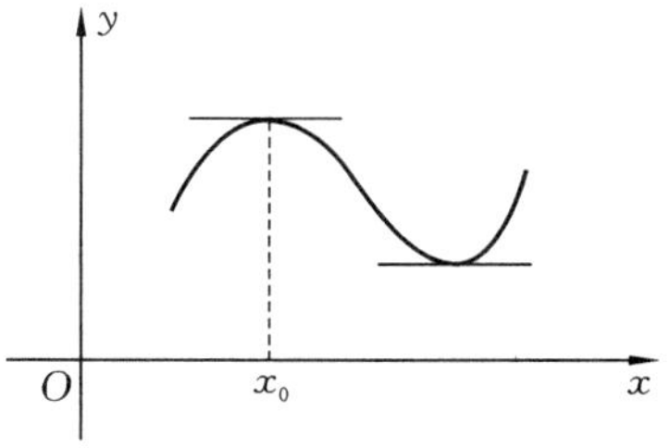

Figure 3-9

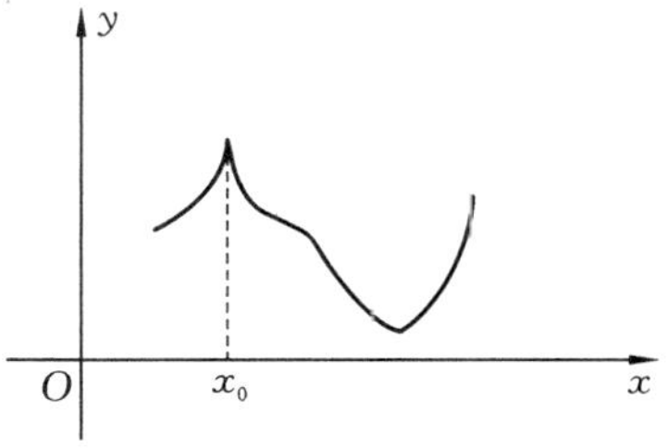

Figure 3-10

If $f'(x_0)$ exists, then $y=f(x)$ has a horizontal tangent at point $(x_0, f(x_0))$ (as shown in Figure 3-9), and by Fermat's Theorem, $f'(x_0)=0$.

If the peak (valley) point is not smooth (Figure 3-11), i.e., $y=f(x)$ has no tangent or only a tangent perpendicular to the x axis at point $(x_0, f(x_0))$, then the derivative $f'(x_0)$ of the function at the extreme point x_0 does not exist.

Thus, the necessary condition for the existence of an extreme value is obtained: The extreme point must be a stationary point or a non-differentiable point.

Theorem 3.3.2 Suppose the function $f(x)$ is differentiable at point x_0, and $f(x_0)$ is an extreme value, then $f'(x_0)=0$.

The converse of this theorem is not true. That is, a stationary point is not necessarily an extreme point. For example, $y=x^3$ has a stationary point $x=0$, and clearly $f(0)$ is not an extreme value of the function $y=x^3$. Similarly, a point where the derivative does not exist is not necessarily an extreme point. For instance, $\sqrt[3]{x}$, $f'(0)$ does not exist, and $f(0)$ is not an extreme value of $y=\sqrt[3]{x}$ (Figure 3-11).

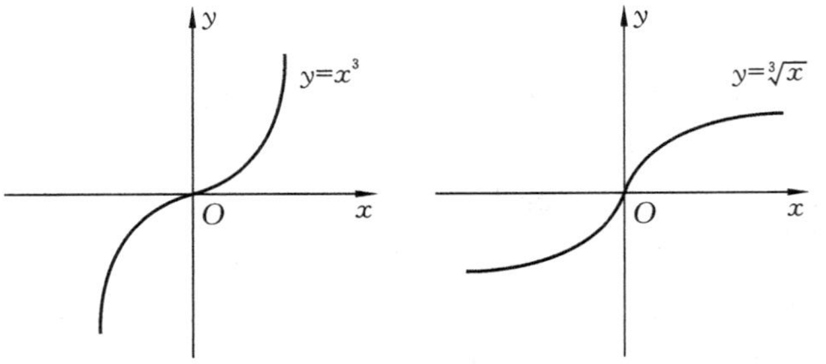

Figure 3-11

What kind of stationary points and non-differentiable points are considered extreme points? Observing Figures 3-10 and 3-11, it can be seen that for a continuous function, the left neighborhood of a peak point (corresponding to a maximum) is rising, and the right neighborhood is falling, i. e., as one passes through the maximum point x_0 from left to right, $f'(x)$ changes from positive to negative. Similarly, when passing through the minimum point x_0, $f'(x)$ changes from negative to positive.

Theorem 3.3.3 (First Sufficient Condition) Suppose the function $f(x)$ is continuous at point x_0 within a neighborhood $U(x_0)$, and differentiable within $\mathring{U}(x_0, \delta)$.

(1) If $f'(x)>0$ when $x\in(x_0-\delta, x_0)$, and $f'(x)<0$ when $x\in(x_0, x_0+\delta)$, then $f(x)$ attains a maximum at point x_0.

(2) If $f'(x)<0$ when $x\in(x_0-\delta, x_0)$, and $f'(x)>0$ when $x\in(x_0, x_0+\delta)$, then $f(x)$ attains a minimum at point x_0.

(3) If the sign of $f'(x)$ remains unchanged when $x\in\mathring{U}(x_0, \delta)$, then $f(x)$ has no extreme value at point x_0.

Proof

(1) It is known from the conditions and the monotonicity criterion that $f(x)$ is increasing on $(x_0-\delta, x_0]$ and decreasing on $[x_0, x_0+\delta)$, hence for any $x\in\mathring{U}(x_0, \delta)$, there is $f(x)<f(x_0)$, so $f(x_0)$ is the maximum value of the function $f(x)$.

The proofs of (2) and (3) can be carried out similarly.

According to Theorem 3.3.2 and Theorem 3.3.3, the method for finding the extreme values of the function $f(x)$ is summarized as follows:

(1) Determine the domain $D(f)$ of the function $y=f(x)$.

(2) Find the derivative $f'(x)$.

(3) Find the stationary points and points of non-differentiability $x_1, x_2, \cdots, x_n$ of the function $f(x)$.

(4) Divide the domain $D(f)$ into several subintervals using $x_1, x_2, \cdots, x_n$, and discuss the sign of $f'(x)$ within each subinterval.

(5) Determine which points are extreme points according to Theorem 3.3.3, and if they are extreme points, determine whether they are maximum or minimum points, and further find the function values at these points—the extreme values.

Example 3.3.6 Find the extreme values of the function $f(x)=(x-1)^3(2x+3)^2$.

Solution $f(x)$ is differentiable everywhere within its domain **R**, and

$f'(x)=3(x-1)^2(2x+3)^2+4(x-1)^3(2x+3)=(x-1)^2(2x+3)(10x+5)$.

Let $f'(x)=0$, yielding the stationary point $x_1=-\frac{3}{2}, x_2=-\frac{1}{2}, x_3=1$. Divide the domain into 4 subintervals using x_1, x_2, x_3, the function $f(x)=(x-1)^3(2x+3)^2$ has a maximum at $x=-\frac{3}{2}$ $f\left(-\frac{3}{2}\right)=0$. At $x=-\frac{1}{2}$, it has a minimum $f\left(-\frac{1}{2}\right)=-\frac{27}{2}$.

Example 3.3.7 Find the intervals of monotonicity and extreme values of the function $f(x)=x-\frac{3}{2}x^{\frac{2}{3}}$.

Solution The domain of definition of $f(x)$ is $D(f)=(-\infty,+\infty)$.

$$f'(x)=1-x^{-\frac{1}{3}}=\frac{\sqrt[3]{x}-1}{\sqrt[3]{x}}.$$

$f(x)$ is not differentiable at point $x=0$, let $f'(x)=0$, obtaining the stationary point $x=1$. Divide the domain into 3 subintervals using $x=0, x=1$, $f(x)=x-\frac{3}{2}x^{\frac{3}{2}}$ increases on $(-\infty,0]$ and $[1,+\infty)$ and decreases on $[0,1]$, and has a maximum $f(0)=1$, a minimum $f(1)=-\frac{1}{2}$.

When the second derivative of the function $f(x)$ exists and is non-zero at the stationary point, the following method can be used to determine whether the stationary point is an extreme point.

Theorem 3.3.4 (Second Sufficient Condition) Suppose the function $f(x)$ is twice differentiable at point x_0, and $f'(x_0)=0, f''(x_0)\neq 0$, then

(1) If $f''(x_0)<0$, then x_0 is a maximum point of $f(x)$.

(2) If $f''(x_0)>0$, then x_0 is a local minimum point of $f(x)$.

Proof

(1) Given $f''(x_0)<0, f'(x_0)=0$, according to the definition of the second derivative,

$$f''(x_0)=\lim_{x\to x_0}\frac{f'(x)-f'(x_0)}{x-x_0}=\lim_{x\to x_0}\frac{f'(x)}{x-x_0}<0.$$

By the local preservation of sign for function limits, there exists a deleted neighborhood $U(x_0, \delta)$ of point x_0 such that when $x \in U(x_0, \delta)$,

$$\frac{f'(x)}{x-x_0}<0.$$

It follows that: When $x \in (x_0-\delta, x_0)$, $f'(x)>0$. When $x \in (x_0, x_0+\delta)$, $f'(x)<0$. According to Theorem 3.3.3, $f(x)$ attains a maximum at point x_0.

(2) Similarly, it can be proven that when $f''(x_0)>0$, $f'(x_0)=0$, $f(x_0)$ is a minimum.

Example 3.3.8 Find the extrema of the function $y=x^3-9x^2+15x+3$.

Solution $$y'=3x^2-18x+15=3(x-1)(x-5).$$

Let $y'=0$, obtaining the stationary point $x_1=1$, $x_2=5$. Also, since

$$y''=6x-18, y''|_{x=1}=-12<0, y''|_{x=5}=12>0.$$

Therefore, $x_1=1$, $x_2=5$ are the points of maximum and minimum for the function, respectively, thus the function $y=f(x)$ has a maximum value $y|_{x=1}=10$ and a minimum value $y|_{x=5}=-22$.

The method of finding extreme values using the second sufficient condition is simple, but its applicable scope is narrow. For instance, at point x_0, $f'(x_0)$ does not exist, or at a stationary point $f''(x_0)$ does not exist, or at a stationary point $f''(x_0)$ exists but equals 0. These situations cannot be addressed by Theorem 3.3.4, and can only be determined using Theorem 3.3.3 (the first sufficient condition) to check whether x_0 is an extreme point.

Exercise 3.3

1. Prove: The function $y=x-\ln(1+x^2)$ is monotonically increasing.
2. Prove: The function $y=\sin x-x$ is monotonically decreasing.
3. Find the intervals of increase and decrease for the following functions:

(1) $y=x^4-2x^2+2$; (2) $y=2x+\frac{8}{x}$;

(3) $y=\sqrt{2x-x^2}$; (4) $y=2x^2-\ln x$;

(5) $y=x^n e^{-x}$ $(n>0, x\geqslant 0)$; (6) $y=2x^2(x-1)^2$.

4. Prove the following inequalities:

(1) When $x\geqslant 0$, $\arctan x\leqslant x$;

(2) When $0<x<\frac{\pi}{2}$, $\frac{2}{\pi}x<\sin x<x$;

(3) When $x\geqslant 0$, $\ln(1+x)\geqslant\frac{\arctan x}{1+x}$;

(4) When $0<x<\frac{\pi}{2}$, $\sin x+\tan x>2x$;

(5) When $b>a>e$, $a^b>b^a$.

5. Prove that the equation $\sin x=x$ has only one real root.
6. Prove: The function $f(x)=\left(1+\frac{1}{x}\right)^x$ is monotonically increasing within $(0,+\infty)$.

7. Suppose $f(x)$ is twice differentiable on $[0,a]$, and $f(0)=0$, $f''(x)<0$, prove that: when $0<x\leqslant a$, $\frac{f(x)}{x}$ is monotonically decreasing.

8. Find the extrema of the following functions:

(1) $y=2x^3-3x^2-12x+21$; (2) $y=x-\ln(1+x)$;

(3) $y=\frac{x}{1+x^2}$; (4) $y=\frac{\ln^2 x}{x}$;

(5) $y=2e^x+e^{-x}$; (6) $y=(x-5)^2\sqrt[3]{(x+1)^2}$;

(7) $y=x^2e^{-x^2}$; (8) $y=|x|e^{-|x-1|}$.

9. For what value of a, does the function $f(x)=a\sin x+\frac{1}{3}\sin 3x$ achieve an extremum at $x=\frac{\pi}{3}$?

10. Suppose $f(x)=a\ln x+bx^2+x$ achieves an extremum at point $x_1=1, x_2=2$, find the value of a, b. At this point, does $f(x)$ achieve a maximum or a minimum at points x_1 and x_2?

11. If the function $f(x)=ax^3+bx^2+cx+d$ satisfies the condition $b^2-3ac<0$, prove that this function has no extrema.

12. Discuss how many real roots the equation $\ln x=ax$ $(a>0)$ has.

13. Consider the function

$$f(x)=\begin{cases} x^4\sin^2\frac{1}{x}, & x\neq 0 \\ 0, & x=0 \end{cases}.$$

(1) Prove that $x=0$ is a point of minimum for the function.

(2) Does the function $f(x)$ satisfy the first and second sufficient conditions for the existence of an extremum at point $x=0$?

14. Suppose $f(x)=(x-x_0)^n g(x)$ $(n\in\mathbf{N})$, and $g(x)$ is continuous at point x_0, $g(x_0)\neq 0$, does $f(x)$ have an extremum at point x_0?

3.4 Maximum and Minimum Values of Functions and Their Applications

In practical problems, what is often required is not the extreme value of a function, but rather its maximum or minimum value. For example, under certain conditions, how to achieve the "most economical use of materials", "maximum profit" or "lowest average cost".

Suppose the function $y=f(x)$ is continuous on $[a,b]$. According to the properties of continuous functions on closed intervals, $f(x)$ must have a maximum value M and a minimum value m, $M(m)$ on $[a,b]$. These values may be attained within (a,b), where they would necessarily be the vertical coordinates at some peak (valley) point on the curve $y=$

of $f(x)$, i.e., some maximum (minimum) value, or they may be attained at the endpoints of $[a,b]$ (Figure 3-9). Therefore, to find the extreme values of the continuous function $f(x)$ on the closed interval $[a,b]$, one only needs to compare the function values at all possible extreme points within (a,b) with those at the endpoints. The specific steps are as follows:

(1) Find the derivative $f'(x)$;

(2) Determine all stationary points and points of non-differentiability of $f(x)$ within (a,b): $x_1, x_2, \cdots, x_n$;

(3) Calculate $f(x_1), f(x_2), \cdots, f(x_n)$ and $f(a), f(b)$;

(4) Compare the function values in (3), where the largest is the maximum value of $f(x)$ on $[a,b]$, and the smallest is the minimum value of $f(x)$ on $[a,b]$.

Example 3.4.1 Find the maximum and minimum values of the function $f(x)=x^5-5x^4+5x^3+1$ on $[-1,2]$.

Solution If $f'(x)=5x^4-20x^3+15x^2=5x^2(x-1)(x-3)$.

Let $f'(x)=0$, obtaining the stationary point $x=0,1,3$. Among them, $x=3\in(-1,2)$ should be discarded. Substituting the stationary points and endpoints into the function expression for calculation, we get

$$f(-1)=-10, f(0)=1, f(1)=2, f(2)=-7.$$

By comparison: The maximum value of $f(x)$ on $[-1,2]$ is 2, and the minimum value is 10.

When finding the extreme values of a function, a special case is worth noting: If a continuous function $f(x)$ has only one extremum $f(x_0)$ within an interval I (finite or infinite, open or closed), then this extremum $f(x_0)$ is the extreme value of $f(x)$ on I. If $f(x_0)$ is a maximum, then $f(x_0)$ is the maximum value of $f(x)$ on I. If $f(x_0)$ is a minimum, then $f(x_0)$ is the minimum value of $f(x)$ on I (Figure 3-12). Practical problems generally fall into this category.

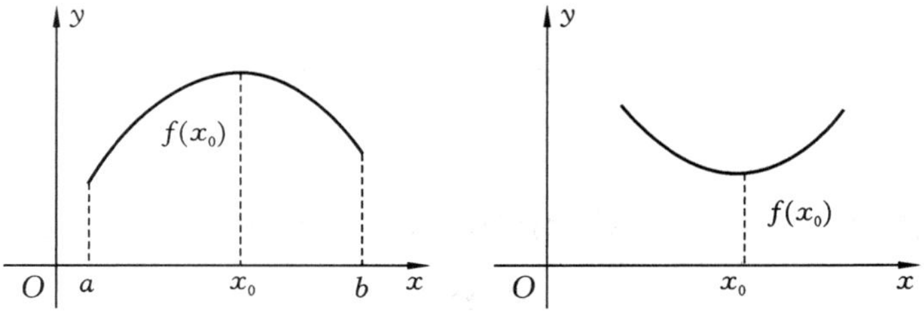

Figure 3-12

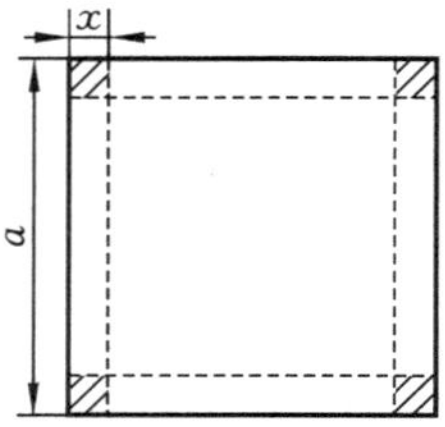

Figure 3-13

Example 3.4.2 A square piece of metal with side length a (Figure 3-13) has each corner cut off to form a small square, and then folded into an open-top box. What should be the side length of the cut-out small squares to maximize the volume of the box?

Solution Let the side length of the cut-out small squares be x. Then the base side length of the box is $a-2x$, and the height is x.

Therefore, the volume of the box is

$$V=x\ (a-2x)^2\ \left(x\in\left(0,\frac{a}{2}\right)\right),$$

$$V'=(a-2x)(a-6x).$$

Let $V'=0$, yielding $x_1=\frac{a}{6}$, $x_2=\frac{a}{2}\notin\left(0,\frac{a}{2}\right)$ which is discarded.

When $x\in\left(0,\frac{a}{6}\right)$, $V'>0$; when $x\in\left(\frac{a}{6},\frac{a}{2}\right)$, $V'<0$. Thus, the function V attains its maximum at $x=\frac{a}{6}$, and this unique maximum is the global maximum of V. That is, when the side length of the cut-out small squares is $\frac{a}{6}$, the volume of the box is maximized.

Example 3.4.3 Draw a tangent to the curve $y=\frac{1}{x^2}$ at a point and find the shortest length of the segment intercepted by the coordinate axes.

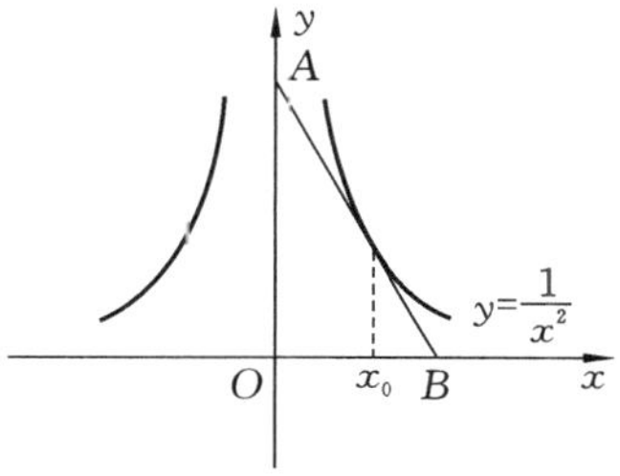

Figure 3-14

Solution As shown in Figure 3-14, the hyperbola $y=\frac{1}{x^2}$ is symmetric about the y axis, so it suffices to discuss the case of $x>0$.

Let the slope of the tangent line at point $\left(x_0,\frac{1}{x_0^2}\right)$ be $x_0>0$, $y=\frac{1}{x^2}$.

$$k=y'|_{x=x_0}=-\frac{2}{x_0^3}.$$

Therefore, the equation of the tangent line to the curve at point $\left(x_0,\frac{1}{x_0^2}\right)$ is

$$y-\frac{1}{x_0^2}=-\frac{2}{x_0^3}(x-x_0).$$

The coordinates of the points where it intersects the two coordinate axes are $A\left(0,\frac{3}{x_0^2}\right)$, $B\left(\frac{3}{2}x_0,0\right)$. Expressing the length d of the intercepted tangent segment AB as a function of x_0, we have

$$d=\sqrt{\frac{9}{x_0^4}+\frac{9}{4}x_0^2}.$$

Obviously, d and d^2 reach their minimum simultaneously, and

$$d^2=f(x_0)=\frac{9}{x_0^4}+\frac{9}{4}x_0^2,$$

$$f'(x_0)=-\frac{36}{x_0^5}+\frac{9}{2}x_0.$$

Let $f'(x_0)=0$, obtaining the stationary point $x_0=\sqrt{2}$, and further

$$f''(x_0)=\frac{180}{x_0^6}+\frac{9}{2}>0.$$

Therefore, when $x_0=\sqrt{2}$, d^2 has a unique minimum value, which is also the minimum value, thus $d^2=\frac{27}{4}$, $d=\frac{3\sqrt{3}}{2}$. Hence, the shortest length of the tangent segment intercepted by the coordinate axes on the curve $y=\frac{1}{x^2}$ is $\frac{3\sqrt{3}}{2}$.

Generally, in solving practical problems, if the nature of the problem allows us to conclude that the function $f(x)$ indeed has a maximum (minimum) value, and this maximum (minimum) value is necessarily attained within the defined interval. On the other hand, $f(x)$ has only one stationary point x_0 within the defined interval, then there is no need to discuss whether $f(x_0)$ is an extreme value. It can be concluded that $f(x_0)$ is the sought maximum (minimum) value.

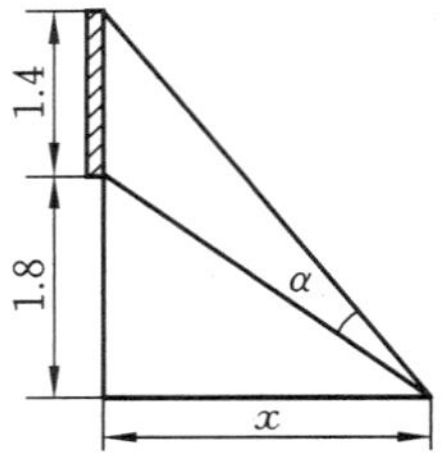

Figure 3-15

Example 3.4.4 A picture of height 1.4 m is hung on the wall, with its bottom edge above the observer's eye level 1.8 m. How far from the wall should the observer stand to view the picture most clearly (i.e., with the maximum viewing angle)?

Solution Let the distance between the observer and the wall be x m, and the viewing angle be α. As shown in Figure 3-15, express α as a function of x, i.e.,

$$\alpha=f(x)=\arctan\frac{1.4+1.8}{x}-\arctan\frac{1.8}{x}=\arctan\frac{3.2}{x}-\arctan\frac{1.8}{x},$$

$$f'(x)=-\frac{3.2}{x^2+3.2^2}+\frac{1.8}{x^2+1.8^2}.$$

Let $f'(x)=0$, yielding the unique stationary point $x=2.4$.

Also, by common knowledge, the observer's view is unclear if they are too close or too far from the wall, thus the maximum value of α should be within $(0,+\infty)$. Therefore, the observer should stand at a distance of 2.4 m from the wall to view the picture most clearly.

Example 3.4.5 Figure 3-16 shows a voltage regulator circuit with an electromotive force of E and internal resistance of r. Assuming the load resistance is R, what should R be for the maximum output power? What is the maximum power?

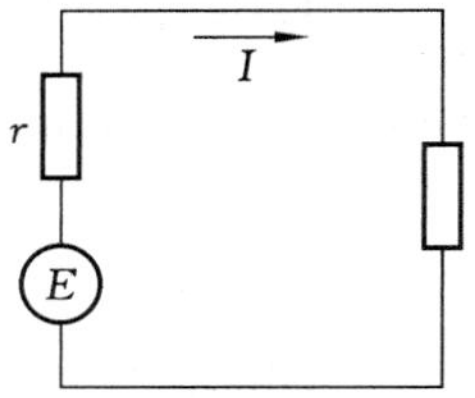

Figure 3-16

Solution According to physics, the output power

$$P=I^2R.$$

Where: I is the current in the circuit.

Also, by Ohm's Law, we have

$$I=\frac{E}{r+R}.$$

Therefore, the output power P is a function of the load resistance R:

$$P=\frac{E^2R}{(r+R)^2}\quad(R>0),$$

$$P'=E^2\cdot\frac{(r+R)^2-2R(r+R)}{(r+R)^4}=\frac{E^2(r-R)}{(r+R)^3}.$$

Let $P'=0$, obtain the unique stationary point $R=r$.

Considering the practical implications, when the load resistance R equals the internal resistance of the power source r, the output power is maximized, with its value being

$$P_{\max}=\frac{E^2r}{(2r)^2}=\frac{E^2}{4r}.$$

Exercise 3.4

1. Find the maximum and minimum values of the following functions over the given intervals.

(1) $y=x^4-2x^2+5, x\in[-2,2]$;

(2) $y=\frac{x-1}{x+1}, x\in[0,4]$;

(3) $y=x+\sqrt{1-x}, x\in[-3,1]$;

(4) $y=\sin 2x-x, x\in\left[-\frac{\pi}{2},\frac{\pi}{2}\right]$.

2. Find a positive number a, such that the sum of the number and its reciprocal is minimized.

3. To construct an open rectangular container with a square base and a volume of 108 m^3, how should it be made to minimize the material used?

4. To transport goods from city A on the canal to city B which is $BC=a$ km away from the canal (Figure 3-17). The unit cost of shipping by boat is m yuan/km, and by train is n yuan/km ($m<n$). Find a point D on the canal to build a railway BD, such that the total shipping cost is minimized.

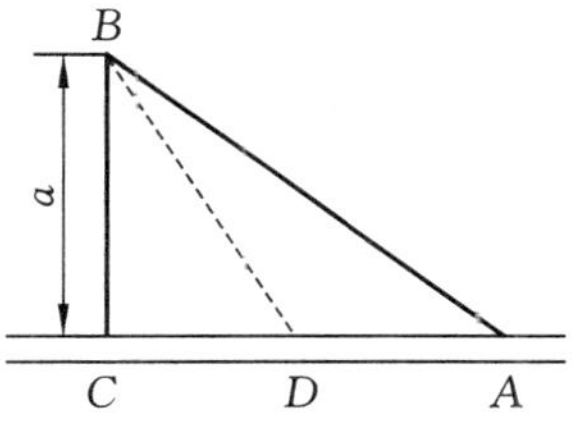

Figure 3-17

5. The cross-section of a certain area's bomb shelter is to be designed as a rectangle topped with a semicircle (Figure 3-18), with a cross-sectional area of 5 m^2. What should the width of the base x be to minimize the perimeter of the cross-section?

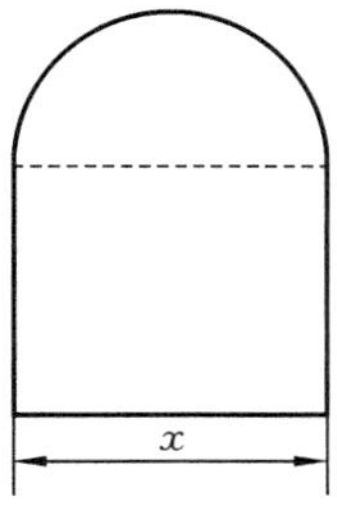

Figure 3-18

6. Ship A travels east at a speed of 20 knots, while at the same time, Ship B travels south at a speed of 16 knots from a position 82 nautical miles directly north of Ship A. How long will it take for the distance between the two ships to be minimized?

7. A real estate company has 50 apartments for rent. When the monthly rent is set at 1000 yuan, all apartments are rented. For every 50 yuan increase in the monthly rent, one more apartment remains unrented, and each rented apartment incurs a maintenance cost of 100 yuan per month. What rent should be set to maximize profit?

8. Find the largest number in the sequence $1, \sqrt{2}, \sqrt[3]{3}, \cdots, \sqrt[n]{n}, \cdots$.

9. Find a point $A(x_0, f(x_0))$ on the parabola $y^2=4x$, such that the normal line at A is intercepted by the parabola at the shortest segment.

10. Let $0\leqslant x\leqslant 1, p>1$, prove:

$$\frac{1}{2^{p-1}}\leqslant x^p+(1-x)^p\leqslant 1.$$

3.5 Concavity and Inflection Points of Curves

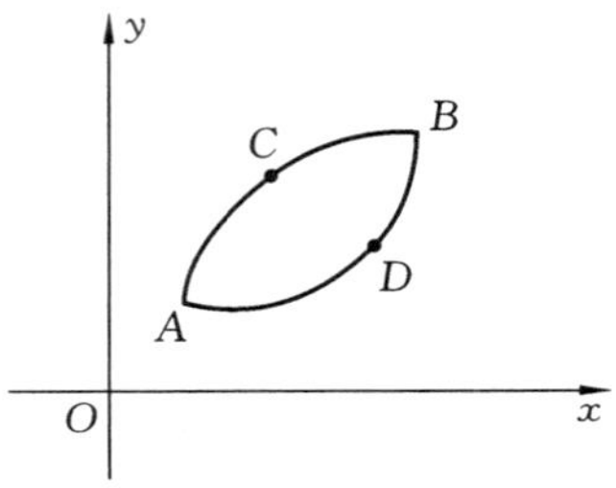

Figure 3-19

The graph of a monotonically increasing function is an ascending curve. However, during the ascent, there is also the issue of the direction of curvature. For example, in Figure 3-19, there are two arcs $\overset{\frown}{ACB}$ and $\overset{\frown}{ADB}$, both of which are ascending, but arc $\overset{\frown}{ACB}$ is a convex (upward) curve arc, while $\overset{\frown}{ADB}$ is concave (downward). To more accurately depict the graph of a function, the concept of the concavity of a curve is given geometrically below, and its determination method is studied.

It can be intuitively seen that in Figure 3-20(a), the curve is concave, characterized by any chord MN always being above the corresponding arc $\overset{\frown}{MN}$, with the midpoint of chord MN being above the corresponding point on the arc. Whereas in Figure 3-20(b), the situation is exactly the opposite. Thus, the following definition is provided:

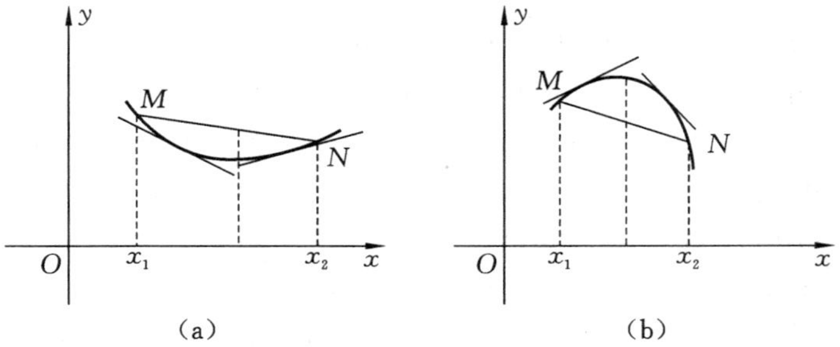

Figure 3-20

Definition 3.5.1 Let the function $f(x)$ be continuous over the interval I, $\forall x_1, x_2\in I, x_1\neq x_2$.

(1) If $f\left(\frac{x_1+x_2}{2}\right)<\frac{1}{2}[f(x_1)+f(x_2)]$, then $f(x)$ is said to be a concave function over I.

(2) If $f\left(\frac{x_1+x_2}{2}\right)>\frac{1}{2}[f(x_1)+f(x_2)]$, then $f(x)$ is said to be a convex function over I.

If the curve arc is smooth, i.e., the tangent exists at each point, then the concavity or convexity can also be defined by the positional relationship between the arc and its tangents.

Definition 3.5.2 If throughout a certain interval, the curve arc has a tangent at every point, and the curve arc lies above its tangent at any point, the curve is said to be concave (or concave arc) in this interval. If the curve arc lies below its tangent at any point, the curve is said to be convex (or convex arc) in this interval.

Observing Figure 3-21 reveals that: As x increases, the tangent slope $k=f'(x)$ at each point on the concave arc is an increasing function, i.e., $f''(x)>0$, whereas the convex arc is the opposite, which indicates that the sign of the second derivative can be used to determine the concavity or convexity of the curve.

Theorem 3.5.1 Let $f(x)$ be continuous on $[a,b]$ and have a second derivative within (a,b).

(1) If $\forall x\in(a,b)$, $f''(x)>0$, then the curve $y=f(x)$ is concave on $[a,b]$.

(2) If $\forall x\in(a,b)$, $f''(x)<0$, then the curve $y=f(x)$ is convex on $[a,b]$.

Proof

(1) Let $\forall x_0\in(a,b)$, then the tangent line equation at point $(x_0, f(x_0))$ is

$$Y=f(x_0)+f'(x_0)(x-x_0).$$

According to the Taylor's Mean Value Theorem, the first-order expansion of $f(x)$ at point x_0 is

$$f(x)=f(x_0)+f'(x_0)(x-x_0)+\frac{f''(\xi)}{2!}(x-x_0)^2 \quad (\xi \text{ is between } x_0, x).$$

Therefore

$$f(x)-Y=\frac{f''(\xi)}{2!}(x-x_0)^2>0.$$

That is, $f(x)>Y$. That is, the curve arc lies above the tangent at any point on it (Figure 3-21). By Definition 3.6.2, the curve is concave on $[a,b]$.

Similarly, it can be proven the (2).

From the above proof process, it can be seen that changing the interval $[a,b]$ to other forms, the theorem still holds.

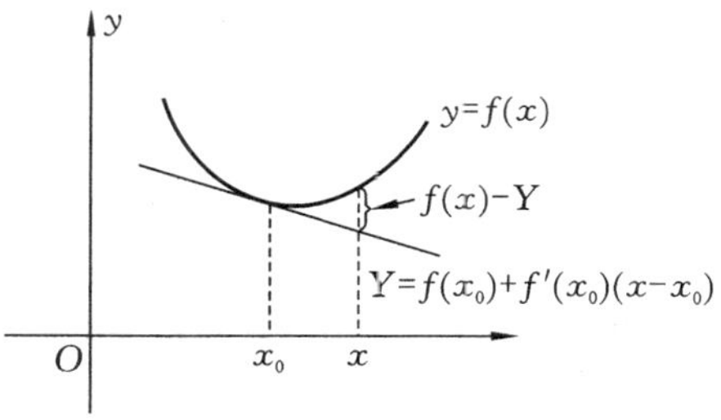

Figure 3-21

Example 3.5.1 Determine the concavity and convexity of the curve $y=\ln x$.

Solution The domain of $y=\ln x$ is $(0,+\infty)$.

$$y'=\frac{1}{x}, y''=-\frac{1}{x^2}<0.$$

Therefore, $y=\ln x$ is convex within $(0,+\infty)$.

Example 3.5.2 Find the intervals of concavity and convexity for the function $y=x^{\frac{1}{3}}$.

Solution The domain of $y=x^{\frac{1}{3}}$ is $(-\infty,+\infty)$.

$$y'=\frac{1}{3}x^{-\frac{2}{3}}, y''=-\frac{2}{9}x^{-\frac{5}{3}}.$$

When $x\in(-\infty,0)$, $y''>0$, so $y=x^{\frac{1}{3}}$ is concave on $(-\infty,0]$. When $x\in(0,+\infty)$, $y''<0$, so $y=x^{\frac{1}{3}}$ is convex on $[0,+\infty)$.

Generally, if a continuous curve arc $y=f(x)$ changes its concavity at a certain point $M(x_0,f(x_0))$, i.e., the curve transitions from convex to concave or vice versa, then point $M(x_0,f(x_0))$ is called an inflection point of the curve. In Example 3.5.2, the origin $(0,0)$ is an inflection point of $y=x^{\frac{1}{3}}$.

Just as the first derivative $f'(x)$ is used to determine the intervals of monotonicity and extreme values, the second derivative $f''(x)$ is used to find the intervals of concavity and inflection points of a function. The key is to determine the sign of $f''(x)$. To do this, one should first find the points where $f''(x)=0$ and the points where $f''(x)$ does not exist (which are potential locations for inflection points), and use these points to divide the domain into several small intervals. Then, discuss the sign of $f''(x)$ within each interval. The specific steps are as follows:

(1) Find the domain $D(f)$ of $f(x)$.

(2) Find the second derivative $f''(x)$ of $f(x)$.

(3) Find the points where $f''(x)=0$ and the points where $f''(x)$ does not exist (if any).

(4) Divide $D(f)$ into subintervals using all points obtained in (3), and determine the sign of $f''(x)$ within each subinterval. If $f''(x)$ has opposite signs in the left and right neighborhoods of point x_0, then $(x_0,f(x_0))$ is an inflection point. If $f''(x)$ has the same sign in the left and right neighborhoods of point x_0, then point $(x_0,f(x_0))$ is not an inflection point.

Example 3.5.3 Discuss the concavity of the function $y=(x-1)\sqrt[3]{x^5}$ and find the inflection points.

Solution The function $y=(x-1)\sqrt[3]{x^5}$ is continuous within its domain **R**.

$$y'=x^{\frac{5}{3}}+(x-1)\frac{5}{3}x^{\frac{2}{3}}=\frac{8}{3}x^{\frac{5}{3}}-\frac{5}{3}x^{\frac{2}{3}},$$

$$y''=\frac{40}{9}x^{\frac{2}{3}}-\frac{10}{9}x^{-\frac{1}{3}}=\frac{10}{9}\frac{4x-1}{\sqrt[3]{x}}.$$

Let $y''=0$, yielding $x=\frac{1}{4}$. When $x=0$, y'' does not exist.

Divide **R** into three subintervals using $x=0, \frac{1}{4}$. Therefore, $y=(x-1)\sqrt[3]{x^2}$ is concave on $(-\infty, 0]$ and $\left[\frac{1}{4}, +\infty\right)$, and convex on $\left[0, \frac{1}{4}\right]$, and there are inflection points at $(0,0)$ and $\left(\frac{1}{4}, -\frac{3}{32\sqrt[3]{2}}\right)$.

Note: Inflection points occur at $(x_0, f(x_0))$ where $f''(x_0)=0$ or $f''(x_0)$ does not exist. However, the converse is not true. For example, at point $x=0$, $y=x^4$ has $y''=0$, but since the sign of y'' does not change as it passes through $x=0$, point $(0,0)$ is not an inflection point of $y=x^4$.

Example 3.5.4 Determine the intervals of monotonicity, extreme values, concavity, and inflection points of the function $f(x)=x^4-2x^3+2$.

Solution

$$f'(x)=4x^3-6x^2=2x^2(2x-3),$$
$$f''(x)=12x^2-12x=12x(x-1).$$

Let $f'(x)=0$, obtaining the stationary point $x=0, \frac{3}{2}$. Let $f''(x)=0$, obtaining $x=0, 1$.

Divide the domain of $f(x)$ into smaller intervals using $x=0, 1, \frac{3}{2}$. Therefore, $f(x)=x^4-2x^3+2$ is decreasing on $\left(-\infty, \frac{3}{2}\right]$, it is increasing within $\left[\frac{3}{2}, +\infty\right)$, with a minimum value $f\left(\frac{3}{2}\right)=\frac{5}{16}$.

The concave intervals are $(-\infty, 0]$ and $[1, +\infty)$, the convex intervals are $[0,1]$, and there are inflection points $(0,2)$ and $(1,1)$.

Depicting the above conclusions in a Cartesian coordinate system yields the graph of $y=x^4-2x^3+2$ (Figure 3-22).

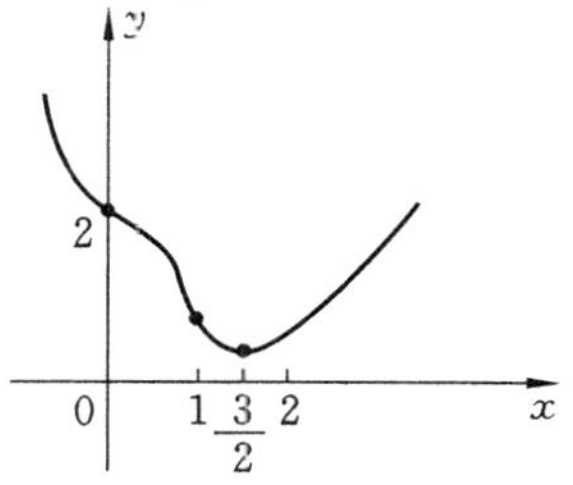

Figure 3-22

Exercise 3.5

1. Prove that the functions $f(x)=x\arctan x$ and $g(x)=x\ln x$ are concave within their domains.

2. Find the intervals of concavity and convexity, and the inflection points of the following functions:

(1) $y=3x^2-x^3$; (2) $y=\ln(1+x^2)$;

(3) $y=x^2+\frac{1}{x}$; (4) $y=\frac{2x}{1+x^2}$;

(5) $y=xe^{-x}$; (6) $y=x+(1-x)^{\frac{5}{3}}$.

3. For what values of a and b, is the point $(1,3)$ an inflection point of the curve $y=ax^3+bx^2$?

4. It is known that $f(x)=x^3+ax^2+bx$ has an extreme value of -2 at point $x=1$. Determine the coefficients a and b, and find all extreme values of $f(x)$ and the inflection points of the curve $y=f(x)$.

5. Prove that the curve $y=\frac{x-1}{x^2+1}$ has three inflection points that lie on the same straight line.

6. Determine the value of k in the function $f(x)=k(x^2-3)^2$, such that the normal at the inflection point passes through the origin.

7. Suppose $y=f(x)$ has a continuous third derivative in a neighborhood of point $x=x_0$, if $f''(x_0)=0, f'''(x_0)\neq 0$, is $(x_0, f(x_0))$ an inflection point? Why?

8. If within (a,b), $f''(x)>0$, and $f(x_0)$ is the minimum value of function $f(x)$ at $(x_0\in(a,b))$, prove that $f(x_0)$ is the minimum value of function $f(x)$ within (a,b).

9. Prove the following inequalities.

(1) When $x,y>0, x\neq y, n>1, \frac{1}{2}(x^n+y^n)>\left(\frac{x+y}{2}\right)^n$;

(2) When $x\neq y, \frac{e^x+e^y}{2}>e^{\frac{x+y}{2}}$;

(3) When $x,y>0, 2\arctan\frac{x+y}{2}\geqslant\arctan x+\arctan y$.

3.6 Sketch Functions

3.6.1 Asymptotes of Curves

1. Horizontal Asymptotes

If $\lim\limits_{x\to+\infty} f(x)=a$ or $\lim\limits_{x\to-\infty} f(x)=a$, then the line $y=a$ is called a horizontal asymptote of the curve $y=f(x)$.

For instance, all three cases shown in Figure 3-23 indicate that $y=a$ is the horizontal asymptote of the curve $y=f(x)$.

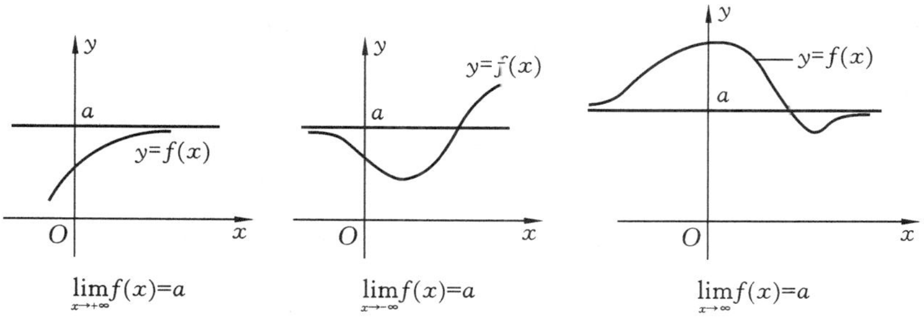

Figure 3-23

2. Vertical Asymptote

If $\lim\limits_{x \to c^+} f(x) = \infty$ or $\lim\limits_{x \to c^-} f(x) = \infty$, the line $x = c$ is called the vertical asymptote of the curve $y = f(x)$.

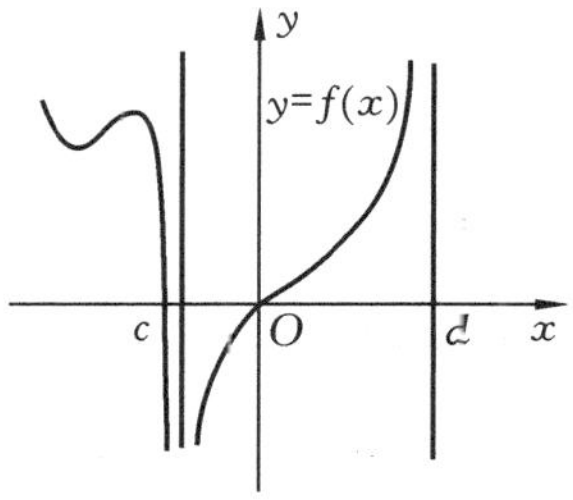

Figure 3-24

As shown in Figure 3-24, the line $x = c$, $x = d$ is the vertical asymptote of the curve $y = f(x)$. In this case, $x = c$, $x = d$ is also an infinite discontinuity of the function $y = f(x)$.

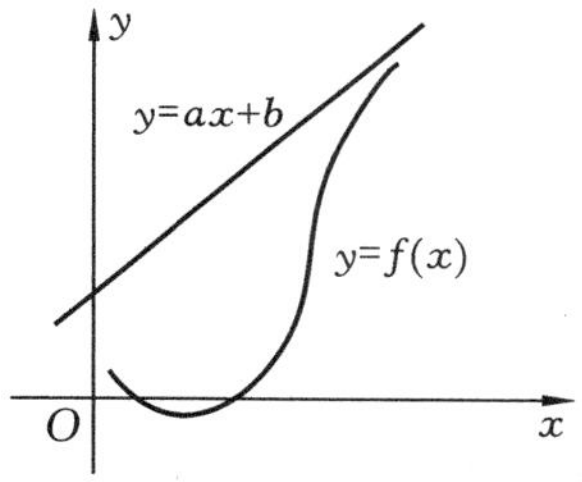

Figure 3-25

3. Oblique Asymptote

For the curve $y = f(x)$ and the line $y = ax + b\ (a \neq 0)$, if $\lim\limits_{x \to +\infty} [f(x) - (ax + b)] = 0$ or $\lim\limits_{x \to -\infty} [f(x) - (ax + b)] = 0$, that is, the difference between the ordinate of the curve and that of the line approaches 0 as $x \to \infty$, then the line $y = ax + b$ is called the oblique asymptote of the curve $y = f(x)$, as shown in Figure 3-25.

How is the a, b determined in the oblique asymptote?

Since

$$\lim_{x \to \infty} [f(x) - (ax + b)] = 0,$$

therefore

$$\lim_{x \to \infty} [f(x) - ax] = b,$$

hence

$$\lim_{x \to \infty} \frac{f(x) - ax}{x} = \lim_{x \to \infty} \left[\frac{f(x)}{x} - a \right] = 0,$$

thus

$$a = \lim_{x \to \infty} \frac{f(x)}{x},$$

$$b = \lim_{x \to \infty} [f(x) - ax].$$

Example 3.6.1 Find the horizontal and vertical asymptotes of the curve $y=\frac{e^x+e^{-x}}{e^x-e^{-x}}$.

Solution Because

$$\lim_{x\to+\infty} y=\lim_{x\to+\infty}\frac{e^x+e^{-x}}{e^x-e^{-x}}=\lim_{x\to+\infty}\frac{e^{2x}+1}{e^{2x}-1}=1,$$

$$\lim_{x\to-\infty} y=\lim_{x\to-\infty}\frac{e^x+e^{-x}}{e^x-e^{-x}}=\lim_{x\to-\infty}\frac{1+e^{-2x}}{1-e^{-2x}}=-1.$$

Therefore, the curve has two horizontal asymptotes $y=1$ and $y=-1$.

On the other hand, the function $y=\frac{e^x+e^{-x}}{e^x-e^{-x}}$ has a discontinuity at $x=0$, and

$$\lim_{x\to 0}\frac{e^x+e^{-x}}{e^x-e^{-x}}=\infty.$$

Thus, the curve has a vertical asymptote $x=0$.

Example 3.6.2 Find the asymptotes of the curve $f(x)=\frac{x^2+2x-1}{x}$.

Solution Since $\lim\limits_{x\to\infty} f(x)=\infty$, the curve $y=f(x)$ has no horizontal asymptotes. Furthermore,

$$\lim_{x\to 0} f(x)=\lim_{x\to 0}\frac{x^2+2x-1}{x}=\infty.$$

Therefore, the curve has a vertical asymptote $x=0$.

Additionally, since

$$a=\lim_{x\to\infty}\frac{f(x)}{x}=\lim_{x\to\infty}\frac{x^2+2x-1}{x^2}=1,$$

$$b=\lim_{x\to\infty}[f(x)-ax]=\lim_{x\to\infty}\left(\frac{x^2+2x-1}{x}-x\right)=\lim_{x\to\infty}\frac{2x-1}{x}=2.$$

Hence, $y=x+2$ is the oblique asymptote of the curve.

3.6.2 Graphical Representation of Functions

Traditionally, the graphical representation of functions was often achieved through the method of plotting points. This method is labor-intensive and yields relatively crude graphs. Nowadays, differential methods can be employed to determine the main features of a function (such as where it increases or decreases, where it is convex or concave, where it has extreme values and what these values are, the coordinates of inflection points, etc.). By plotting a few key points based on these characteristics, a more accurate representation of the function's graph can be drawn. This process generally follows these steps:

(1) Determine the domain of the function $y=f(x)$ and examine its parity (symmetry) and periodicity.

(2) Find the first derivative $f'(x)$ and the second derivative $f''(x)$ of the function, and determine the points where $f'(x)$ and $f''(x)$ are equal to 0 and the points where they

do not exist.

(3) List discussion to determine the intervals of monotonicity, extreme values, intervals of concavity and inflection points.

(4) Examine the asymptotes of the function.

(5) Find certain special points of the function, such as intersections with the coordinate axes, points of non-differentiability, or add a few points as needed.

(6) Synthesize the above discussion to draw the graph of the function.

Example 3.6.3 Sketch the graph of the function $y=\frac{1}{\sqrt{2\pi}}e^{-\frac{x^2}{2}}$.

Solution The domain of $y=\frac{1}{\sqrt{2\pi}}e^{-\frac{x^2}{2}}$ is $(-\infty,+\infty)$, and the range is $y>0$. Clearly, the function is not periodic, but its graph is symmetric about the y axis, allowing us to discuss only the $x\geqslant 0$ part.

$$y'=\frac{-x}{\sqrt{2\pi}}e^{-\frac{x^2}{2}},\ y''=\frac{1}{\sqrt{2\pi}}(x^2-1)e^{-\frac{x^2}{2}}.$$

Let $y'=0$, obtaining the stationary point $x=0$. Let $y''=0$, obtaining $x=\pm 1$. The curve has one peak point $\left(0,\frac{1}{\sqrt{2\pi}}\right)$ and two inflection points $\left(\pm 1,\frac{1}{\sqrt{2\pi}}e^{-\frac{1}{2}}\right)$.

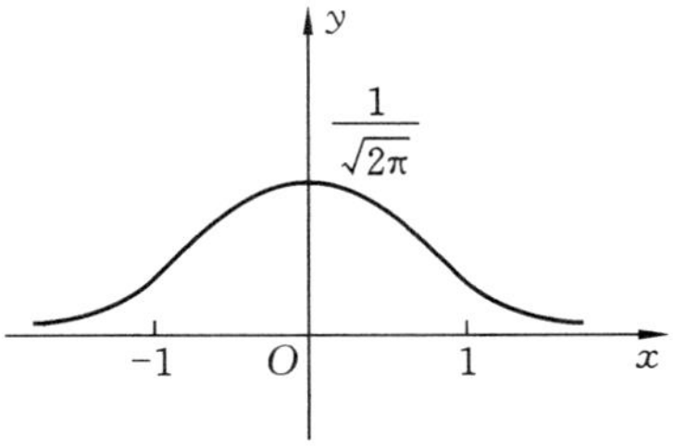

Figure 3-26

Clearly, the function has no discontinuities, no vertical asymptotes, but due to

$$\lim_{x\to\infty}f(x)=\lim_{x\to\infty}\frac{1}{\sqrt{2\pi}}e^{-\frac{x^2}{2}}=0.$$

Therefore, there is a horizontal asymptote $y=0$.

Depicting the above results in a rectangular coordinate system yields the graph of $y=\frac{1}{\sqrt{2\pi}}e^{-\frac{x^2}{2}}$ (Figure 3-26).

Example 3.6.4 Sketch the graph of $f(x)=\frac{(x-3)^2}{4(x-1)}$.

Solution The domain of $f(x)$ is $(-\infty,1)\cup(1,+\infty)$. Because

$$\lim_{x\to 1}\frac{(x-3)^2}{4(x-1)}=\infty,$$

therefore, $x=1$ is a vertical asymptote of $y=f(x)$. Also,

$$a=\lim_{x\to\infty}\frac{f(x)}{x}=\lim_{x\to\infty}\frac{(x-1)^2}{4x(x-1)}=\frac{1}{4},$$

$$b=\lim_{x\to\infty}(f(x)-ax)=\lim_{x\to\infty}\left[\frac{(x-3)^2}{4(x-1)}-\frac{1}{4}x\right]=\lim_{x\to\infty}\frac{-5x+9}{4(x-1)}=-\frac{5}{4}.$$

Hence, the curve has a slant asymptote $y=\frac{1}{4}x-\frac{5}{4}$. Also, because

$$f'(x)=\frac{(x+1)(x-3)}{4(x-1)^2},f''(x)=\frac{2}{(x-1)^3}.$$

Let $f'(x)=0$, obtaining the stationary point $x=-1$, $x=3$, within the domain of $f(x)$, $f''(x)$ exists, and $f''(x)\neq 0$, thus the curve has no inflection points.

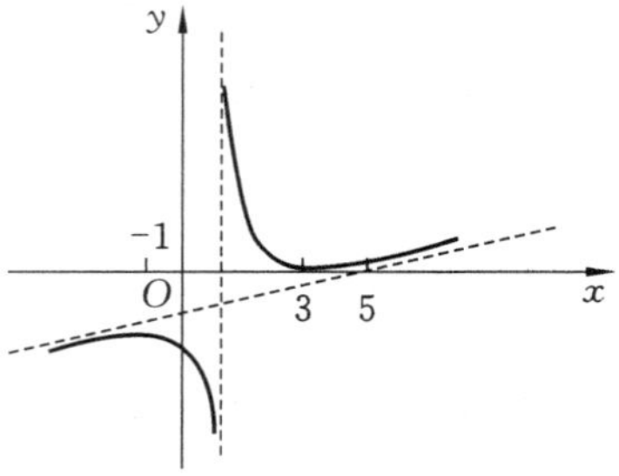

Figure 3-27

The function has a maximum value $f(-1)=-2$ and a minimum value $f(3)=0$.

Additionally, the intersection points of the curve with the coordinate axes are $\left(0,-\frac{9}{4}\right)$, $(3,0)$. Further, let $x=2$, obtaining $f(2)=\frac{1}{4}$.

Plotting points and connecting them yields the graph of this function (Figure 3-27).

Exercise 3.6

1. Find the asymptotes of the following functions:

(1) $y=e^{\frac{1}{x}}$; (2) $y=\frac{1}{x^2-4x+5}$;

(3) $y=x+e^{-x}$; (4) $y=\frac{e^x}{1+x}$;

(5) $y=xe^{-2x}$; (6) $y=x-\ln(x+1)$.

2. Draw the graphs of the following functions:

(1) $y=3x-x^3$; (2) $y=xe^{-x}$;

(3) $y=\frac{2x}{1+x^2}$; (4) $y=\ln(1+x^2)$.

Comprehensive Review 3

1. Choose one correct term among "sufficient", "necessary" and "sufficient and necessary" to fill in the blanks below.

(1) If $f(x)$ is differentiable on $[a,b]$, then within (a,b), $f'(x)\equiv 0$ is the ________ condition for $[a,b]$ to be $f(x)\equiv f(a)$.

(2) The non-existence of $f'(x_0)=0$ or $f'(x_0)$ is a ________ condition for the function $f(x)$ to attain an extremum at the point x_0.

2. Multiple Choice Questions.

(1) Let $y=f(x)$ satisfy $f(0)=1$, $f'(0)=0$, when $x\neq 0$ then $f'(x)>0$, when $x<0$ then $f''(x)<0$, when $x>0$ then $f''(x)>0$, its graph is (   ).

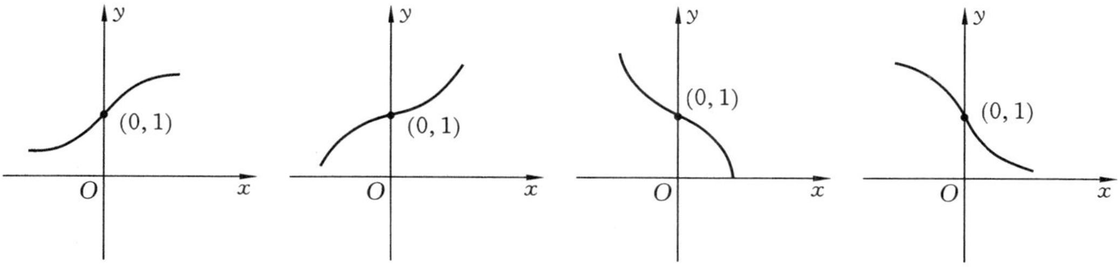

A. B. C. D.

(2) Let $[0,1]$ be on $f''(x)>0$, then the order of magnitude of the numbers $f'(0)$, $f'(1)$, $f(1)-f(0)$ and $f(0)-f(1)$ is(   ).

A. $f'(1)>f'(0)>f(1)-f(0)$ B. $f'(1)>f(1)-f(0)>f'(0)$

C. $f(1)-f(0)>f'(1)>f'(0)$ D. $f'(1)>f(0)-f(1)>f'(0)$

3. Let the function $f(x)=\begin{cases}3-x^2, & 0\leqslant x\leqslant 1\\ \dfrac{x}{2}, & 1<x\leqslant 2\end{cases}$, be on the interval $[0,2]$, does $f(x)$ satisfy the conditions of the Mean Value Theorem? How many solutions satisfy the equation $f(2)-f(0)=f'(\xi)(2-0)\xi$ are there?

4. Let $k>0$, what value of k makes the equation $\arctan x-kx=0$ have a positive root?

5. Find the following limits:

(1) $\lim\limits_{x\to\pi}\dfrac{\sin 3x}{\tan 5x}$;   (2) $\lim\limits_{x\to\frac{\pi}{2}^+}\dfrac{\tan x}{\ln\left(x-\dfrac{\pi}{2}\right)}$;

(3) $\lim\limits_{x\to\pi}(\pi-x)\tan\dfrac{x}{2}$;   (4) $\lim\limits_{x\to 0}\dfrac{(1+x)^{\frac{1}{x}}-\mathrm{e}}{x}$;

(5) $\lim\limits_{x\to\infty}\left[x-x^2\ln\left(1+\dfrac{1}{x}\right)\right]$;   (6) $\lim\limits_{x\to\infty}\left(\sin\dfrac{2}{x}+\cos\dfrac{1}{x}\right)^x$.

6. Determine the intervals of monotonicity and extreme values for the following functions:

(1) $y=\arctan x-\dfrac{1}{2}\ln(1+x^2)$;   (2) $y=x^{\frac{1}{x}}$.

7. Prove the following inequalities:

(1) When $x>0$, $\dfrac{x}{1+x^2}<\arctan x$;

(2) When $x>0$, $\ln\left(1+\dfrac{1}{x}\right)>\dfrac{1}{1+x}$.

8. Determine the value of a, b, c, such that the function $y=x^3+3ax^2+3bx+c$ has a maximum at point $x=-1$, and point $(0,3)$ is an inflection point.

9. At which point on the curve arc $y=\sin x\ (0<x<\pi)$ is the radius of curvature the smallest? Find the radius of curvature at that point.

10. Given $a_0+\frac{a_1}{2}+\frac{a_2}{3}+\cdots+\frac{a_n}{n+1}=0$, prove that the equation

$$a_0+a_1x+a_2x^2+\cdots+a_nx^n=0$$

has a root in $(0,1)$.

11. Suppose $f(x)$ is continuous on $[a,b]$, differentiable within (a,b), and a,b has the same sign. Prove that there exists $\xi\in(a,b)$, such that

$$2\xi[f(b)-f(a)]=(b^2-a^2)f'(\xi).$$

12. Given the derivable function $y=f(x)$ determined by the equation $x^3-3xy^2+2y^3=32$, discuss and find the extreme values of $f(x)$.

13. Find the extreme values of $y=x^m(1-x)^n$ $(m,n\in\mathbf{Z}^+)$.

14. Given $f(x)=nx(1-x)^n$ $(n\in\mathbf{Z}^+)$, find:

(1) The maximum value $M(n)$ of $f(x)$ on $0\leqslant x\leqslant 1$;

(2) Find $\lim\limits_{n\to\infty}M(n)$.

15. Find the following limits:

(1) $\lim\limits_{n\to\infty}\frac{n^2}{a^n}$ $(a>1)$; (2) $\lim\limits_{n\to\infty}\left(\cos\frac{t}{n}\right)^n$.

16. Determine the order of infinitesimal of $\frac{2}{3}(\cos x-\cos 2x)$ with respect to x when $x\to 0$.

17. Suppose $f(x)$ has a second derivative on $[a,b]$, and

$$f(a)=f(b)=0, f'(a)f'(b)>0.$$

Prove: There exist $\xi\in(a,b)$ and $\tau\in(a,b)$, such that $f(\xi)=0, f''(\tau)=0$.

18. Suppose $f(x), g(x)$ has a second derivative on $[a,b]$, and

$$g''(x)\neq 0, f(a)=f(b)=g(a)=g(b)=0.$$

Prove: (1) Within (a,b), $g(x)\neq 0$;

(2) There exists $\xi\in(a,b)$, such that $\frac{f(\xi)}{g(\xi)}=\frac{f''(\xi)}{g''(\xi)}$.

19. Suppose the function $f(x)$ has a second continuous derivative in some neighborhood of point $x=0$, and $\lim\limits_{x\to 0}\left(1+x+\frac{f(x)}{x}\right)^{\frac{1}{x}}=e^3$. Find $f(0)$, $f'(0)$, $f''(0)$ and $\lim\limits_{x\to 0}\left(1+\frac{f(x)}{x}\right)^{\frac{1}{x}}$.

20. Suppose

$$f(x)=\begin{cases}\frac{\varphi(x)-\cos x}{x}, & x\neq 0\\ a, & x=0\end{cases},$$

where: $\varphi(x)$ has a second continuous derivative, and $\varphi(0)=1, \varphi'(0)=0$.

(1) Determine a, such that $f(x)$ is continuous at $x=0$;

(2) Find $f'(x)$;

(3) Discuss the continuity of $f'(x)$ at point $x=0$.

Chapter 4 Indefinite Integrals

The *I Ching* emphasizes the mutual transformation and harmony between Yin and Yang, and holds that all things in the world are in a state of balance between Yin and Yang. The harmony of Yin and Yang is not only reflected in the static balance, but also reflected in the dynamic process of change. The indefinite integral can be seen as the inverse of the derivative operation, which describes how the original function can be restored or constructed through "cumulative effects" as the function changes. This cumulative effect reflects the harmonization and balance of the two forces (i.e., the increase and decrease of the function) in the process of change.

The idea of "accumulation" in the *I Ching* is reflected in the gradual accumulation of things into a new state or form through the constant harmonization of Yin and Yang. The "realization of cumulative effect" in indefinite integrals is reflected in that by integrating the derivatives of functions, we can get a series of original functions, which are related to each other by the integral constant, forming a family of original functions. This original function family can be regarded as the different states or forms formed by different accumulation methods (that is, different integral constants) in the process of function change. Imagining a winding river, the indefinite integral is like telling us that as long as we know the flow rate of the river at a certain point (i.e., the derivative), we can infer the water level of the river at different locations (i.e., the original function) through the integral operation. Each function in the original function family represents the possible state of the river under different initial water level conditions. This "cumulative effect realization" gives us a deeper understanding of how functions change.

Chapter 2 discussed how to find the derivative or differential of a known function. Now we need to consider the inverse problem of the differential operation: Given a function $f(x)$, we need to find another function $F(x)$, such that the derivative of $F(x)$ exactly equals $f(x)$, i.e., $F'(x) = f(x)$. This is the problem of indefinite integrals that this chapter will discuss.

This chapter first introduces the concept and properties of indefinite integrals, then discusses the methods for finding indefinite integrals. Finding indefinite integrals is one of the fundamental problems in integral calculus.

4.1 Concept and Properties of Indefinite Integrals

4.1.1 Primitive Function and Indefinite Integral

From differential calculus, it is known that if the equation of a curve is given as $y=f(x)$, then the slope of the tangent line (if it exists) at any point $(x, f(x))$ on the curve is $K=f'(x)$. For example, the slope of the tangent line to the curve $y=x^2$ at point $(x, f(x))$ is $K=2x$. Now, we need to solve the opposite problem: Given the slope of the tangent line at any point on the curve, how to find the equation of the curve. To this end, the concept of a primitive function is introduced.

Definition 4.1.1 If $F(x)$ is a differentiable function on the interval I, and for any $x\in I$, there exists

$$F'(x)=f(x) \text{ or } \mathrm{d}F(x)=f(x)\mathrm{d}x.$$

It is said that $F(x)$ is an antiderivative of $f(x)$ over the interval I.

For example, $\forall x\in(-\infty,+\infty)$, $(\sin x)'=\cos x$, hence $\sin x$ is an antiderivative of $\cos x$ over the interval $(-\infty,+\infty)$.

For instance, when $x\in(-\infty,+\infty)$,

$$[\ln(x+\sqrt{1+x^2})]'=\frac{1}{x+\sqrt{1+x^2}}\left(1+\frac{x}{\sqrt{1+x^2}}\right)=\frac{1}{\sqrt{1+x^2}}.$$

hence, $\ln(x+\sqrt{1+x^2})$ is an antiderivative of $\frac{1}{\sqrt{1+x^2}}$ over the interval $(-\infty,+\infty)$.

Studying antiderivatives involves addressing the following two issues:

(1) Under what conditions does the function $f(x)$ necessarily have an antiderivative? If an antiderivative exists, is it unique?

(2) If the antiderivative of a given function exists, how can it be found?

Regarding the first issue, the two following theorems provide the answer. As for the second issue, the answer lies in the various integration methods to be introduced in this chapter.

Theorem 4.1.1 (Existence Theorem of Antiderivative) If the function $f(x)$ is continuous over the interval I, then $f(x)$ must have an antiderivative over the interval I, i.e., there exists a differentiable function $F(x)$, such that

$$F'(x)=f(x) \quad (x\in I).$$

Simply put, a continuous function always has an antiderivative. Since elementary functions are continuous within their defined intervals, they must have antiderivatives within their defined intervals.

Theorem 4.1.2 Suppose $F(x)$ is an antiderivative of $f(x)$ over the interval I, then

(1) $F(x)+C$ is also the primitive function of $f(x)$ over I, where C is an arbitrary constant;

(2) Any two primitive functions of $f(x)$ over I differ by only a constant.

Proof (1) from $F'(x)=f(x)$, we get

$$[F(x)+C]'=F'(x)+C'=f(x) \quad (x\in I).$$

Therefore, $F(x)+C$ is also a primitive function of $f(x)$ over I.

(2) Let $F(x)$ and $\Phi(x)$ be any two primitive functions of $f(x)$ over I, then we have

$$[\Phi(x)-F(x)]'=\Phi'(x)-F'(x)=f(x)-f(x)=0.$$

By the corollary of the Mean Value Theorem, it is known that

$$\Phi(x)-F(x)=C \quad (C \text{ is a constant}).$$

That is

$$\Phi(x)=F(x)+C.$$

Theorem 4.1.2 indicates that if a function has a primitive function, then it has infinitely many primitive functions. If $F(x)$ is a primitive function of $f(x)$, then the family of functions $F(x)+C$ constitutes all primitive functions of $f(x)$.

Definition 4.1.2 Let $F(x)$ be a primitive function of $f(x)$ over the interval I, then all primitive functions $F(x)+C$ of $f(x)$ are called the indefinite integral of $f(x)$ over the interval I, denoted as $\int f(x)\mathrm{d}x$, namely

$$\int f(x)\mathrm{d}x = F(x)+C.$$

Where: the notation $\int$ is the integral sign, $f(x)$ is the integrand, $f(x)\mathrm{d}x$ is the integrand expression, x is the variable of integration, and C is the integration constant.

From Definition 4.1.2, the two examples given at the beginning of this section can be expressed as

$$\int \cos x\,\mathrm{d}x = \sin x + C,$$

$$\int \frac{1}{\sqrt{1+x^2}}\mathrm{d}x = \ln\left(x+\sqrt{1+x^2}\right)+C.$$

Example 4.1.1 Find $\int \frac{1}{1+x^2}\mathrm{d}x$.

Solution Since $(\arctan x)'=\frac{1}{1+x^2}$, $\arctan x$ is an antiderivative of $\frac{1}{1+x^2}$, therefore

$$\int \frac{1}{1+x^2}\mathrm{d}x = \arctan x + C.$$

Example 4.1.2 Find $\int \frac{1}{x}\mathrm{d}x$.

Solution When $x>0$, $(\ln x)'=\frac{1}{x}$, so within $(0,+\infty)$, there is

$$\int \frac{1}{x}\mathrm{d}x = \ln x + C.$$

When $x<0$, $[\ln(-x)]' = \frac{1}{-x} \cdot (-x)' = \frac{1}{x}$, so within $(-\infty, 0)$, there is

$$\int \frac{1}{x}\mathrm{d}x = \ln(-x) + C.$$

Summarizing the above results, it can be written as

$$\int \frac{1}{x}\mathrm{d}x = \ln|x| + C.$$

Example 4.1.3 Suppose a curve passes through the point $(1,0)$, and the tangent slope at any point (x, y) on the curve is equal to twice the abscissa of that point, find the equation of this curve.

Solution Let the equation of the curve be $y = f(x)$, according to the problem, there is

$$\frac{\mathrm{d}y}{\mathrm{d}x} = 2x.$$

That is, $f(x)$ is an antiderivative of $2x$.

Also, because $\int 2x\mathrm{d}x = x^2 + C$, the equation of the curve is

$$y = x^2 + C.$$

Since the curve passes through the point $(1,0)$, therefore $0 = 1 + C$, thus $C = -1$. Hence, the equation of the curve is

$$y = x^2 - 1.$$

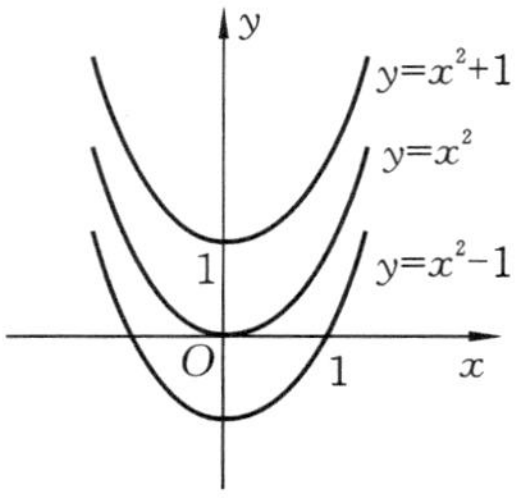

Figure 4-1

The graph of the antiderivative of the function $f(x)$ is called the integral curve of $f(x)$. The curve sought in Example 4.1.3 is the integral curve of the function $2x$ passing through the point $(1,0)$. Obviously, this integral curve can be obtained by translating another integral curve (such as $y = x^2$) along the y axis (Figure 4-1).

According to the definition of indefinite integrals, the following relationship can be obtained:

Since $\int f(x)\mathrm{d}x$ is the antiderivative of $f(x)$, therefore

$$\frac{\mathrm{d}}{\mathrm{d}x}\int f(x)\mathrm{d}x = f(x) \text{ or } \mathrm{d}\int f(x)\mathrm{d}x = f(x)\mathrm{d}x.$$

That is, integrating first and then differentiating cancels out the two operations.

Again, since $F(x)$ is the antiderivative of $F'(x)$, therefore

$$\int F'(x)\mathrm{d}x = F(x) + C \text{ or } \int \mathrm{d}F(x) = F(x) + C.$$

That is, differentiating first and then integrating cancels out the two operations but also adds an arbitrary constant.

Therefore, the operation of finding an indefinite integral, denoted as "$\int$", is the inverse of the differentiation operation "d". When the notations "$\int$" and "d" are combined, they either cancel out or differ by a constant.

4.1.2 Properties of Indefinite Integrals

Using the rules of differentiation and the definition of indefinite integrals, the following operational properties can be derived:

Property 4.1.1 Suppose the functions $f(x)$ and $g(x)$ have antiderivatives, then

$$\int [f(x) \pm g(x)]\,dx = \int f(x)\,dx \pm \int g(x)\,dx.$$

Proof Differentiate the right-hand side of equation, yielding

$$\left[\int f(x)\,dx \pm \int g(x)\,dx\right]' = \left[\int f(x)\,dx\right]' \pm \left[\int g(x)\,dx\right]' = f(x) \pm g(x).$$

This indicates that the right-hand side of equation is the antiderivative of $f(x) \pm g(x)$. Additionally, the right-hand side of equation contains two integral signs, which formally include two arbitrary constants. Since the sum or difference of arbitrary constants is still an arbitrary constant, it actually contains only one arbitrary constant. Therefore, the right-hand side of equation is the indefinite integral of $f(x) \pm g(x)$.

Property 4.1.1 can be generalized to the case of the sum of a finite number of functions.

Similarly, it can be proven that when finding the indefinite integral, a non-zero constant factor can be taken outside the integral sign.

Property 4.1.2 Suppose the function $f(x)$ has an antiderivative, and k is a non-zero constant, then

$$\int k f(x)\,dx = k \int f(x)\,dx.$$

4.1.3 Basic Integration Formulas

Since integration is the inverse operation of differentiation, it is natural to derive corresponding integration formulas from derivative formulas. A table of some basic integration formulas, known as the basic integration table, is compiled. Readers are strongly advised to memorize the basic integration table, as the computation of indefinite integrals will ultimately rely on applying these basic integration formulas.

(1) $\int k\,dx = kx + C$ (where k is a constant);

(2) $\int x^{\mu}\,dx = \dfrac{x^{\mu+1}}{\mu+1} + C$ $(\mu \neq -1)$; (3) $\int \dfrac{dx}{x} = \ln|x| + C$;

(4) $\int \dfrac{dx}{1+x^2} = \arctan x + C$; (5) $\int \dfrac{dx}{\sqrt{1-x^2}} = \arcsin x + C$;

(6) $\int \cos x\,dx = \sin x + C$;

(7) $\int \sin x\,dx = -\cos x + C$;

(8) $\int \frac{dx}{\cos^2 x} = \int \sec^2 x\,dx = \tan x + C$;

(9) $\int \frac{dx}{\sin^2 x} = \int \csc^2 x\,dx = -\cot x + C$;

(10) $\int \sec x \tan x\,dx = \sec x + C$;

(11) $\int \csc x \cot x\,dx = -\csc x + C$;

(12) $\int e^x\,dx = e^x + C$;

(13) $\int a^x\,dx = \frac{a^x}{\ln a} + C$;

(14) $\int \operatorname{sh} x\,dx = \operatorname{ch} x + C$;

(15) $\int \operatorname{ch} x\,dx = \operatorname{sh} x + C$.

4.1.4 Direct Integration Method

Utilizing the properties of indefinite integrals and basic integral formulas, one can directly compute the indefinite integrals of some functions.

Example 4.1.4 Compute $\int \frac{1}{x\sqrt[3]{x}}dx$.

Solution Although the integrand is expressed in fractional and radical forms, it is essentially a power function.

$$\int \frac{dx}{x^3\sqrt{x}} = \int x^{-\frac{4}{3}}dx = \frac{1}{-\frac{4}{3}+1}x^{-\frac{4}{3}+1} + C = -3x^{-\frac{1}{3}} + C = -\frac{3}{\sqrt[3]{x}} + C.$$

Note: To verify the correctness of the integral result, one can differentiate the result. If its derivative equals the integrand, the integral result is correct, otherwise, it is incorrect.

Example 4.1.5 Compute $\int \frac{(x-1)^3}{x^2}dx$.

Solution

$$\int \frac{(x-1)^3}{x^2}dx = \int \frac{x^3 - 3x^2 + 3x - 1}{x^2}dx = \int\left(x - 3 + \frac{3}{x} - \frac{1}{x^2}\right)dx$$

$$= \int x\,dx - 3\int dx + 3\int \frac{1}{x}dx - \int \frac{1}{x^2}dx = \frac{x^2}{2} - 3x + 3\ln|x| + \frac{1}{x} + C.$$

Example 4.1.6 Compute $\int (2^x e^x - 3\sin x)\,dx$.

Solution

$$\int (2^x e^x - 3\sin x)\,dx = \int (2e)^x dx - 3\int \sin x\,dx$$

$$= \frac{(2e)^x}{\ln(2e)} + 3\cos x + C = \frac{2^x e^x}{1+\ln 2} + 3\cos x + C.$$

Example 4.1.7 Compute $\int \frac{x^4+1}{x^2+1}dx$.

Solution This type of integral is not found in the basic integral table. One can first transform the integrand algebraically to fit the formulas in the integral table.

$$\int \frac{x^4+1}{x^2+1}\mathrm{d}x = \int \frac{x^4-1+2}{x^2+1}\mathrm{d}x = \int\left(x^2-1+\frac{2}{x^2+1}\right)\mathrm{d}x = \frac{1}{3}x^3 - x + 2\arctan x + C.$$

Example 4.1.8 Compute $\int \frac{1}{\sin^2 x\cos^2 x}\mathrm{d}x$.

Solution

$$\int \frac{1}{\sin^2 x\cos^2 x}\mathrm{d}x = \int \frac{\sin^2 x+\cos^2 x}{\sin^2 x\cos^2 x}\mathrm{d}x = \int \frac{1}{\cos^2 x}\mathrm{d}x + \int \frac{1}{\sin^2 x}\mathrm{d}x = \tan x - \cot x + C.$$

Example 4.1.9 Compute the following indefinite integrals:

(1) $\int \tan^2 x\,\mathrm{d}x$; (2) $\int \sin^2\frac{x}{2}\mathrm{d}x$.

Solution (1) $\int \tan^2 x\,\mathrm{d}x = \int(\sec^2 x - 1)\,\mathrm{d}x = \int \sec^2 x\,\mathrm{d}x - \int \mathrm{d}x = \tan x - x + C$.

(2) $\int \sin^2\frac{x}{2}\mathrm{d}x = \int \frac{1}{2}(1-\cos x)\,\mathrm{d}x = \frac{1}{2}\int \mathrm{d}x - \frac{1}{2}\int \cos x\,\mathrm{d}x = \frac{1}{2}x - \frac{1}{2}\sin x + C$.

Example 4.1.10 Compute $\int f(x)\mathrm{d}x$, where $f(x) = \begin{cases} x^2, & x \leqslant 0 \\ \sin x, & x > 0 \end{cases}$.

Solution

$$F(x) = \int f(x)\mathrm{d}x = \begin{cases} \frac{x^3}{3} + C_1, & x \leqslant 0 \\ -\cos x + C_2, & x > 0 \end{cases}.$$

Since $F(x)$ is continuous at point $x=0$, therefore

$$\lim_{x\to 0^-} F(x) = \lim_{x\to 0^+} F(x) = F(0).$$

We obtain $-1+C_2 = C_1$. Taking $C = C_2$, thus,

$$\int f(x)\mathrm{d}x = \begin{cases} \frac{x^3}{3} - 1 + C, & x \leqslant 0 \\ -\cos x - C, & x > 0 \end{cases}.$$

Example 4.1.11 Given $f'(\sin x) = \cos 2x$, compute $f(x)$.

Solution Because $f'(\sin x) = \cos 2x = 1 - 2\sin^2 x$, therefore,

$$f'(x) = 1 - 2x^2.$$

Hence

$$f(x) = \int f'(x)\mathrm{d}x = \int(1-2x^2)\,\mathrm{d}x = x - \frac{2}{3}x^3 + C.$$

Exercise 4.1

1. Find the following indefinite integrals:

(1) $\int 3x^3\sqrt{x}\,\mathrm{d}x$;

(2) $\int \frac{1}{x^3\sqrt{x}}\mathrm{d}x$;

(3) $\int (x^3+2x-1)\,\mathrm{d}x$;

(4) $\int \frac{(1-x)^2}{\sqrt{x}}\mathrm{d}x$;

(5) $\int (\sqrt{x}+1)(\sqrt[3]{x^2}-1)\,dx$;

(6) $\int \frac{3x^4+3x^2+1}{x^2+1}dx$;

(7) $\int \sec x(\sec x-\tan x)\,dx$;

(8) $\int \cos\theta(\tan\theta+\sec\theta)\,d\theta$;

(9) $\int e^x\left(1-\frac{e^{-x}}{\sqrt{x}}\right)dx$;

(10) $\int \frac{x^2}{1+x^2}dx$;

(11) $\int \left(\frac{3}{1+x^2}+\frac{2}{\sqrt{1-x^2}}\right)dx$;

(12) $\int \frac{2\cdot 3^x-3\cdot 4^x}{5^x}dx$;

(13) $\int (t^2-2\sin t)\,dt$;

(14) $\int \cos^2\frac{x}{2}dx$;

(15) $\int \cot^2 x\,dx$;

(16) $\int \frac{dx}{1+\cos 2x}$;

(17) $\int \frac{\cos 2x}{\cos x-\sin x}dx$;

(18) $\int \frac{\cos 2x}{\cos^2 x\sin^2 x}dx$.

2. A curve passes through the point $(e^2,3)$ and the slope of the tangent at any point is equal to the reciprocal of the abscissa of that point. Find the equation of the curve.

3. Prove:

$$\int [f(x)g'(x)+g(x)f'(x)]\,dx=f(x)g(x)+C.$$

4. By observation, find the following integrals:

(1) $\int (e^x\sin x+e^x\cos x)\,dx$;

(2) $\int (2x\cos x-x^2\sin x)\,dx$.

5. An object starts from rest and after t (s), its velocity is $3t^2$ (m/s). Question:

(1) How far has the object traveled from the starting point after 3 s ?

(2) How long does it take for the object to travel 360 m ?

6. Prove: The functions $\arcsin(2x-1)$, $\arccos(1-2x)$ and $2\arctan\sqrt{\frac{x}{1-x}}$ are both antiderivatives of $\frac{1}{\sqrt{x-x^2}}$.

4.2 Integration by Substitution

In Section 4.1, when calculating indefinite integrals, the final step always involves applying basic integration formulas, and the number of indefinite integrals that can be calculated directly is very limited. The question arises: How can we find the indefinite integral $\int \cos 2x\,dx$? If we write it as

$$\int \cos 2x\,dx=\sin 2x+C,$$

the result is incorrect. Because when verifying by differentiation, we have $(\sin 2x + C)' = 2\cos 2x \neq \cos 2x$. However, if we let $u = 2x$, that is, $x = \frac{u}{2}$, then

$$\int \cos 2x \, dx = \int \cos u \, d\frac{u}{2} = \frac{1}{2}\int \cos u \, du = \frac{1}{2}\sin u + C = \frac{1}{2}\sin 2x + C.$$

This result is undoubtedly correct. This method, which transforms certain indefinite integrals into forms that can be integrated using basic integration formulas through appropriate variable substitution (substitution), is called the method of integration by substitution, or simply the substitution method. Below, we introduce two different usages, known as the first type of substitution method and the second type of substitution method.

4.2.1 The First Type of Substitution Method

If the indefinite integral $\int g(x)dx$ is not easy to find directly, but the integrand can be decomposed as

$$g(x) = f[\varphi(x)]\varphi'(x).$$

Make the substitution $u = \varphi(x)$, and note that $\varphi'(x)dx = d\varphi(x)$, then the integral of the variable x can be transformed into the integral of the variable u, thus we have

$$\int g(x)dx = \int f[\varphi(x)]\varphi'(x)dx = \int f(u)\,du.$$

If $\int f(u)\,du$ can be integrated, the problem of computing the indefinite integral $\int g(x)dx$ is solved, which is the first type of substitution method. The first type of substitution method is essentially the reverse application of the differential method for composite functions to find indefinite integrals.

Theorem 4.2.1 Let $F(u)$ be an antiderivative of $f(u)$, and $u = \varphi(x)$ be differentiable, then there is a substitution formula

$$\int f[\varphi(x)]\varphi'(x)dx = \int f(u)\,du = F(u) + C = F[\varphi(x)] + C.$$

The integration process of the first type of substitution method can be decomposed into

$$\int g(x)dx \longrightarrow \int f[\varphi(x)]\varphi'(x)dx \longrightarrow \int f[\varphi(x)]\,d\varphi(x)$$

$$\xrightarrow[u=\varphi(x)]{\text{substitution}} \int f(u)\,du = F(u) + C \xrightarrow{\text{return}} F[\varphi(x)] + C.$$

The key to the first type of substitution method is to match the integrand to the differential of some function, hence, it is also known as the method of matching differentials.

Example 4.2.1 Find $\int \frac{1}{3x+2}\mathrm{d}x$.

Solution By matching differentials, $\int \frac{1}{3x+2} \cdot \frac{1}{3}\mathrm{d}(3x+2)$, then let $u=3x+2$, and we have

$$\int \frac{1}{3x+2}\mathrm{d}x = \frac{1}{3}\int \frac{1}{u}\mathrm{d}u = \frac{1}{3}\ln|u|+C = \frac{1}{3}\ln|3x+2|+C.$$

In general, there is

$$\int f(ax+b)\,\mathrm{d}x = \int \frac{1}{a}f(ax+b)\,\mathrm{d}(ax+b) = \frac{1}{a}\left[\int f(u)\,\mathrm{d}u\right]_{u=\varphi(x)}.$$

Example 4.2.2 Find $\int x\mathrm{e}^{x^2}\mathrm{d}x$.

Solution Let $u=x^2$, then $\mathrm{d}u=\mathrm{d}x^2=2x\,\mathrm{d}x$, and we have

$$\int x\mathrm{e}^{x^2}\mathrm{d}x = \frac{1}{2}\int \mathrm{e}^{x^2}\cdot 2x\,\mathrm{d}x = \frac{1}{2}\int \mathrm{e}^u\,\mathrm{d}u = \frac{1}{2}\mathrm{e}^u + C = \frac{1}{2}\mathrm{e}^{x^2}+C.$$

In general, there is

$$\int x^{n-1}f(x^n)\,\mathrm{d}x = \frac{1}{n}\int f(u)\,\mathrm{d}u\bigg|_{u=x^n}.$$

Example 4.2.3 Find $\int \frac{x^2}{(x-4)^3}\mathrm{d}x$.

Solution Let $u=x-4$, then $x=u+4$, $\mathrm{d}x=\mathrm{d}u$. Thus,

$$\int \frac{x^2}{(x-4)^3}\mathrm{d}x = \int \frac{(u+4)^2}{u^3}\mathrm{d}u = \int (u^2+8u+16)u^{-3}\mathrm{d}u = \int (u^{-1}+8u^{-2}+16u^{-3})\,\mathrm{d}u$$

$$= \ln|u| - 8u^{-1} - 8u^{-2} + C = \ln|x-4| - \frac{8}{x-4} - \frac{8}{(x-4)^2} + C.$$

Example 4.2.4 Find the indefinite integral:

(1) $\int \frac{1}{a^2+x^2}\mathrm{d}x$; (2) $\int \frac{1}{\sqrt{a^2-x^2}}\mathrm{d}x$ $(a>0)$.

Solution

(1) $$\int \frac{1}{a^2+x^2}\mathrm{d}x = \int \frac{1}{a^2}\frac{1}{1+\left(\frac{x}{a}\right)^2}\mathrm{d}x = \frac{1}{a}\int \frac{1}{1+\left(\frac{x}{a}\right)^2}\mathrm{d}\left(\frac{x}{a}\right)$$

$$\xrightarrow{u=\frac{x}{a}} \frac{1}{a}\int \frac{1}{1+u^2}\mathrm{d}u = \frac{1}{a}\arctan u + C = \frac{1}{a}\arctan\frac{x}{a}+C.$$

(2) The solution is similar to (1), and is left to the reader. The result is given below:

$$\int \frac{1}{\sqrt{a^2-x^2}}\mathrm{d}x = \arcsin\frac{x}{a}+C.$$

After becoming proficient in variable substitution, one can omit the process of writing intermediate variables for substitution and back-substitution. Commonly used integration by parts formulas are given in Table 4-1.

Table 4-1

Integral type	Substitution formulas
$\int f(ax+b)\,dx = \frac{1}{a}\int f(ax+b)\,d(ax+b)\quad(a\neq 0)$	$u=ax+b$
$\int f(x^{\mu})x^{\mu-1}\,dx = \frac{1}{\mu}\int f(x^{\mu})\,d(x^{\mu})\quad(\mu\neq 0)$	$u=x^{\mu}$
$\int f(\ln x)\cdot\frac{1}{x}\,dx = \int f(\ln x)\,d(\ln x)$	$u=\ln x$
$\int f(e^x)\cdot e^x\,dx = \int f(e^x)\,d(e^x)$	$u=e^x$
$\int f(a^x)\cdot a^x\,dx = \frac{1}{\ln a}\int f(a^x)\,d(a^x)$	$u=a^x$
$\int f(\sin x)\cdot\cos x\,dx = \int f(\sin x)\,d(\sin x)$	$u=\sin x$
$\int f(\cos x)\cdot\sin x\,dx = -\int f(\cos x)\,d(\cos x)$	$u=\cos x$
$\int f(\tan x)\sec^2 x\,dx = \int f(\tan x)\,d(\tan x)$	$u=\tan x$
$\int f(\cot x)\csc^2 x\,dx = -\int f(\cot x)\,d(\cot x)$	$u=\cot x$
$\int f(\arctan x)\frac{1}{1+x^2}\,dx = \int f(\arctan x)\,d(\arctan x)$	$u=\arctan x$
$\int f(\arcsin x)\frac{1}{\sqrt{1-x^2}}\,dx = \int f(\arcsin x)\,d(\arcsin x)$	$u=\arcsin x$

Example 4.2.5 Find the indefinite integral:

(1) $\int \tan x\,dx$; (2) $\int \sin 2x\,dx$.

Solution

(1)
$$\int \tan x\,dx = \int \frac{\sin x}{\cos x}dx = -\int \frac{d(\cos x)}{\cos x} = -\ln|\cos x| + C.$$

Similarly, we can obtain

$$\int \cot x\,dx = \ln|\sin x| + C.$$

(2)
$$\int \sin 2x\,dx = \frac{1}{2}\int \sin 2x\,d(2x) = -\frac{1}{2}\cos 2x + C.$$

Another solution

$$\int \sin 2x\,dx = 2\int \sin x\cos x\,dx = 2\int \sin x\,d(\sin x) = \sin^2 x + C.$$

Another solution

$$\int \sin 2x\,dx = 2\int \sin x\cos x\,dx = -2\int \cos x\,d(\cos x) = -\cos^2 x + C.$$

The above results indicate that $-\frac{1}{2}\cos 2x$, $\sin^2 x$, $-\cos^2 x$ are all primitive functions of $\sin 2x$.

Example 4.2.6 Find $\int \frac{e^{3\sqrt{x}}}{\sqrt{x}} dx$.

Solution

$$\int \frac{e^{3\sqrt{x}}}{\sqrt{x}} dx = 2\int e^{3\sqrt{x}} d(\sqrt{x}) = \frac{2}{3}\int e^{3\sqrt{x}} d(3\sqrt{x}) = \frac{2}{3}e^{3\sqrt{x}} + C.$$

Example 4.2.7 Find $\int \frac{1}{x(1+2\ln x)} dx$.

Solution

$$\int \frac{1}{x(1+2\ln x)} dx = \int \frac{1}{1+2\ln x} d(\ln x) = \frac{1}{2}\int \frac{1}{1+2\ln x} d(1+2\ln x) = \frac{1}{2}\ln|1+2\ln x| + C.$$

Example 4.2.8 Find $\int \frac{1}{x^2-8x+25} dx$.

Solution $$\int \frac{1}{x^2-8x+25} dx = \int \frac{1}{(x-4)^2+9} dx = \frac{1}{9}\int \frac{dx}{1+\left(\frac{x-4}{3}\right)^2}$$

$$= \frac{1}{3}\int \frac{1}{1+\left(\frac{x-4}{3}\right)^2} d\left(\frac{x-4}{3}\right) = \frac{1}{3}\arctan\frac{x-4}{3} + C.$$

Example 4.2.9 Find $\int \frac{1}{x^2-a^2} dx$.

Solution Since $\frac{1}{x^2-a^2} = \frac{1}{2a}\left(\frac{1}{x-a} - \frac{1}{x+a}\right)$, therefore

$$\int \frac{1}{x^2-a^2} dx = \frac{1}{2a}\int \left(\frac{1}{x-a} - \frac{1}{x+a}\right) dx = \frac{1}{2a}\left(\int \frac{1}{x-a} dx - \int \frac{1}{x+a} dx\right)$$

$$= \frac{1}{2a}\left[\int \frac{1}{x-a} d(x-a) - \int \frac{1}{x+a} d(x+a)\right]$$

$$= \frac{1}{2a}(\ln|x-a| - \ln|x+a|) + C = \frac{1}{2a}\ln\left|\frac{x-a}{x+a}\right| + C.$$

Example 4.2.10 Find $\int \sin^4 x \cos^5 x \, dx$.

Solution

$$\int \sin^4 x \cos^5 x \, dx = \int \sin^4 x \cos^4 x \, d(\sin x) = \int \sin^4 x \, (1-\sin^2 x)^2 d(\sin x)$$

$$= \int (\sin^4 x - 2\sin^6 x + \sin^8 x) \, d(\sin x) = \frac{1}{5}\sin^5 x - \frac{2}{7}\sin^7 x + \frac{1}{9}\sin^9 x + C.$$

Example 4.2.11 Find $\int \cos^2 x \, dx$.

Solution

$$\int \cos^2 x \, dx = \int \frac{1+\cos 2x}{2} dx = \frac{1}{2}\left(\int dx + \int \cos 2x \, dx\right)$$

$$= \frac{1}{2}\int dx + \frac{1}{4}\int \cos 2x \, d(2x) = \frac{x}{2} + \frac{\sin 2x}{4} + C.$$

Example 4.2.12 Find $\int \csc x \, dx$.

Solution

$$\int \csc x \, dx = \int \frac{1}{\sin x} dx = \int \frac{1}{2\sin\frac{x}{2}\cos\frac{x}{2}} dx = \int \frac{1}{\tan\frac{x}{2}\cos^2\frac{x}{2}} d\left(\frac{x}{2}\right)$$

$$= \int \frac{1}{\tan\frac{x}{2}} d\left(\tan\frac{x}{2}\right) = \ln\left|\tan\frac{x}{2}\right| + C.$$

Because

$$\tan\frac{x}{2} = \frac{\sin\frac{x}{2}}{\cos\frac{x}{2}} = \frac{2\sin^2\frac{x}{2}}{\sin x} = \frac{1-\cos x}{\sin x} = \csc x - \cot x,$$

therefore

$$\int \csc x \, dx = \ln|\csc x - \cot x| + C.$$

Example 4.2.13 Find $\int \sec x \, dx$.

Solution

$$\int \sec x \, dx = \int \frac{dx}{\cos x} = \int \frac{d\left(x+\frac{\pi}{2}\right)}{\sin\left(x+\frac{\pi}{2}\right)}$$

$$= \ln\left|\csc\left(x+\frac{\pi}{2}\right) - \cot\left(x+\frac{\pi}{2}\right)\right| + C = \ln|\sec x + \tan x| + C.$$

Another solution

$$\int \sec x \, dx = \int \frac{\sec x(\sec x + \tan x)}{\sec x + \tan x} dx = \int \frac{d(\sec x + \tan x)}{\sec x + \tan x} = \ln|\sec x + \tan x| + C.$$

Example 4.2.14 Find $\int \sec^6 x \, dx$.

Solution

$$\int \sec^6 x \, dx = \int (\sec^2 x)^2 \sec^2 x \, dx = \int (1+\tan^2 x)^2 d(\tan x)$$

$$= \int (1 + 2\tan^2 x + \tan^4 x) d(\tan x) = \tan x + \frac{2}{3}\tan^3 x + \frac{1}{5}\tan^5 x + C.$$

Example 4.2.15 Find $\int \tan^5 x \sec^3 x \, dx$.

Solution

$$\int \tan^5 x \sec^3 x \, dx = \int \tan^4 x \sec^2 x \sec x \tan x \, dx = \int (\sec^2 x - 1)^2 \sec^2 x \, d(\sec x)$$

$$= \int (\sec^6 x - 2\sec^4 x + \sec^2 x) d(\sec x) = \frac{1}{7}\sec^7 x - \frac{2}{5}\sec^5 x + \frac{1}{3}\sec^3 x + C.$$

Example 4.2.16 Find $\int \frac{1}{1+\sin x}\mathrm{d}x$.

Solution

$$\int \frac{1}{1+\sin x}\mathrm{d}x = \int \frac{1-\sin x}{1-\sin^2 x}\mathrm{d}x = \int \frac{1}{\cos^2 x}\mathrm{d}x + \int \frac{1}{\cos^2 x}\mathrm{d}(\cos x) = \tan x - \frac{1}{\cos x} + C.$$

Another solution

$$\int \frac{1}{1+\sin x}\mathrm{d}x = \int \frac{\mathrm{d}x}{1+\cos\left(\frac{\pi}{2}-x\right)} = \int \frac{\mathrm{d}x}{2\cos^2\left(\frac{\pi}{4}-\frac{x}{2}\right)}$$

$$= -\int \frac{\mathrm{d}\left(\frac{\pi}{4}-\frac{x}{2}\right)}{\cos^2\left(\frac{\pi}{4}-\frac{x}{2}\right)} = -\tan\left(\frac{\pi}{4}-\frac{x}{2}\right) + C.$$

Example 4.2.17 Find $\int \sin 3x \cos 2x\,\mathrm{d}x$.

Solution By

$$\sin\alpha\cos\beta = \frac{1}{2}[\sin(\alpha+\beta) + \sin(\alpha-\beta)],$$

we get

$$\sin 3x \cos 2x = \frac{1}{2}(\sin 5x + \sin x).$$

Thus

$$\int \sin 3x \cos 2x\,\mathrm{d}x = \frac{1}{2}\int(\sin 5x + \sin x)\,\mathrm{d}x = \frac{1}{2}\left[\int \frac{1}{5}\sin 5x\,\mathrm{d}(5x) + \int \sin x\,\mathrm{d}x\right]$$

$$= -\frac{1}{10}\cos 5x - \frac{1}{2}\cos x + C.$$

Example 4.2.18 Find $\int \frac{x^2-1}{x^4+1}\mathrm{d}x$.

Solution

$$\frac{x^2-1}{x^4+1} = \frac{1-\frac{1}{x^2}}{x^2+\frac{1}{x^2}} = \frac{1-\frac{1}{x^2}}{\left(x+\frac{1}{x}\right)^2-2},$$

so

$$\int \frac{x^2-1}{x^4+1}\mathrm{d}x = \int \frac{1-\frac{1}{x^2}}{\left(x+\frac{1}{x}\right)^2-2}\mathrm{d}x = \int \frac{\mathrm{d}\left(x+\frac{1}{x}\right)}{\left(x+\frac{1}{x}\right)^2-2} \xrightarrow{t=x+\frac{1}{x}} \int \frac{\mathrm{d}t}{t^2-2} = \frac{1}{2\sqrt{2}}\ln\left|\frac{t-\sqrt{2}}{t+\sqrt{2}}\right| + C$$

$$= \frac{1}{2\sqrt{2}}\ln\left|\frac{x+\frac{1}{x}-\sqrt{2}}{x+\frac{1}{x}+\sqrt{2}}\right| + C = \frac{1}{2\sqrt{2}}\ln\left|\frac{x^2-\sqrt{2}x+1}{x^2+\sqrt{2}x+1}\right| + C.$$

Although the first type of substitution method is the inverse process of differentiating composite functions, it is generally much more difficult than differentiating composite functions. This is because it requires certain skills, and there is no general rule for how to appropriately choose the variable substitution $u=\varphi(x)$. Therefore, to master the substitution method, one must be familiar with some typical examples and accumulate experience by doing more exercises, so as to apply it skillfully.

4.2.2 The Second Type of Substitution Method

The first type of substitution method involves a variable substitution $u=\varphi(x)$, transforming the integral $\int f[\varphi(x)]\varphi'(x)\mathrm{d}x$ into $\int f(u)\,\mathrm{d}u$. However, some integrals require a variable substitution of the form $x=\psi(t)$, transforming the integral $\int f(x)\mathrm{d}x$ into $\int f[\psi(t)]\psi'(t)\mathrm{d}t$. After solving the latter integral, the inverse function $t=\psi^{-1}(x)$ of $x=\psi(t)$ is substituted back, which is the second type of substitution method.

The formula for the second type of substitution method can be expressed as

$$\int f(x)\mathrm{d}x=\left\{\int f[\psi(t)]\psi'(t)\mathrm{d}t\right\}_{t=\psi^{-1}(x)}.$$

To ensure the validity of the above formula, in addition to the integrand having an antiderivative, the condition for the existence of the inverse function $t=\psi^{-1}(x)$ must also be met. The following theorem is provided.

Theorem 4.2.2 Let $x=\psi(t)$ be a monotonic and differentiable function, and $\psi'(t)\neq 0$. Supposing $f[\psi(t)]\psi'(t)$ has an antiderivative $F(t)$, then

$$\int f(x)\mathrm{d}x=\int f[\psi(t)]\psi'(t)\mathrm{d}t=F(t)+C=F[\psi^{-1}(x)]+C,$$

where: $\psi^{-1}(x)$ is the inverse function of $x=\psi(t)$.

Proof Since $F(t)$ is the primitive function of $f[\psi(t)]\psi'(t)$, let $G(x)=F[\psi^{-1}(x)]$, then by the rules of differentiation for composite and inverse functions, we obtain

$$G'(x)=\frac{\mathrm{d}F}{\mathrm{d}t}\frac{\mathrm{d}t}{\mathrm{d}x}=f[\psi(t)]\psi'(t)\frac{1}{\psi'(t)}=f[\psi(t)]=f(x).$$

That is, $G(x)$ is the primitive function of $f(x)$. Thus, the conclusion is proven.

As seen from Theorem 4.2.2, the substitution and back-substitution processes of the second type of substitution method are exactly opposite to those of the first type of substitution method.

Example 4.2.19 Find $\int\sqrt{a^2-x^2}\,\mathrm{d}x\ (a>0)$.

Solution The difficulty in solving this integral lies in the presence of the radical $\sqrt{a^2-x^2}$, so one can consider using the trigonometric identity $\sin^2 t+\cos^2 t=1$ to eliminate the radical.

Let $x = a\sin t, t \in \left(-\frac{\pi}{2}, \frac{\pi}{2}\right)$, then $dx = a\cos t\,dt$, and

$$\sqrt{a^2 - x^2} = \sqrt{a^2 - a^2\sin^2 t} = a\cos t.$$

Thus, the integral becomes

$$\int \sqrt{a^2 - x^2}\,dx = \int a\cos t \cdot a\cos t\,dt = a^2\int \cos^2 t\,dt = \frac{a^2}{2}\int (1 + \cos 2t)\,dt$$

$$= \frac{a^2}{2}\left(t + \frac{1}{2}\sin 2t\right) + C = \frac{a^2}{2}(t + \sin t\cos t) + C.$$

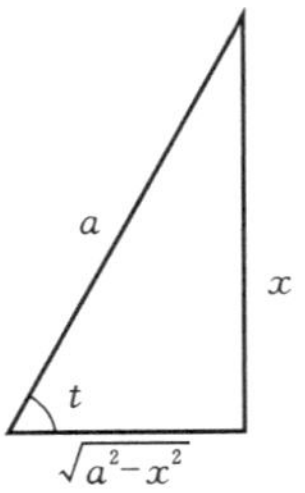

Figure 4-2

To revert the variable t back to the original variable x, one can construct an auxiliary triangle (Figure 4-2) based on $\sin t = \frac{x}{a}$, and thus obtain

$$\cos t = \frac{\sqrt{a^2 - x^2}}{a}.$$

Hence

$$\int \sqrt{a^2 - x^2}\,dx = \frac{a^2}{2}\left(\arcsin\frac{x}{a} + \frac{x}{a}\,\frac{\sqrt{a^2 - x^2}}{a}\right) + C = \frac{a^2}{2}\arcsin\frac{x}{a} + \frac{x}{2}\sqrt{a^2 - x^2} + C.$$

Note: If one lets $x = a\cos t$, the result can be calculated similarly.

Example 4.2.20 Find $\int \frac{1}{\sqrt{x^2 + a^2}}\,dx \ (a > 0)$.

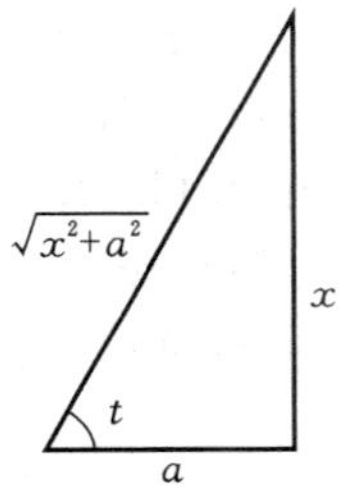

Figure 4-3

Solution To eliminate the radical, let $x = a\tan t, t \in \left(-\frac{\pi}{2}, \frac{\pi}{2}\right)$, as shown in Figure 4-3, then

$$dx = a\sec^2 t\,dt, \sqrt{x^2 + a^2} = a\sec t.$$

Thus

$$\int \frac{1}{\sqrt{x^2 + a^2}}\,dx = \int \frac{a\sec^2 t\,dt}{a\sec t} = \int \sec t\,dt = \ln|\sec t + \tan t| + C$$

$$= \ln\left|\frac{\sqrt{x^2 + a^2}}{a} + \frac{x}{a}\right| + C = \ln\left(x + \sqrt{x^2 + a^2}\right) + C_1,$$

where: $C_1 = C - \ln a$.

The substitutions used in the above two examples are both trigonometric substitutions, with the purpose of eliminating the radical. The general rules are

(1) If the integrand contains $\sqrt{a^2 - x^2}$, let $x = a\sin t$ or $x = a\cos t$;

(2) If the integrand contains $\sqrt{a^2 + x^2}$, let $x = a\tan t$ or $x = a\cot t$;

(3) If the integrand contains $\sqrt{x^2 - a^2}$, let $x = a\sec t$ or $x = a\csc t$.

Some indefinite integrals can be calculated using two different substitution methods.

Example 4.2.21 Find $\int \frac{dx}{x^2\sqrt{x^2-1}}(x>1)$.

Solution Using the first type of substitution method:

$$\int \frac{dx}{x^2\sqrt{x^2-1}}=\int \frac{dx}{x^3\sqrt{1-\frac{1}{x^2}}}=\int \frac{1}{x}\frac{-1}{\sqrt{1-\frac{1}{x^2}}}d\left(\frac{1}{x}\right)$$

$$\xrightarrow{u=\frac{1}{x}}\int \frac{-u}{\sqrt{1-u^2}}du=\sqrt{1-u^2}+C=\frac{1}{x}\sqrt{x^2-1}+C.$$

Another solution

Using the second type of substitution method: Let $x=\sec t\left(0<t<\frac{\pi}{2}\right)$, then

$$\int \frac{dx}{x^2\sqrt{x^2-1}}=\int \frac{\sec t\tan t}{\sec^2 t\tan t}dt=\int \cos t\,dt=\sin t+C=\frac{1}{x}\sqrt{x^2-1}+C.$$

When the degree of x in the denominator of the integrand is higher than that in the numerator (generally at least 2 degrees higher), consider using the reciprocal substitution $x=\frac{1}{t}$, which is used to eliminate the variable factor x in the denominator of the integrand.

Example 4.2.22 Find $\int \frac{dx}{x\sqrt{x^{2n}-1}}\ (x>1, n\in \mathbf{N}^+)$.

Solution Let $x=\frac{1}{t}$, then $dx=-\frac{1}{t^2}dt$.

$$\int \frac{dx}{x\sqrt{x^{2n}-1}}=\int \frac{1}{\frac{1}{t}\sqrt{\frac{1}{t^{2n}}-1}}\left(-\frac{1}{t^2}dt\right)=-\int \frac{t^{n-1}}{\sqrt{1-t^{2n}}}dt=-\frac{1}{n}\int \frac{1}{\sqrt{1-(t^n)^2}}dt^n$$

$$=-\frac{1}{n}\arcsin t^n+C=-\frac{1}{n}\arcsin\frac{1}{x^n}+C.$$

Using the tangent substitution $x=a\tan t$, it is also possible to eliminate the high power of (x^2+a^2) in the denominator of the integrand. See the following example.

Example 4.2.23 Find $\int \frac{x^3}{(x^2-2x+2)^2}dx$.

Solution The denominator is the square of a quadratic irreducible factor, complete the square of the quadratic irreducible factor to $(x-1)^2+1$, let $x-1=\tan t\left(-\frac{\pi}{2}<t<\frac{\pi}{2}\right)$, then $x^2-2x+2=\sec^2 t$, $dx=\sec^2 t\,dt$. Thus,

$$\int \frac{x^3}{(x^2-2x+2)^2}dx=\int \frac{(\tan t+1)^3}{\sec^4 t}\cdot \sec^2 t\,dt$$

$$=\int (\sin^3 t\cos^{-1} t+3\sin^2 t+3\sin t\cos t+\cos^2 t)\,dt$$

$$=\int(\sin^2 t\cos^{-1}t+3\cos t)\sin t\,dt+\int(3\sin^2 t+\cos^2 t)\,dt$$

$$=\int[(1-\cos^2 t)\cos^{-1}t+3\cos t][-d(\cos t)]+\int(2-\cos 2t)\,dt$$

$$=-\int(\cos^{-1}t+2\cos t)\,d(\cos t)+2t-\frac{1}{2}\sin 2t$$

$$=-\ln(\cos t)-\cos^2 t+2t-\sin t\cos t+C.$$

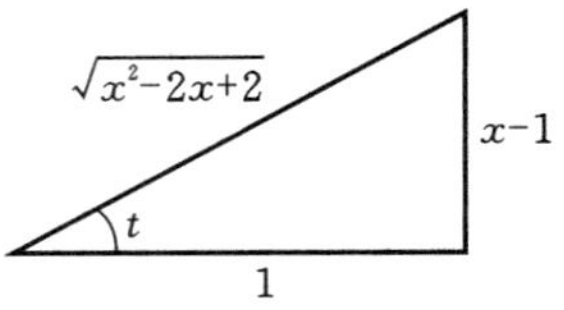

Figure 4-4

According to $\tan t=x-1$, construct an auxiliary triangle (Figure 4-4), then

$$\cos t=\frac{1}{\sqrt{x^2-2x+2}},\ \sin t=\frac{x-1}{\sqrt{x^2-2x+2}},$$

thus

$$\int\frac{x^3}{(x^2-2x+2)^2}dx=\frac{1}{2}\ln(x^2-2x+2)+2\arctan(x-1)-\frac{x}{x^2-2x+2}+C.$$

Some results from the examples in this section will be frequently encountered later, so they are often used as formulas. Here, in addition to the formulas in the basic integral table in Section 4.1.3, the following commonly used integral formulas are supplemented (where the constant $a>0$):

(16) $\int\tan x\,dx=-\ln|\cos x|+C$;

(17) $\int\cot x\,dx=\ln|\sin x|+C$;

(18) $\int\sec x\,dx=\ln|\sec x+\tan x|+C$;

(19) $\int\csc x\,dx=\ln|\csc x-\cot x|+C$;

(20) $\int\frac{dx}{a^2+x^2}=\frac{1}{a}\arctan\frac{x}{a}+C$;

(21) $\int\frac{dx}{x^2-a^2}=\frac{1}{2a}\ln\left|\frac{x-a}{x+a}\right|+C$;

(22) $\int\frac{dx}{\sqrt{a^2-x^2}}=\arcsin\frac{x}{a}+C$;

(23) $\int\frac{dx}{\sqrt{x^2+a^2}}=\ln(x+\sqrt{x^2+a^2})+C$;

(24) $\int\frac{dx}{\sqrt{x^2-a^2}}=\ln\left|x+\sqrt{x^2-a^2}\right|+C$.

Example 4.2.24 Find $\int\frac{dx}{\sqrt{4x^2-9}}$.

Solution $\int\frac{dx}{\sqrt{4x^2-9}}=\int\frac{dx}{\sqrt{(2x)^2-3^2}}=\frac{1}{2}\int\frac{d(2x)}{\sqrt{(2x)^2-3^2}}$.

Using formula (24), we get

$$\int \frac{dx}{\sqrt{4x^2-9}} = \frac{1}{2}\ln\left|2x+\sqrt{4x^2-9}\right| + C.$$

Example 4.2.25 Find $\int \frac{dx}{\sqrt{1+x-x^2}}$.

Solution $$\int \frac{dx}{\sqrt{1+x-x^2}} = \int \frac{d\left(x-\frac{1}{2}\right)}{\sqrt{\left(\frac{\sqrt{5}}{2}\right)^2-\left(x-\frac{1}{2}\right)^2}}.$$

Using integral formula (22), we get

$$\int \frac{dx}{\sqrt{1+x-x^2}} = \arcsin\frac{x-\frac{1}{2}}{\frac{\sqrt{5}}{2}} + C = \arcsin\frac{2x-1}{\sqrt{5}} + C.$$

Exercise 4.2

1. Fill in the appropriate coefficients in the blanks on the right-hand side of each equation to make the equation hold. For example, $dx = \frac{1}{3}d(3x-5)$.

(1) $dx =$ ______ $d(ax)$; (2) $dx =$ ______ $d(7x-3)$;

(3) $x\,dx =$ ______ $d(x^2)$; (4) $x\,dx =$ ______ $d(5x^2)$;

(5) $x\,dx =$ ______ $d(1-x^2)$; (6) $x^3dx =$ ______ $d(3x^4-2)$;

(7) $e^{2x}dx =$ ______ $d(e^{2x})$; (8) $e^{-\frac{x}{2}}dx =$ ______ $d(1+e^{-\frac{x}{2}})$;

(9) $\sin\frac{3}{2}x\,dx =$ ______ $d\left(\cos\frac{3}{2}x\right)$; (10) $\frac{dx}{x} =$ ______ $d(5\ln|x|)$;

(11) $\frac{dx}{x} =$ ______ $d(3-5\ln|x|)$; (12) $\frac{dx}{1+9x^2} =$ ______ $d(\arctan 3x)$;

(13) $\frac{dx}{\sqrt{1-x^2}} =$ ______ $d(1-\arcsin x)$; (14) $\frac{x\,dx}{\sqrt{1-x^2}} =$ ______ $d(\sqrt{1-x^2})$.

2. Find the following indefinite integrals:

(1) $\int (3-2x)^3dx$; (2) $\int \sqrt{4+3x}\,dx$;

(3) $\int \frac{dx}{\sqrt{2x-1}}$; (4) $\int (6-2x)^{\alpha}dx \ (\alpha \neq -1)$;

(5) $\int \frac{dx}{3x+1}$; (6) $\int \frac{dx}{1-3x}$;

(7) $\int \sin\frac{1}{2}x\,dx$; (8) $\int \sin^2 2x\,dx$;

(9) $\int \cos^2 \alpha x \, dx \quad (\alpha \neq 0)$;

(10) $\int e^{\sqrt{2}x} dx$;

(11) $\int \frac{dx}{\sqrt{4-x^2}}$;

(12) $\int \frac{dx}{4+x^2}$;

(13) $\int (\sin \alpha x - e^{2x}) dx$;

(14) $\int \cos^2 (\omega t + \varphi) dt$;

(15) $\int \frac{\sin x}{\cos^3 x} dx$;

(16) $\int \frac{\sin x + \cos x}{\sqrt[3]{\sin x - \cos x}} dx$;

(17) $\int \tan^{10} x \sec^2 x \, dx$;

(18) $\int \frac{dx}{x \ln x \ln(\ln x)}$;

(19) $\int \frac{dx}{(\arcsin x)^2 \sqrt{1-x^2}}$;

(20) $\int \frac{10^{2\arccos x}}{\sqrt{1-x^2}} dx$;

(21) $\int \tan \sqrt{1+x^2} \cdot \frac{x \, dx}{\sqrt{1+x^2}}$;

(22) $\int \frac{\arctan \sqrt{x}}{\sqrt{x}(1+x)} dx$;

(23) $\int \frac{1+\ln x}{(x \ln x)^2} dx$;

(24) $\int \frac{1}{\sin x \cos x} dx$;

(25) $\int \frac{\ln(\tan x)}{\cos x \sin x} dx$;

(26) $\int \frac{dx}{e^x + e^{-x}}$;

(27) $\int \sin 2x \cos 3x \, dx$;

(28) $\int \sin 5x \sin 7x \, dx$;

(29) $\int \tan^3 x \sec x \, dx$;

(30) $\int \frac{x \ln(1+x^2)}{1+x^2} dx$;

(31) $\int \frac{1-x}{\sqrt{9-4x^2}} dx$;

(32) $\int \frac{x^3}{9+x^2} dx$;

(33) $\int \frac{dx}{(x+1)(x-2)}$;

(34) $\int \frac{x}{x^2-x-2} dx$;

(35) $\int \frac{x-1}{x^2+2x+3}$;

(36) $\int \frac{3\sin x + 2\cos x}{2\sin x + 3\cos x} dx$;

(37) $\int \frac{\sin x}{\sin x - \cos x} dx$;

(38) $\int \frac{1}{1+\sqrt{1-x^2}} dx$;

(39) $\int \frac{dx}{\sqrt{(x^2+1)^3}}$;

(40) $\int \frac{\sqrt{x^2-9}}{x} dx \quad (x > 3)$;

(41) $\int \frac{x^3+1}{(x^2+1)^2} dx$;

(42) $\int \frac{dx}{1+\sqrt{2x}}$;

(43) $\int \frac{1-x}{\sqrt{9-4x^2}} dx$;

(44) $\int \frac{dx}{x^4(1+x^2)}$;

(45) $\int \sqrt{e^x - 2} \, dx$;

(46) $\int \frac{dx}{1+\sqrt{1+x}}$.

4.3 Integration by Parts

Based on the method of differentiating composite functions, the substitution method for integration was obtained. Now, using the method of differentiating the product of two functions, another fundamental method for finding indefinite integrals, the integration by parts is derived.

Suppose the function $u=u(x), v=v(x)$ has continuous derivatives. From the formula for the derivative of the product of two functions, we have

$$(uv)'=u'v+uv'.$$

Rearranging the terms, we get

$$uv'=(uv)'-vu'.$$

Taking the indefinite integral on both sides of the equation, we obtain

$$\int uv'\mathrm{d}x = uv - \int vu'\mathrm{d}x \text{ or } \int u\,\mathrm{d}v = uv - \int v\,\mathrm{d}u.$$

They are called the integration by parts formula. If finding $\int u\,\mathrm{d}v$ is difficult, but finding $\int v\,\mathrm{d}u$ is relatively easy, then the integration by parts formula can demonstrate its effectiveness.

Example 4.3.1 Find $\int x\cos x\,\mathrm{d}x$.

Solution We now attempt to find this integral using the integration by parts method. First, we split the integrand into two parts, u and $\mathrm{d}v$. If we set $u=x$, $\mathrm{d}v=\cos x\,\mathrm{d}x$, then $\mathrm{d}u=\mathrm{d}x$, $v=\sin x$. Substituting into equation $\int u\,\mathrm{d}v = uv - \int v\,\mathrm{d}u$, we get

$$\int x\cos x\,\mathrm{d}x = x\sin x - \int \sin x\,\mathrm{d}x.$$

Since $\int \sin x\,\mathrm{d}x$ can be easily integrated, we have

$$\int x\cos x\,\mathrm{d}x = x\sin x + \cos x + C.$$

When finding this integral, if we set $u=\cos x$, $\mathrm{d}v=x\,\mathrm{d}x$, then $\mathrm{d}u=-\sin x\,\mathrm{d}x$, $v=\frac{x^2}{2}$. Thus,

$$\int x\cos x\,\mathrm{d}x = \frac{x^2}{2}\cos x + \int \frac{x^2}{2}\sin x\,\mathrm{d}x.$$

The integral on the right-hand side of the equation is even more difficult to find than the original integral.

Conclusion: When calculating indefinite integrals using the method of integration by parts, appropriately selecting u and $\mathrm{d}v$ is crucial. The principles for selecting u and $\mathrm{d}v$ are:

(1) It is easy to obtain v from $\mathrm{d}v$;

(2) $\int v\mathrm{d}u$ is easier to integrate than $\int u\mathrm{d}v$.

Example 4.3.2 Find $\int x\mathrm{e}^{-x}\mathrm{d}x$.

Solution Let $u=x$, $\mathrm{d}v=\mathrm{e}^{-x}\mathrm{d}x$, then $\mathrm{d}u=\mathrm{d}x$, $v=-\mathrm{e}^{-x}$. Thus,

$$\int x\mathrm{e}^{-x}\mathrm{d}x=-x\mathrm{e}^{-x}+\int \mathrm{e}^{-x}\mathrm{d}x=-x\mathrm{e}^{-x}-\mathrm{e}^{-x}+C.$$

Some integrals require the continuous application of the method of integration by parts.

Example 4.3.3 Find $\int x^2\mathrm{e}^x\mathrm{d}x$.

Solution $\int x^2\mathrm{e}^x\mathrm{d}x=\int x^2\mathrm{d}(\mathrm{e}^x)=x^2\mathrm{e}^x-2\int x\mathrm{e}^x\mathrm{d}x=x^2\mathrm{e}^x-2\int x\mathrm{d}(\mathrm{e}^x)$

$$=x^2\mathrm{e}^x-2\left(x\mathrm{e}^x-\int \mathrm{e}^x\mathrm{d}x\right)=\mathrm{e}^x(x^2-2x+2)+C.$$

Note: If the integrand is a product of a power function (with a positive integer exponent) and an exponential or sine (cosine) function, one can set the power function as u and treat the rest of the integrand as $\mathrm{d}v$. After applying the integration by parts formula, the power of the power function is reduced by one.

Example 4.3.4 Find $\int x\arctan x\,\mathrm{d}x$.

Solution Let $u=\arctan x$, $\mathrm{d}v=x\mathrm{d}x=\mathrm{d}\left(\frac{x^2}{2}\right)$, then

$$\begin{aligned}\int x\arctan x\,\mathrm{d}x&=\frac{x^2}{2}\arctan x-\int\frac{x^2}{2}\mathrm{d}(\arctan x)=\frac{x^2}{2}\arctan x-\frac{1}{2}\int\frac{x^2}{1+x^2}\mathrm{d}x\\&=\frac{x^2}{2}\arctan x-\frac{1}{2}\int\left(1-\frac{1}{1+x^2}\right)\mathrm{d}x\\&=\frac{x^2}{2}\arctan x-\frac{1}{2}(x-\arctan x)+C.\end{aligned}$$

Example 4.3.5 Find $\int x^3\ln x\,\mathrm{d}x$.

Solution Let $u=\ln x$, $\mathrm{d}v=x^3\mathrm{d}x=\mathrm{d}\left(\frac{x^4}{4}\right)$, then

$$\int x^3\ln x\,\mathrm{d}x=\frac{1}{4}x^4\ln x-\frac{1}{4}\int x^4\cdot\frac{1}{x}\mathrm{d}x=\frac{1}{4}x^4\ln x-\frac{1}{16}x^4+C.$$

Note: If the integrand is a product of a power function and a logarithmic or inverse trigonometric function, one can set the logarithmic or inverse trigonometric function as u, and the remaining integrand expression as $\mathrm{d}v$. After applying the integration by parts formula, the logarithmic or inverse trigonometric function will disappear.

Example 4.3.6 Find $\int e^x \sin x\, dx$.

Solution

$$\int e^x \sin x\, dx = \int \sin x\, d(e^x) = e^x \sin x - \int e^x d(\sin x) = e^x \sin x - \int e^x \cos x\, dx = e^x \sin x - \int \cos x\, d(e^x)$$

$$= e^x \sin x - e^x \cos x + \int e^x d(\cos x) = e^x \sin x - e^x \cos x - \int e^x \sin x\, dx$$

Solving for the variable

$$\int e^x \sin x\, dx = \frac{e^x}{2}(\sin x - \cos x) + C.$$

Below are some more examples for the reader to carefully understand the problem-solving methods.

Example 4.3.7 Find $\int \sec^3 x\, dx$.

Solution

$$\int \sec^3 x\, dx = \int \sec x \sec^2 x\, dx = \int \sec x\, d(\tan x) = \sec x \tan x - \int \sec x \tan^2 x\, dx$$

$$= \sec x \tan x - \int \sec x\,(\sec^2 x - 1)\, dx = \sec x \tan x - \int \sec^3 x\, dx + \int \sec x\, dx$$

$$= \sec x \tan x + \ln|\sec x + \tan x| - \int \sec^3 x\, dx.$$

Solving for the variable

$$\int \sec^3 x\, dx = \frac{1}{2}(\sec x \tan x + \ln|\sec x + \tan x|) + C.$$

Example 4.3.8 Find $\int \ln(1+\sqrt{x})\, dx$.

Solution Let $t = \sqrt{x}$, then $x = t^2$, we get

$$\int \ln(1+\sqrt{x})\, dx = \int \ln(1+t)\, d(t^2) = t^2 \ln(1+t) - \int t^2 d[\ln(1+t)]$$

$$= t^2 \ln(1+t) - \int \frac{t^2}{1+t} dt = t^2 \ln(1+t) - \int \frac{t^2-1}{1+t} dt - \int \frac{1}{1+t} dt$$

$$= t^2 \ln(1+t) - \int (t-1)\, dt - \int \frac{1}{1+t} dt = t^2 \ln(1+t) - \frac{t^2}{2} + t - \ln(1+t) + C$$

$$= (x-1)\ln(1+\sqrt{x}) + \sqrt{x} - \frac{x}{2} + C.$$

Example 4.3.9 Given that an antiderivative of $f(x)$ is e^{-x^2}, find $\int x f'(x) dx$.

Solution Using the integration by parts formula, we get

$$\int x f'(x) dx = \int x\, d[f(x)] = x f(x) - \int f(x) dx,$$

according to the problem

$$\int f(x) dx = e^{-x^2} + C,$$

differentiating both sides of the equation with respect to x, we get

$$f(x)=-2xe^{-x^2},$$

therefore

$$\int xf'(x)dx=xf(x)-\int f(x)dx=-2x^2e^{-x^2}-e^{-x^2}-C=-e^{-x^2}(2x^2+1)-C.$$

Example 4.3.10 Find $\int\frac{x^2e^x}{(x+2)^2}dx$.

Solution Choose $u=x^2e^x$, then

$$dv=\frac{1}{(x+2)^2}dx=d\left(\frac{-1}{x+2}\right),$$

$$\int\frac{x^2e^x}{(x+2)^2}dx=\int x^2e^x d\left(\frac{-1}{x+2}\right)=x^2e^x\left(\frac{-1}{x+2}\right)-\int\frac{-1}{x+2}d(x^2e^x)$$

$$=-\frac{x^2e^x}{x+2}+\int\frac{2xe^x+x^2e^x}{x+2}dx=-\frac{x^2e^x}{x+2}+\int xe^x dx$$

$$=-\frac{x^2e^x}{x+2}+xe^x-\int e^x dx=-\frac{x^2e^x}{x+2}+xe^x-e^x+C.$$

Exercise 4.3

1. Find the following indefinite integrals:

(1) $\int x\sin x\,dx$; (2) $\int x^2\cos x\,dx$;

(3) $\int x\,2^{-x}dx$; (4) $\int x^2\ln x\,dx$;

(5) $\int(x+1)\ln x\,dx$; (6) $\int\arcsin x\,dx$;

(7) $\int\frac{\ln x}{\sqrt{x}}dx$; (8) $\int x\ln(x-1)dx$;

(9) $\int\ln^2 x\,dx$; (10) $\int e^{\sqrt{x}}dx$;

(11) $\int x\sin x\cos x\,dx$; (12) $\int x\tan^2 x\,dx$;

(13) $\int\frac{\ln(\sin x)}{\cos^2 x}dx$; (14) $\int\frac{\arcsin x}{x^2}dx$;

(15) $\int\frac{x}{\cos^2 x}dx$; (16) $\int\frac{\arctan x}{x^2}dx$;

(17) $\int\cos(\ln x)dx$; (18) $\int\sin x\ln(\tan x)dx$;

(19) $\int\frac{\ln(\cos x)}{\cos^2 x}dx$; (20) $\int\ln\left(x+\sqrt{1+x^2}\right)dx$.

2. Given $f'(e^x)=2x$, find $f(x)$.

3. Given that an antiderivative of $f(x)$ is $\dfrac{\sin x}{x}$, find $\int x f'(x)\mathrm{d}x$.

4. Given that an antiderivative of $f'(x)$ is $x\ln x$, find $\int f'(\sqrt{x})\,\mathrm{d}x$.

5. Proof: If $I_n=\int \tan^n x\,\mathrm{d}x\ (n=2,3,\cdots)$, then

$$I_n=\frac{1}{n-1}\tan^{n-1}x-I_{n-2}.$$

4.4 Integration of Rational Functions and Functions Reducible to Rational Functions

This section will briefly introduce the indefinite integrals of some special types of functions, including the integrals of rational functions and functions reducible to rational functions (such as rational expressions of trigonometric functions, integrals of simple irrational functions, etc.), and finally introduce the use of integral tables.

4.4.1 Integration of Rational Functions

A function represented by the quotient of two polynomials $P(x)$ and $Q(x)$, namely

$$\frac{P(x)}{Q(x)}=\frac{a_0x^n+a_1x^{n-1}-\cdots+a_{n-1}x+a_n}{b_0x^m+b_1x^{m-1}-\cdots+b_{m-1}x+b_m}$$

is called a rational function, also known as a rational fraction, where m and n are non-negative integers, the coefficients of the terms in the numerator and denominator are real numbers, $a_0\neq 0$, $b_0\neq 0$, and there are no common factors between $P(x)$ and $Q(x)$.

When $n<m$, it is called a proper fraction, when $n\geqslant m$, it is called an improper fraction. An improper fraction can always be converted into the sum of a polynomial and a proper fraction using polynomial division. For example,

$$\frac{x^4}{x^2+1}=x^2-1+\frac{1}{x^2+1}.$$

Since the integral of a polynomial is easy to obtain, the problem of integrating rational functions only needs to be discussed for proper rational fractions.

For a proper fraction $\dfrac{P(x)}{Q(x)}$, if the denominator can be factored into the product of two polynomials

$$Q(x)=Q_1(x)Q_2(x),$$

$Q_1(x)$ and $Q_2(x)$ have no common factors, then it can be decomposed into the sum of two proper fractions:

$$\frac{P(x)}{Q(x)}=\frac{P_1(x)}{Q_1(x)}-\frac{P_2(x)}{Q_2(x)}.$$

The above steps are referred to as decomposing a proper fraction into a sum of partial fractions. If $Q_1(x)$ or $Q_2(x)$ can be further decomposed into the product of two polynomials with no common factors, then it can be further split into simpler partial fractions. It follows that the decomposition of a proper fraction will only yield partial fractions of the form $\frac{P_1(x)}{(x-a)^k}$ and $\frac{P_2(x)}{(x^2+px+q)^l}$, where $p^2-4q<0$, $P_1(x)$ is a polynomial of degree less than k, and $P_2(x)$ is a polynomial of degree less than 2. Below are a few examples of integrating proper fractions.

Example 4.4.1 Find $\int \frac{3x+1}{x^2-x-6}\mathrm{d}x$.

Solution

$$Q(x)=x^2-x-6=(x+2)(x-3),$$

thus, we can assume

$$\frac{3x+1}{x^2-x-6}=\frac{A}{x+2}+\frac{B}{x-3}.$$

Where: A, B are the coefficients to be determined. Multiplying both sides of the equation by the denominator, we get

$$A(x-3)+B(x+2)=3x+1.$$

Comparing the coefficients of like powers on both sides of the equation, we have

$$\left.\begin{array}{r}A+B=3\\-3A+2B=1\end{array}\right\}\Rightarrow A=1, B=2,$$

thus

$$\int \frac{3x+1}{x^2-x-6}\mathrm{d}x=\int\left(\frac{1}{x+2}+\frac{2}{x-3}\right)\mathrm{d}x=\ln|x+2|+2\ln|x-3|+C.$$

Example 4.4.2 Find $\int \frac{x^3+2x^2-3x+2}{(x-1)^2(1+x^2)}\mathrm{d}x$.

Solution The proper fraction can be decomposed into the following form:

$$\frac{x^3+2x^2-3x+2}{(x-1)^2(1+x^2)}=\frac{Ax+B}{(x-1)^2}+\frac{Cx+D}{1+x^2}.$$

Multiplying both sides by the denominator, we get

$$(Ax+B)(1+x^2)+(Cx+D)(x-1)^2=x^3+2x^2-3x+2.$$

That is

$$(A+C)x^3+(B-2C+D)x^2+(A+C-2D)x+B+D=x^3+2x^2-3x+2.$$

Comparing coefficients on both ends, we obtain the system of equations

$$\left.\begin{array}{r}A+C=1\\B-2C+D=2\\A+C-2D=-3\\B+D=2\end{array}\right\}\Rightarrow A=1, B=0, C=0, D=2.$$

Thus

$$\frac{x^3+2x^2-3x+2}{(x-1)^2(1+x^2)}=\frac{x}{(x-1)^2}+\frac{2}{1+x^2}=\frac{x-1+1}{(x-1)^2}+\frac{2}{1+x^2}=\frac{1}{x-1}+\frac{1}{(x-1)^2}+\frac{2}{1+x^2}.$$

Therefore

$$\int\frac{x^3+2x^2-3x+2}{(x-1)^2(1+x^2)}\mathrm{d}x=\int\frac{\mathrm{d}x}{x-1}+\int\frac{\mathrm{d}x}{(x-1)^2}+2\int\frac{\mathrm{d}x}{1+x^2}=\ln|x-1|-\frac{1}{x-1}+2\arctan x+C.$$

Example 4.4.3 Find $\int\frac{x-2}{x^2+2x+3}\mathrm{d}x$.

Solution The denominator is already a quadratic irreducible factor and cannot be further factored. Since $(x^2+2x+3)'=2x+2$ and the numerator $x-2$ are both linear expressions, we can consider splitting the numerator into the following two terms:

$$x-2=\frac{1}{2}(x^2+2x+3)'-3=\frac{1}{2}(2x+2)-3.$$

Hence

$$\int\frac{x-2}{x^2+2x+3}\mathrm{d}x=\int\frac{\frac{1}{2}(2x+2)-3}{x^2+2x+3}\mathrm{d}x=\frac{1}{2}\int\frac{2x+2}{x^2+2x+3}\mathrm{d}x-3\int\frac{\mathrm{d}x}{x^2+2x+3}$$

$$=\frac{1}{2}\int\frac{\mathrm{d}(x^2+2x+3)}{x^2+2x+3}-3\int\frac{\mathrm{d}(x+1)}{(x+1)^2+(\sqrt{2})^2}$$

$$=\frac{1}{2}\ln(x^2+2x+3)-\frac{3}{\sqrt{2}}\arctan\frac{x+1}{\sqrt{2}}+C.$$

Example 4.4.4 Find $\int\frac{x-3}{(x-1)(x^2-1)}\mathrm{d}x$.

Solution The two factors $x-1$ and x^2-1 in the denominator of the integrand have a common factor, so they need to be further factored into $(x-1)^2(x+1)$. Thus, we assume

$$\frac{x-3}{(x-1)^2(x+1)}=\frac{Ax+B}{(x-1)^2}+\frac{C}{x+1}.$$

Multiplying both sides by the denominator, we get

$$x-3=(Ax+B)(x+1)+C(x-1)^2,$$

that is

$$x-3=(A+C)x^2+(A+B-2C)x+B+C,$$

we have

$$\left.\begin{aligned}A+C&=0\\A+B-2C&=1\\B+C&=-3\end{aligned}\right\}\Rightarrow A=1,B=-2,C=-1.$$

Therefore

$$\int\frac{x-3}{(x-1)(x^2-1)}\mathrm{d}x=\int\frac{x-3}{(x-1)^2(x+1)}\mathrm{d}x=\int\left[\frac{x-2}{(x-1)^2}-\frac{1}{x+1}\right]\mathrm{d}x$$

$$=\int\frac{x-1-1}{(x-1)^2}\mathrm{d}x-\ln|x+1|=\ln|x-1|+\frac{1}{x-1}-\ln|x+1|+C.$$

4.4.2 Integration of Rational Trigonometric Functions

A function formed by a finite number of arithmetic operations on $\sin x$, $\cos x$ and constants is called a rational trigonometric function, denoted as $R(\sin x, \cos x)$.

The integration of rational trigonometric functions is quite flexible, with many methods available. Some methods were introduced in the context of substitution and integration by parts. The method to be introduced here involves transforming the integral of a rational trigonometric function into the integral of a rational function through appropriate substitutions.

From trigonometric identities, it is known that $\sin x$ and $\cos x$ can both be expressed as rational functions of $\tan\frac{x}{2}$, i.e.,

$$\sin x = 2\sin\frac{x}{2}\cos\frac{x}{2} = \frac{2\tan\frac{x}{2}}{\sec^2\frac{x}{2}} = \frac{2\tan\frac{x}{2}}{1+\tan^2\frac{x}{2}},$$

$$\cos x = \cos^2\frac{x}{2} - \sin^2\frac{x}{2} = \frac{1-\tan^2\frac{x}{2}}{\sec^2\frac{x}{2}} = \frac{1-\tan^2\frac{x}{2}}{1+\tan^2\frac{x}{2}}.$$

Therefore, if the substitution $u=\tan\frac{x}{2}$ is made, then $x=2\arctan u$, and thus we have

$$\sin x = \frac{2u}{1+u^2}, \cos x = \frac{1-u^2}{1+u^2}, \mathrm{d}x = \frac{2\mathrm{d}u}{1+u^2}.$$

The integral of a rational trigonometric function can then be transformed into the integral of a rational function, i.e.,

$$\int R(\sin x, \cos x)\,\mathrm{d}x = \int R\left(\frac{2u}{1+u^2}, \frac{1-u^2}{1+u^2}\right)\frac{2}{1+u^2}\mathrm{d}u.$$

This substitution is known as the universal substitution.

In some cases (such as when the powers of $\sin x$ and $\cos x$ in the rational trigonometric function are both even), the substitution $u=\tan x$ is also commonly used, from which it is easy to deduce

$$\sin x = \frac{u}{\sqrt{1+u^2}}, \cos x = \frac{1}{\sqrt{1+u^2}}, \mathrm{d}x = \frac{1}{1+u^2}\mathrm{d}u.$$

This substitution is often referred to as the modified universal substitution.

Example 4.4.5 Find $\int \frac{\cot x}{1+\sin x+\cos x}\mathrm{d}x$.

Solution Let $u=\tan\frac{x}{2}$, then we have

$$\mathrm{d}x=\frac{2}{1+u^2}\mathrm{d}u,\sin x=\frac{2u}{1+u^2},\cos x=\frac{1-u^2}{1+u^2},\cot x=\frac{1-u^2}{2u}.$$

Thus

$$\int\frac{\cot x}{1+\sin x+\cos x}\mathrm{d}x=\int\frac{\frac{1-u^2}{2u}}{1+\frac{2u}{1+u^2}+\frac{1-u^2}{1+u^2}}\frac{2}{1+u^2}\mathrm{d}u=\int\frac{1-u}{2u}\mathrm{d}u$$

$$=\frac{1}{2}\int\frac{\mathrm{d}u}{u}-\frac{1}{2}\int\mathrm{d}u=\frac{1}{2}\ln|u|-\frac{1}{2}u+C=\frac{1}{2}\ln\left|\tan\frac{x}{2}\right|-\frac{1}{2}\tan\frac{x}{2}+C.$$

Example 4.4.6 Find $\int\frac{1}{\sin^4 x}\mathrm{d}x$.

Solution Method 1: Using the universal substitution formula, let $u=\tan\frac{x}{2}$, then

$$\int\frac{1}{\sin^4 x}\mathrm{d}x=\int\frac{1}{\left(\frac{2u}{1+u^2}\right)^4}\frac{2}{1+u^2}\mathrm{d}u=\int\frac{1-3u^2+3u^4+u^6}{8u^4}\mathrm{d}u$$

$$=\frac{1}{8}\int(u^{-4}+3u^{-2}+3+u^2)\,\mathrm{d}u=\frac{1}{8}\left(-\frac{1}{3u^3}-\frac{3}{u}+3u+\frac{u^3}{3}\right)+C$$

$$=-\frac{1}{24}\cot^3\frac{x}{2}-\frac{3}{8}\cot\frac{x}{2}+\frac{3}{8}\tan\frac{x}{2}+\frac{1}{24}\tan^3\frac{x}{2}+C.$$

Method 2: Using the modified universal substitution formula, let $u=\tan x$, then

$$\int\frac{1}{\sin^4 x}\mathrm{d}x=\int\frac{1}{\left(\frac{u}{\sqrt{1+u^2}}\right)^4}\frac{1}{1+u^2}\mathrm{d}u=\int\frac{1+u^2}{u^4}\mathrm{d}u$$

$$=-\frac{1}{3u^3}-\frac{1}{u}+C=-\frac{1}{3}\cot^3 x-\cot x+C.$$

Method 3: Without using the universal substitution formula:

$$\int\frac{1}{\sin^4 x}\mathrm{d}x=\int\csc^2 x\,(1+\cot^2 x)\,\mathrm{d}x=\int\csc^2 x\,\mathrm{d}x+\int\cot^2 x\csc^2 x\,\mathrm{d}x$$

$$=-\cot x-\frac{1}{3}\cot^3 x+C.$$

Note: Comparing the above three methods, it is clear that the universal substitution is not necessarily the best method. Therefore, when calculating integrals of rational trigonometric functions, one should first consider other means and resort to the universal substitution only when necessary.

4.4.3 Integration of Simple Irrational Functions

To find the integral of a simple irrational function, the basic idea is to use appropriate substitutions to rationalize it, transforming it into the integral of a rational function. The following examples illustrate this.

Example 4.4.7 Find $\int \frac{x}{\sqrt[3]{3x+1}}\mathrm{d}x$.

Solution To remove the square root, let $u=\sqrt[3]{3x+1}$, then $x=\frac{u^3-1}{3}$, $\mathrm{d}x=u^2\mathrm{d}u$, so

$$\int \frac{x}{\sqrt[3]{3x+1}}\mathrm{d}x=\int \frac{u^3-1}{3u}\cdot u^2\mathrm{d}u=\frac{1}{3}\int (u^4-u)\,\mathrm{d}u=\frac{1}{3}\left(\frac{u^5}{5}-\frac{u^2}{2}\right)+C$$

$$=\frac{1}{15}(3x+1)^{\frac{5}{3}}-\frac{1}{6}(3x+1)^{\frac{2}{3}}+C.$$

Example 4.4.8 Find $\int \frac{1}{\sqrt{x}\,(1+\sqrt[3]{x})}\mathrm{d}x$.

Solution To remove both radicals simultaneously, let $\sqrt[6]{x}=t$, i.e., $x=t^6$, then $\mathrm{d}x=6t^5\mathrm{d}t$, thus

$$\int \frac{1}{\sqrt{x}\,(1+\sqrt[3]{x})}\mathrm{d}x=\int \frac{6t^5}{t^3(1+t^2)}\mathrm{d}t=\int \frac{6t^2}{1+t^2}\mathrm{d}t$$

$$=6\int \frac{t^2+1-1}{1+t^2}\mathrm{d}t=6\int\left(1-\frac{1}{1+t^2}\right)\mathrm{d}t$$

$$=6(t-\arctan t)+C=6(\sqrt[6]{x}-\arctan^6\sqrt{x})+C.$$

Example 4.4.9 Find $\int \frac{1}{x}\sqrt{\frac{1+x}{x}}\,\mathrm{d}x$.

Solution To remove the square root, let $\sqrt{\frac{1+x}{x}}=t$, then

$$\frac{1+x}{x}=t^2,\ x=\frac{1}{t^2-1},\ \mathrm{d}x=-\frac{2t\,\mathrm{d}t}{(t^2-1)^2}.$$

Therefore

$$\int \frac{1}{x}\sqrt{\frac{1+x}{x}}\,\mathrm{d}x=\int (t^2-1)\,t\cdot\frac{-2t}{(t^2-1)^2}\mathrm{d}t=-2\int\frac{t^2}{t^2-1}\mathrm{d}t$$

$$=-2\int\left(1+\frac{1}{t^2-1}\right)\mathrm{d}t=-2t-\ln\left|\frac{t-1}{t+1}\right|+C$$

$$=-2t+2\ln(t+1)-\ln|t^2-1|+C$$

$$=-2\sqrt{\frac{1+x}{x}}+2\ln\left(\sqrt{\frac{1+x}{x}}+1\right)+\ln|x|+C.$$

The above three examples show that if the integrand contains simple radicals $\sqrt[n]{ax+b}$ or $\sqrt[n]{\frac{ax+b}{cx+d}}$, one can set this simple radical to u. Since such a transformation has an inverse function, and the inverse function is a rational function of u, the original integral can be transformed into an integral of a rational function.

4.4.4 Use of Integration Tables

From the previous discussion, it can be seen that the calculation of integrals is more flexible and complex than that of derivatives. For practical convenience, commonly used integral formulas are often compiled into tables, which are called integral tables. In the integral tables, all integral formulas are arranged according to the type of integrand. One only needs to determine the type of the integrand or transform it into a type listed in the table, and then consult the formula to obtain the integral. Some calculators and computer software also have practical functions for finding indefinite integrals. However, for beginners, it is essential to first master various basic integration methods.

Below are two examples of using the integral table to find integral results.

Example 4.4.10 Find the integral $\int \frac{1}{5-4\cos x}dx$ using the table.

Solution The formula can be found in the integral table:

$$\int \frac{dx}{a+b\cos x} = \frac{2}{a+b}\sqrt{\frac{a+b}{a-b}}\arctan\left(\sqrt{\frac{a-b}{a+b}}\tan\frac{x}{2}\right)+C \quad (a^2>b^2).$$

Substituting $a=5, b=-4$ into the formula, we get

$$\int \frac{1}{5-4\cos x}dx = \frac{2}{3}\arctan\left(3\tan\frac{x}{2}\right)+C.$$

Example 4.4.11 Find the integral $\int \frac{dx}{x\sqrt{4x^2+9}}$ using the table.

Solution This integral cannot be found directly in the table and requires a change of variables.

Let $2x=u$, then $\sqrt{4x^2+9}=\sqrt{u^2+3^2}$, thus

$$\int \frac{dx}{x\sqrt{4x^2+9}} = \int \frac{\frac{1}{2}du}{\frac{u}{2}\sqrt{u^2+3^2}} = \int \frac{du}{u\sqrt{u^2+3^2}}.$$

The formula can be found in the integral table:

$$\int \frac{dx}{x\sqrt{x^2+a^2}} = \frac{1}{a}\ln\frac{\sqrt{x^2+a^2}-a}{|x|}+C.$$

Therefore

$$\int \frac{du}{u\sqrt{u^2+3^2}} = \frac{1}{3}\ln\frac{\sqrt{u^2+3^2}-3}{|u|}+C.$$

Substituting $u=2x$ into the formula, we get

$$\int \frac{dx}{x\sqrt{4x^2+9}} = \frac{1}{3}\ln\frac{\sqrt{4x^2+9}-3}{2|x|}+C.$$

Before concluding this chapter, it should be noted that since elementary functions are continuous within their definition intervals, their indefinite integrals must exist. However, the existence of an indefinite integral is not the same as whether the indefinite integral can be expressed using elementary functions. In fact, there are many elementary functions whose indefinite integrals exist but cannot be expressed using elementary functions, such as

$$\int e^{-x^2}\,dx,\ \int \frac{\sin x}{x}\,dx,\ \int \frac{dx}{\ln x},\ \int \frac{dx}{\sqrt{1+x^4}}$$

and so on, which are not elementary functions.

Exercise 4.4

1. Find the indefinite integrals of the following rational functions:

(1) $\int \frac{2x+3}{x^2+3x-10}\,dx$; (2) $\int \frac{x^3}{9+x^2}\,dx$;

(3) $\int \frac{dx}{x(x^2+1)}$; (4) $\int \frac{x^2+1}{(x+1)^2(x-1)}\,dx$;

(5) $\int \frac{x}{x^2+2x+2}\,dx$; (6) $\int \frac{dx}{(x^2+1)(x^2+x)}$.

2. Find the indefinite integrals of the following functions:

(1) $\int \frac{dx}{3+\cos x}$; (2) $\int \frac{dx}{1+\sin x+\cos x}$;

(3) $\int \tan^3 x\,dx$; (4) $\int \frac{\cos x}{2\sin x+3\cos x}\,dx$;

(5) $\int \frac{1}{x\sqrt{2x+1}}\,dx$; (6) $\int \frac{\sqrt{x+1}-1}{\sqrt{x+1}+1}\,dx$;

(7) $\int \frac{dx}{\sqrt{x}+\sqrt[4]{x}}$; (8) $\int \frac{1}{\sqrt{1+e^x}}\,dx$;

(9) $\int \frac{dx}{1+\sqrt[3]{x+1}}$; (10) $\int \sqrt{\frac{1-x}{1+x}}\cdot\frac{1}{x}\,dx$.

Comprehensive Review 4

1. Find the following indefinite integrals (where a is a constant):

(1) $\int \frac{dx}{e^{-x}-e^{x}}$; (2) $\int \frac{x}{(1-x)^3}\,dx$;

(3) $\int \frac{x^2}{a^6-x^6}\,dx\ (a>0)$; (4) $\int \frac{1+\cos x}{x+\sin x}\,dx$;

(5) $\int \frac{\ln(\ln x)}{x} dx$;

(6) $\int \frac{\sin x \cos x}{1+\sin^4 x} dx$;

(7) $\int \tan^4 x \, dx$;

(8) $\int \frac{dx}{x(x^6+4)}$;

(9) $\int \sqrt{\frac{a+x}{a-x}} dx \quad (a>0)$;

(10) $\int \frac{dx}{\sqrt{x(1+x)}}$;

(11) $\int x \cos^2 x \, dx$;

(12) $\int \frac{dx}{(a^2-x^2)^{5/2}}$;

(13) $\int \frac{dx}{x^4 \sqrt{1+x^2}}$;

(14) $\int \sqrt{x} \sin \sqrt{x} \, dx$;

(15) $\int \ln(1+x^2) \, dx$;

(16) $\int \frac{\sin^2 x}{\cos^3 x} dx$;

(17) $\int \arctan \sqrt{x} \, dx$;

(18) $\int \frac{\sqrt{1+\cos x}}{\sin x} dx$;

(19) $\int \frac{x^3}{(1+x^8)^2} dx$;

(20) $\int \frac{\sin x}{1+\sin x} dx$;

(21) $\int \frac{x+\sin x}{1+\cos x} dx$;

(22) $\int \frac{\sqrt[3]{x}}{x(\sqrt{x}+\sqrt[3]{x})} dx$;

(23) $\int \frac{dx}{(1+e^x)^2}$;

(24) $\int \frac{e^{3x}+e^x}{e^{4x}-e^{2x}+1} dx$;

(25) $\int \frac{x e^x}{(e^x+1)^2} dx$;

(26) $\int \ln^2 (x+\sqrt{1+x^2}) \, dx$;

(27) $\int \frac{\ln x}{(1+x^2)^{3/2}} dx$;

(28) $\int \sqrt{1-x^2} \arcsin x \, dx$;

(29) $\int \frac{x^3 \arccos x}{\sqrt{1-x^2}} dx$;

(30) $\int \frac{\cot x}{1+\sin x} dx$;

(31) $\int \frac{dx}{\sin^3 x \cos x}$;

(32) $\int \frac{dx}{(2+\cos x)\sin x}$.

2. Given $f'(\sin^2 x)=\cos 2x+\tan^2 x$, find $f(x)$ $(0<x<1)$.

3. Given $f'\left(x\tan\frac{x}{2}\right)=(x+\sin x)\tan\frac{x}{2}+\cos x$, find $f(x)$.

4. The derivative function $f'(x)$ of the function $f(x)$ is an upward-opening quadratic parabola that intersects the x axis at points $x=0$ and $x=2$. If the maximum value of $f(x)$ is 4 and the minimum value is 0, find $f(x)$.

5. Given that $F(x)>0$ is the antiderivative of $f(x)$ and $F(0)=1$, when $x\geqslant 0$, we have $f(x)F(x)=\sin^2 2x$. Solve for $f(x)$.

Chapter 5 Definite Integrals and Their Applications

The *I Ching* emphasizes the mutual transformation and harmony between Yin and Yang, and holds that all things in the world are in a state of balance between Yin and Yang. The harmony of Yin and Yang is not only reflected in the static balance, but also reflected in the dynamic process of change. Through the continuous accumulation of Yin and Yang, things gradually develop into a new state or form. A definite integral can be seen as a quantitative description of the "cumulative effect" of a function over a certain interval. This cumulative effect reflects the harmonization and balance of the two forces (such as the positive and negative values of the function) in the process of change. The result of the definite integral is the ultimate embodiment of this Yin and Yang harmony and accumulation.

The idea of "division" in the *I Ching* is embodied in the careful observation and analysis of things, by dividing into smaller parts to better understand the whole. The process of calculating a definite integral is to divide the interval into infinitesimal parts and sum the function values on each small interval. This "art of division and summation" embodies the fine grasp of the change of the function in the interval, and approximates the real cumulative effect by infinitely fine division and summation. Imagine a meandering river, the integral is like dividing the river into numerous small sections, then calculating the flow of each section (i.e., the value of the function), and finally summing the flow of all sections to get the total flow of the whole river in the interval. This "art of dividing and summing" allows us to accurately calculate the cumulative effect of a function over an interval.

The *I Ching* emphasizes "flexibility" and "application", and holds that only when theory is applied to practice can it exert its value. As an important part of calculus, definite integral is widely used. Definite integrals are widely used in geometry, physics, engineering, economics and other fields. It can be used to calculate not only geometric quantities such as area and volume, but also physical quantities such as work done by variable force and position of center of mass, as well as economic quantities such as marginal cost and marginal benefit. This wide application lies in the value of definite integral as a mathematical tool.

This chapter discusses the second fundamental problem of integral calculus—the definite integral. Many problems in natural science and production practice, such as the area of plane figures, the arc length of curves, water pressure, and the work done by a variable force along a straight line, can be reduced to definite integral problems. We will introduce the concept of the definite integral by considering two practical problems from geometry and physics, followed by a discussion of its properties and calculation methods. Applications of the definite integral will be introduced in Sections 5.5 to 5.8.

5.1 Concept and Properties of Definite Integrals

5.1.1 Examples of Definite Integral Problems

1. Area of a Curved Trapezoid

Let $y=f(x)$ be a continuous function on the interval $[a,b]$, and $f(x)\geqslant 0$. The figure enclosed by the lines $x=a$, $x=b$, $y=0$ and the curve $y=f(x)$ is called a curved trapezoid (Figure 5-1), where the curved arc is called the curved side, and the segment on the x axis corresponding to the interval $[a,b]$ is called the base.

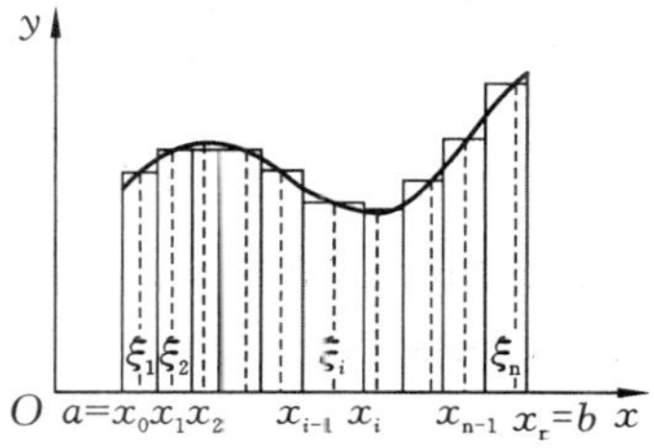

Figure 5-1

As we know, the height of a rectangle is constant, and its area can be calculated using the formula

$$\text{Rectangle Area}=\text{Base}\times\text{Height}.$$

To define and calculate, the height $f(x)$ of the curvilinear trapezoid at each point on the base varies over $[a,b]$, so its area cannot be directly defined and calculated using the formula for the area of a rectangle. To resolve the contradiction between variable height and constant height, the basic idea is: Divide the interval $[a,b]$ into many small subintervals, correspondingly, the curvilinear trapezoid is divided into many narrow curvilinear trapezoids. Since the height $f(x)$ varies continuously over the interval $[a,b]$, its variation is very small within a very small interval and can be approximately considered constant. Thus, the height at a certain point in each subinterval is used to approximate the variable height in the same subinterval, i.e., a narrow rectangle is used to approximate the corresponding narrow curvilinear trapezoid. In this way, the sum of the areas of all narrow rectangles is an approximation of the area of the curvilinear trapezoid. Obviously, the finer the division, the closer the sum of the areas of the narrow rectangles is to the area of the curvilinear trapezoid. If the division is infinitely refined, so that the length of each subinterval tends to zero, the limit of the sum of the areas of all narrow rectangles can be defined as the area of the curvilinear trapezoid. This also provides a method for calculating the area of the curvilinear trapezoid, with the specific steps detailed as follows:

(1) Partition. Insert any number of points in the interval $[a, b]$

$$a=x_0<x_1<x_2<\cdots<x_{n-1}<x_n=b.$$

Divide $[a, b]$ into n small intervals $[x_0, x_1]$, $[x_1, x_2]$, $\cdots$, $[x_{n-1}, x_n]$, their lengths being

$$\Delta x_1=x_1-x_0, \Delta x_2=x_2-x_1, \cdots, \Delta x_n=x_n-x_{n-1}.$$

Through each point x_i, draw a line segment parallel to the y axis, dividing the curved trapezoid into n small curved trapezoids (Figure 5-1).

(2) Approximation. Take any point ξ_i in the small interval $[x_{i-1}, x_i]$, use a rectangle with $[x_{i-1}, x_i]$ as the base and $f(\xi_i)$ as the height to approximate the area ΔA_i of the ith small curved trapezoid, then

$$\Delta A_i \approx f(\xi_i)\Delta x_i \quad (i=1,2,\cdots,n).$$

(3) Summation. The sum of the areas of all these n small rectangles is an approximation of the area A of the curved trapezoid, i.e.,

$$A \approx f(\xi_1)\Delta x_1 + f(\xi_2)\Delta x_2 + \cdots + f(\xi_n)\Delta x_n = \sum_{i=1}^{n} f(\xi_i)\Delta x_i.$$

(4) Limit. To ensure that the lengths of all small intervals tend to zero, it is required that the maximum length of the small intervals tends to zero. If we denote

$$\lambda=\max\{\Delta x_1, \Delta x_2, \cdots, \Delta x_n\},$$

then the above condition can be expressed as $\lambda \to 0$. When $\lambda \to 0$ (at this time, the number of small intervals n increases infinitely, i.e., $n \to \infty$), taking the limit of the above sum gives the area of the curved trapezoid

$$A=\lim_{\lambda \to 0}\sum_{i=1}^{n} f(\xi_i)\Delta x_i.$$

2. Distance of Variable Speed Linear Motion

As everyone knows, the formula for calculating the distance of uniform rectilinear motion is

$$\text{Distance} = \text{Speed} \times \text{Time}.$$

Now let's consider uniformly accelerated rectilinear motion: Suppose an object is moving in a straight line, and the velocity $v=v(t)$ is known to be a continuous function over the time interval $[T_1, T_2]$ with t. How can we calculate the distance s traveled by the object during this time?

In this problem, the velocity changes with time t, so the distance cannot be directly calculated using the formula for uniform rectilinear motion. However, since $v(t)$ is continuous, the change in velocity is small over very short time intervals, and it can be approximated as uniform motion. Therefore, if the time interval is divided into many small intervals, and uniform motion is used to approximate variable motion within each small interval, the approximate distance for each small interval can be calculated, then, by summing these values, an approximate total distance can be obtained. Finally, the exact distance can be calculated using the method of limits. The specific steps are as follows:

(1) Partition. Insert $n-1$ points within the time interval $[T_1, T_2]$

$$T_1 = t_0 < t_1 < t_2 < \cdots < t_{n-1} < t_n = T_2.$$

Divide $[T_1, T_2]$ into n small time intervals

$$[t_0, t_1], [t_1, t_2], \cdots, [t_{n-1}, t_n].$$

The lengths of each small time interval are respectively

$$\Delta t_1 = t_1 - t_0, \cdots, \Delta t_i = t_i - t_{i-1}, \cdots, \Delta t_n = t_n - t_{n-1}.$$

The distances traveled by the object in each small time interval are $\Delta s_1, \Delta s_2, \cdots, \Delta s_n$.

(2) Approximation. For each small time interval, select a point τ_i, and approximate the velocity at each moment in $[t_{i-1}, t_i]$ with the velocity $v(\tau_i)$ at time τ_i. This yields an approximate value for the distance traveled by the object in the small time interval $[t_{i-1}, t_i]$, Δs_i, i.e.,

$$\Delta s_i \approx v(\tau_i)\Delta t_i \quad (i = 1, 2, \cdots, n).$$

(3) Summation. Sum the approximate values of the distances obtained in n small time intervals to approximate the total distance traveled by the object in the variable-speed linear motion, i.e.,

$$s = \Delta s_1 + \Delta s_2 + \cdots + \Delta s_n \approx v(\tau_1)\Delta t_1 + v(\tau_2)\Delta t_2 + \cdots + v(\tau_n)\Delta t_n = \sum_{i=1}^{n} v(\tau_i)\Delta t_i.$$

(4) Limit. Denote $\lambda = \max\{\Delta t_1, \Delta t_2, \cdots, \Delta t_n\}$, and when $\lambda \to 0$, take the limit of the above sum to obtain the exact value of the distance traveled in the variable-speed linear motion.

$$s = \lim_{\lambda \to 0} \sum_{i=1}^{n} v(\tau_i)\Delta t_i.$$

5.1.2 Definition of the Definite Integral

The two problems discussed above, despite their different practical meanings, both involve calculating a quantity that depends on a function and its variable's interval of change. From the perspective of the method used, both are completed through the steps of partition, approximation, summation, and limit. From the mathematical structure of the results obtained, they both represent the limit of a specific type of sum with the same structure.

Setting aside the specific significance of these problems, by grasping their common essence and characteristics in terms of quantitative relationships and generalizing them, one can abstract the definition of the definite integral.

Definition 5.1.1 Let the function $f(x)$ be bounded on the interval $[a, b]$. Insert $n-1$ partition points arbitrarily within $[a, b]$.

$$a = x_0 < x_1 < x_2 < \cdots < x_{n-1} < x_n = b.$$

Divide the interval $[a, b]$ into n subintervals,

$$[x_0,x_1],[x_1,x_2],\cdots,[x_{n-1},x_n].$$

The lengths of these subintervals are, respectively,

$$\Delta x_1=x_1-x_0,\Delta x_2=x_2-x_1,\cdots,\Delta x_n=x_n-x_{n-1}.$$

$\forall \xi_i \in [x_{i-1},x_i]$, form the product $f(\xi_i)\Delta x_i\ (i=1,2,\cdots,n)$, and sum $S=\sum_{i=1}^{n} f(\xi_i)\Delta x_i$. Denote $\lambda=\max\{\Delta x_1,\Delta x_2,\cdots,\Delta x_n\}$, if, regardless of how $[a,b]$ is partitioned and regardless of how the point ξ_i is chosen in the subinterval $[x_{i-1},x_i]$, the sum S always tends to a definite limit I as $\lambda\to 0$, then the limit I is called the definite integral of the function $f(x)$ over the interval $[a,b]$. It is denoted as $\int_a^b f(x)\mathrm{d}x$, i.e.,

$$\int_a^b f(x)\mathrm{d}x = I = \lim_{\lambda\to 0}\sum_{i=1}^{n} f(\xi_i)\Delta x_i.$$

Where: $f(x)$ is the integrand, $f(x)\mathrm{d}x$ is the integrand expression, x is the variable of integration, a is the lower limit of integration, b is the upper limit of integration, and $[a,b]$ is the interval of integration.

Regarding the definition of the definite integral, the following points are further explained:

(1) The method of partitioning the interval $[a,b]$ and the choice of the point ξ_i are arbitrary in the definition.

(2) The definite integral represents the limit of a sum and is a definite number. When the limit of the sum exists, the definite integral of $f(x)$ over $[a,b]$ is said to exist, also known as $f(x)$ being integrable over $[a,b]$, otherwise, it is called non-integrable.

(3) If the integrand function is not changed, nor is the interval of integration, but only the integration variable x is rewritten as another letter, such as t or u, then the limit of the sum I remains unchanged, which means the value of the definite integral remains unchanged, i.e.,

$$\int_a^b f(x)\mathrm{d}x = \int_a^b f(t)\mathrm{d}t = \int_a^b f(u)\,\mathrm{d}u.$$

That is, the value of the definite integral depends only on the integrand function and the interval of integration, and is independent of the notation of the integration variable.

According to the definition of the definite integral, the two examples discussed at the beginning of this section can be expressed as follows:

(1) The area A of the curvilinear trapezoid formed by the continuous curve $y=f(x)$ $(f(x)\geqslant 0)$, the straight lines $x=a$, $x=b$, and the x axis is equal to the definite integral of the curvilinear function $f(x)$ over the interval $[a,b]$, i.e.,

$$A=\int_a^b f(x)\mathrm{d}x.$$

(2) For an object moving in a straight line with variable speed $v=v(t)\ (v(t)\geqslant 0)$, the distance s traveled from time $t=T_1$ to time $t=T_2$ is equal to the definite integral of the velocity function over the interval $[T_1,T_2]$, i.e.,

$$s = \int_{T_1}^{T_2} v(t)\,\mathrm{d}t.$$

Under what conditions does the function $f(x)$ satisfy to ensure that $f(x)$ is definitely integrable over $[a,b]$? This question will not be discussed in depth here, but only the following two sufficient conditions are provided.

Theorem 5.1.1 If $f(x)$ is continuous over the interval $[a,b]$, then $f(x)$ is integrable over $[a,b]$.

Theorem 5.1.2 If $f(x)$ is bounded over the interval $[a,b]$ and has only a finite number of discontinuities, then $f(x)$ is integrable over $[a,b]$. The geometric meaning of the definite integral is discussed below.

Over the interval $[a,b]$, when $f(x) \geqslant 0$, the definite integral geometrically represents the area of the curved trapezoid, i.e., $\int_a^b f(x)\,\mathrm{d}x = A$. If $f(x) < 0$ over $[a,b]$, the curved trapezoid is below the x axis, and the value of the definite integral is negative, geometrically representing the negative value of the area of this curved trapezoid, i. e., $\int_a^b f(x)\,\mathrm{d}x = -A$. Generally, if the function $y = f(x)$ takes both positive and negative values over $[a,b]$, the definite integral represents the area of the graph above the x axis minus the area of the graph below the x axis (Figure 5-2).

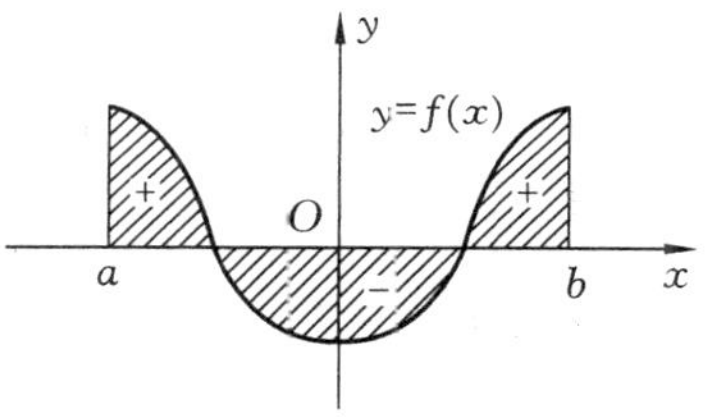

Figure 5-2

From the geometric meaning of the definite integral, the following results can be easily obtained (readers can verify by drawing the graph themselves):

$$\int_{-\pi}^{\pi} \sin x\,\mathrm{d}x = 0,$$

$$\int_{-a}^{a} \sqrt{a^2 - x^2}\,\mathrm{d}x = \frac{1}{2}\pi a^2.$$

Example 5.1.1 Calculate $\int_0^1 x^2\,\mathrm{d}x$ using the definition of the definite integral.

Solution Since the integrand $f(x) = x^2$ is continuous over the interval $[0,1]$ and continuous functions are integrable, the value of the definite integral is independent of the division of the interval $[0,1]$ and the choice of points ξ_i. To simplify the problem, let's divide the interval $[0,1]$ into n equal parts, with division points $x_i = \frac{i}{n}(i = 1, 2, \cdots, n-1)$, $\Delta x_i = \frac{1}{n}$, and take any ξ_i in the small interval $[x_{i-1}, x_i]$.

The right endpoint of the interval $\xi_i = x_i$. Thus, the integral sum is obtained.

$$\sum_{i=1}^{n} f(\xi_i)\Delta x_i = \sum_{i=1}^{n} \xi_i^2 \Delta x_i = \sum_{i=1}^{n} x_i^2 \Delta x_i = \sum_{i=1}^{n} \left(\frac{i}{n}\right)^2 \cdot \frac{1}{n} = \frac{1}{n^3}\sum_{i=1}^{n} i^2$$

$$= \frac{1}{n^3}(1^2 + 2^2 + \cdots + n^2) = \frac{n(n+1)(2n+1)}{6n^3} = \frac{1}{6}\left(1+\frac{1}{n}\right)\left(2+\frac{1}{n}\right).$$

When $\lambda \to 0$, i.e., $n \to \infty$, we obtain

$$\int_0^1 x^2 \mathrm{d}x = \lim_{\lambda \to 0} \sum_{i=1}^{n} f(\xi_i) \Delta x_i = \lim_{n \to \infty} \frac{1}{6}\left(1+\frac{1}{n}\right)\left(2+\frac{1}{n}\right) = \frac{1}{3}.$$

5.1.3 Properties of Definite Integrals

To further discuss the theory and computation of definite integrals, we introduce some properties of definite integrals below. For the convenience of calculation and application, we first make two supplementary stipulations regarding definite integrals:

(1) When $a = b$, $\int_a^b f(x)\mathrm{d}x = 0$;

(2) When $a > b$, $\int_a^b f(x)\mathrm{d}x = -\int_b^a f(x)\mathrm{d}x$.

According to the above stipulations, exchanging the upper and lower limits of a definite integral does not change its absolute value but reverses its sign. Therefore, in the following discussion, unless otherwise specified, there is no restriction on the size of the upper and lower limits of the definite integral, and the integrand is assumed to be integrable.

Property 5.1.1 $\int_a^b [f(x) \pm g(x)]\mathrm{d}x = \int_a^b f(x)\mathrm{d}x \pm \int_a^b g(x)\mathrm{d}x$.

Proof

$$\begin{aligned}\int_a^b [f(x) \pm g(x)]\mathrm{d}x &= \lim_{\lambda \to 0} \sum_{i=1}^{n} [f(\xi_i) \pm g(\xi_i)] \Delta x_i \\ &= \lim_{\lambda \to 0} \sum_{i=1}^{n} f(\xi_i) \Delta x_i \pm \lim_{\lambda \to 0} \sum_{i=1}^{n} g(\xi_i) \Delta x_i = \int_a^b f(x)\mathrm{d}x \pm \int_a^b g(x)\mathrm{d}x.\end{aligned}$$

Property 5.1.1 can be generalized to the case of finitely many functions.

Similarly, it can be shown that:

Property 5.1.2 $\int_a^b kf(x)\mathrm{d}x = k\int_a^b f(x)\mathrm{d}x$ (where k is a constant).

Property 5.1.3 $\int_a^b f(x)\mathrm{d}x = \int_a^c f(x)\mathrm{d}x + \int_c^b f(x)\mathrm{d}x$.

Proof First, consider the case of $a<c<b$.

Since the function is integrable over $[a, b]$, the limit of the integral sum remains unchanged regardless of how $[a, b]$ is partitioned. Therefore, c can always be taken as a partition point, and thus the integral sum over $[a, b]$ equals the integral sum over $[a, c]$ plus the integral sum over $[c, b]$, i.e.,

$$\sum_{[a,b]} f(\xi_i) \Delta x_i = \sum_{[a,c]} f(\xi_i) \Delta x_i + \sum_{[c,b]} f(\xi_i) \Delta x_i.$$

Let $\lambda \to 0$, taking the limit on both sides of the equation, we get

$$\int_a^b f(x)\mathrm{d}x = \int_a^c f(x)\mathrm{d}x + \int_c^b f(x)\mathrm{d}x.$$

Next, consider the case of $a<b<c$.

In this case, the point b lies between a, c, and from the above proof, we know that

$$\int_a^c f(x)\mathrm{d}x = \int_a^b f(x)\mathrm{d}x + \int_b^c f(x)\mathrm{d}x.$$

That is, $\int_a^b f(x)\mathrm{d}x = \int_a^c f(x)\mathrm{d}x - \int_b^c f(x)\mathrm{d}x = \int_a^c f(x)\mathrm{d}x + \int_c^b f(x)\mathrm{d}x.$

The same reasoning applies to the case $c<a<b$. Therefore, regardless of the relative position of a, b, c, Property 5.1.3 always holds.

Property 5.1.3 indicates that the definite integral is additive with respect to the interval of integration.

Property 5.1.4 $\int_a^b 1\mathrm{d}x = \int_a^b \mathrm{d}x = b-a.$

The proof of this property is left to the reader to complete based on the definition of the definite integral.

Property 5.1.5 If $f(x)\geqslant 0$ on $[a,b]$, then $\int_a^b f(x)\mathrm{d}x \geqslant 0$ $(a<b)$.

Proof From $f(x)\geqslant 0$, it follows that $f(\xi_i)\geqslant 0 (i=1,2,\cdots,n)$. Also, it is known that $\Delta x_i \geqslant 0 (i=1,2,\cdots,n)$. Therefore,

$$\sum_{i=1}^{n} f(\xi_i)\Delta x_i \geqslant 0.$$

By the corollary of the limit sign-preserving property, we have

$$\int_a^b f(x)\mathrm{d}x = \lim_{\lambda\to 0}\sum_{i=1}^{n} f(\xi_i)\Delta x_i \geqslant 0.$$

Corollary 5.1.1 If $f(x)\leqslant g(x)$ on the interval $[a,b]$, then

$$\int_a^b f(x)\mathrm{d}x \leqslant \int_a^b g(x)\mathrm{d}x \quad (a<b).$$

Corollary 5.1.2 $\left|\int_a^b f(x)\mathrm{d}x\right| \leqslant \int_a^b |f(x)|\mathrm{d}x \quad (a<b).$

The proofs of the above two corollaries are left to the reader and are omitted here.

Property 5.1.6 (Estimation Theorem) Let M and m be the maximum and minimum values of the function $f(x)$ on the interval $[a,b]$, respectively, then

$$m(b-a)\leqslant \int_a^b f(x)\mathrm{d}x \leqslant M(b-a).$$

It is easy to prove Property 5.1.6 using Properties 5.1.4 and 5.1.5.

Note: Property 5.1.6 has a clear geometric meaning, i.e., the area $\int_a^b f(x)\mathrm{d}x$ of the curved trapezoid with $[a,b]$ as the base and $y=f(x)$ $(\geqslant 0)$ as the curved edge lies between the areas $m(b-a)$ and $M(b-a)$ of rectangles with the same base and heights m and M respectively (Figure 5-3).

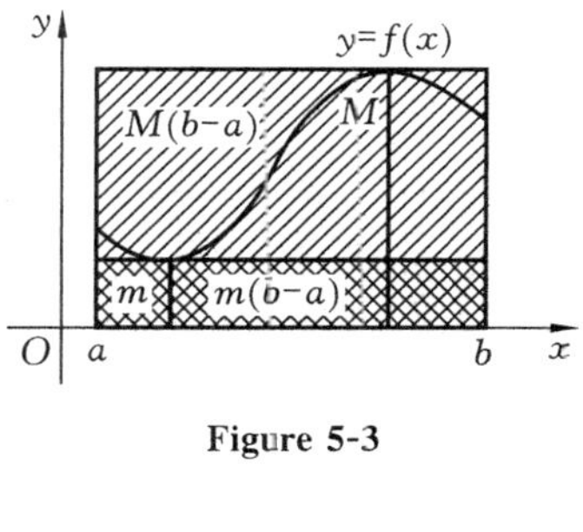

Figure 5-3

Example 5.1.2 Estimate the value of the integral $\int_{\frac{1}{\sqrt{3}}}^{\sqrt{3}} x\arctan x\,\mathrm{d}x$.

Solution Let $f(x)=x\arctan x, x\in\left[\frac{1}{\sqrt{3}},\sqrt{3}\right]$, from $f'(x)=\arctan x+\frac{x}{1+x^2}>0$, it is known that $f(x)$ is monotonically increasing on $\left[\frac{1}{\sqrt{3}},\sqrt{3}\right]$, hence maximum value $M=f(\sqrt{3})=\frac{\sqrt{3}}{3}\pi$, minimum value $m=f\left(\frac{1}{\sqrt{3}}\right)=\frac{\sqrt{3}}{18}\pi$.

Therefore

$$\frac{\sqrt{3}}{18}\pi\left(\sqrt{3}-\frac{1}{\sqrt{3}}\right)\leqslant\int_{\frac{1}{\sqrt{3}}}^{\sqrt{3}} x\arctan x\,\mathrm{d}x\leqslant\frac{\sqrt{3}}{3}\pi\left(\sqrt{3}-\frac{1}{\sqrt{3}}\right),$$

that is

$$\frac{\pi}{9}\leqslant\int_{\frac{1}{\sqrt{3}}}^{\sqrt{3}} x\arctan x\,\mathrm{d}x\leqslant\frac{2\pi}{3}.$$

Property 5.1.7 (Mean Value Theorem for Definite Integrals) If the function $f(x)$ is continuous on the closed interval $[a,b]$, then there exists at least one point ξ in $[a,b]$, such that

$$\int_a^b f(x)\,\mathrm{d}x=f(\xi)(b-a)\quad(a\leqslant\xi\leqslant b).$$

This formula is called the integral mean value formula.

Dividing the inequality in Property 5.1.6 by the interval length $b-a$, we obtain

$$m\leqslant\frac{1}{b-a}\int_a^b f(x)\,\mathrm{d}x\leqslant M.$$

This indicates that the value $\frac{1}{b-a}\int_a^b f(x)\,\mathrm{d}x$ lies between the minimum and maximum values of the function $f(x)$. By the Intermediate Value Theorem for continuous functions on a closed interval, there exists at least one point ξ in the interval $[a,b]$, such that

$$\frac{1}{b-a}\int_a^b f(x)\,\mathrm{d}x=f(\xi).$$

That is

$$\int_a^b f(x)\,\mathrm{d}x=f(\xi)(b-a)\quad(a\leqslant\xi\leqslant b).$$

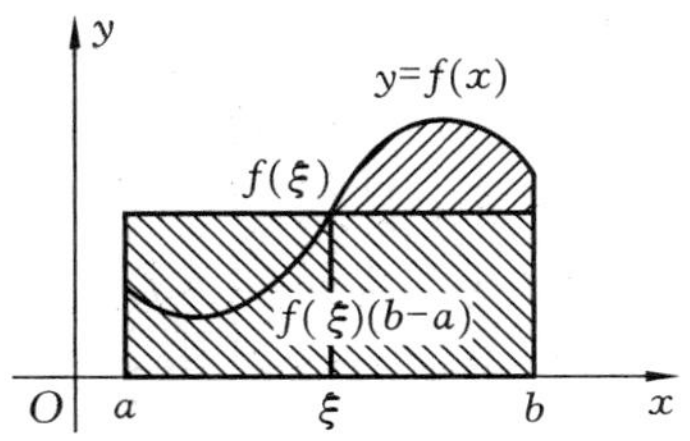

Figure 5-4

Note: The mean value theorem for definite integrals geometrically implies that there exists at least one point ξ in the interval $[a,b]$, such that the area $\int_a^b f(x)\,\mathrm{d}x$ of the curvilinear trapezoid with base $[a,b]$ and curved side $y=f(x)$ is equal to the area $f(\xi)(b-a)$ of a rectangle with the same base and height $f(\xi)$ (Figure 5-4).

According to the geometric interpretation, the value $\frac{1}{b-a}\int_a^b f(x)\mathrm{d}x$ represents the average height of the continuous curve $f(x)$ over the interval $[a,b]$, and it is called the average value of the function $f(x)$ over the interval $[a,b]$. This concept extends the notion of the average value of a finite number of values. For example,

$$v(\xi)=\frac{1}{T_2-T_1}\int_{T_1}^{T_2} v(t)\mathrm{d}t \quad (T_1 \leqslant \xi \leqslant T_2)$$

represents the average velocity of rectilinear motion over the time period $[T_1, T_2]$.

Example 5.1.3 Let $f(x)$ be differentiable, and $\lim_{x\to+\infty} f(x)=1$, find $\lim_{x\to+\infty}\int_x^{x+2} t\sin\frac{3}{t}f(t)\mathrm{d}t$.

Solution By the mean value theorem for definite integrals, we know that there exists $\exists\xi\in[x,x+2]$, such that

$$\int_x^{x+2} t\sin\frac{3}{t}f(t)\mathrm{d}t=\xi\sin\frac{3}{\xi}f(\xi)(x+2-x),$$

thus

$$\lim_{x\to+\infty}\int_x^{x+2} t\sin\frac{3}{t}f(t)\mathrm{d}t=2\lim_{\xi\to+\infty}\xi\sin\frac{3}{\xi}f(\xi)=2\lim_{\xi\to+\infty}\frac{\xi}{3}\sin\frac{3}{\xi}\lim_{\xi\to+\infty}3f(\xi)=2\lim_{\xi\to+\infty}3f(\xi)=6.$$

Exercise 5.1

1. Using the geometric meaning of definite integrals, prove the following equalities:

(1) $\int_0^1 2x\,\mathrm{d}x=1$; (2) $\int_0^1\sqrt{1-x^2}\,\mathrm{d}x=\frac{\pi}{4}$;

(3) $\int_{-\pi}^{\pi}\sin x\,\mathrm{d}x=0$; (4) $\int_{-\frac{\pi}{2}}^{\frac{\pi}{2}}\cos x\,\mathrm{d}x=2\int_0^{\frac{\pi}{2}}\cos x\,\mathrm{d}x$.

2. Prove the properties of definite integrals:

(1) $\int_a^b kf(x)\mathrm{d}x=k\int_a^b f(x)\mathrm{d}x$ (k is a constant);

(2) $\int_a^b 1\mathrm{d}x=\int_a^b \mathrm{d}x=b-a$.

3. Estimate the values of the following integrals:

(1) $\int_1^4 (x^2+1)\,\mathrm{d}x$; (2) $\int_{\frac{\pi}{4}}^{\frac{5\pi}{4}}(1+\sin^2 x)\,\mathrm{d}x$;

(3) $\int_1^2 \frac{x}{1+x^2}\mathrm{d}x$; (4) $\int_2^0 \mathrm{e}^{x^2-x}\,\mathrm{d}x$.

4. Use the mean value theorem of integrals to find the following limits:

(1) $\lim_{n\to\infty}\int_n^{n+p}\frac{\sin x}{x}\mathrm{d}x$; (2) $\lim_{n\to\infty}\int_0^{\frac{1}{2}}\frac{x^n}{1+x}\mathrm{d}x$.

5. Compare the sizes of the following pairs of integrals:

(1) $\int_0^1 x^2\mathrm{d}x$ and $\int_0^1 x^3\mathrm{d}x$; (2) $\int_1^2 x^2\mathrm{d}x$ and $\int_1^2 x^3\mathrm{d}x$;

(3) $\int_1^2 \ln x \, dx$ and $\int_1^2 (\ln x)^2 dx$; (4) $\int_0^1 x \, dx$ and $\int_0^1 \ln(1+x) dx$;

(5) $\int_0^1 e^x \, dx$ and $\int_0^1 (x+1) \, dx$; (6) $\int_0^{\frac{\pi}{2}} x \, dx$ and $\int_0^{\frac{\pi}{2}} \sin x \, dx$.

6. Given that an object moves in a straight line with velocity $v(t)=3t+5$ (m/s), use definite integrals to represent the distance L traveled by the object during the interval $T_1=1$ s, $T_2=3$ s, and use the geometric meaning of definite integrals to find the value of L.

7. Find the limit $\lim\limits_{n\to\infty} \dfrac{1^p+2^p+\cdots+n^p}{n^{p+1}}$ $(p>0)$ $\left(\text{given} \int_0^1 x^{\mu} \, dx = \dfrac{1}{\mu+1}\right)$.

8. Suppose $f(x)$ is continuous at $[0,1]$, differentiable on $(0,1)$, and $3\int_0^{\frac{1}{4}} f(x) dx = \int_{\frac{1}{4}}^1 f(x) dx$. Prove that there must exist $\xi \in (0,1)$ such that $f'(\xi)=0$.

5.2 Fundamental Theorem of Calculus

In integral calculus, two main problems need to be addressed: The first is the determination of the primitive function, which has been discussed in Chapter 4. The second is the calculation of definite integrals. Calculating definite integrals by their definition would be extremely difficult. Therefore, finding an effective method for calculating definite integrals became crucial for the development of integral calculus. It is well known that the concept of an indefinite integral as a primitive function and the concept of a definite integral as the limit of a sum are entirely unrelated. However, Newton and Leibniz not only discovered but also established the profound intrinsic connection between these two concepts, thereby ingeniously opening a new path for calculating definite integrals—the Newton-Leibniz formula. Newton and Leibniz are thus recorded in history as the founders of calculus.

5.2.1 The Relationship Between Position Function and Velocity Function

Suppose an object is moving along a straight line. On this line, a fixed origin, positive direction, and unit length are chosen to form a number axis. Let the position of the object at time t be $s(t)$ and its velocity be $v(t)$ $(v(t)\geqslant 0)$. According to Section 5.1, the distance traveled by the object during the time interval $[T_1, T_2]$ is

$$s=\int_{T_1}^{T_2} v(t) \, dt.$$

On the other hand, this distance can also be expressed as the increment of the position function $s(t)$ over the interval $[T_1, T_2]$:

$$s(T_2)-s(T_1).$$

It follows that there is the following relationship between the position function $s(t)$ and the velocity function $v(t)$:

$$\int_{T_1}^{T_2} v(t)\mathrm{d}t = s(T_2) - s(T_1).$$

Since $s'(t) = v(t)$, that is, the position function $s(t)$ is the antiderivative of the velocity function $v(t)$, finding the distance traveled by the object during the time interval $[T_1, T_2]$ is transformed into finding the increment of the antiderivative $s(t)$ of $v(t)$ over the interval $[T_1, T_2]$. Is this conclusion universally valid? The answer is affirmative. Generally, the definite integral $\int_a^b f(x)\mathrm{d}x$ of the function $f(x)$ over the interval $[a,b]$ equals the increment $F(b) - F(a)$ of the antiderivative $F(x)$ of $f(x)$ over the interval $[a,b]$. The following discussion will unfold step by step.

5.2.2 Function of the Upper Limit of Integration and Its Derivative

Suppose the function $f(x)$ is continuous over the interval $[a,b]$, and x is a point on $[a,b]$. Then the function

$$\Phi(x) = \int_a^x f(x)\mathrm{d}x$$

is called the integral upper limit (or definite integral with variable upper limit).

In the equation, both the integral variable and the integral upper limit are represented by the letter x, but it is important to note that their meanings are different. To distinguish them, the integral variable is usually changed to t (because the definite integral is independent of the notation of the integral variable), i.e.,

$$\Phi(x) = \int_a^x f(x)\mathrm{d}x = \int_a^x f(t)\mathrm{d}t.$$

The geometric meaning of $\Phi(x)$ is the area of the trapezoid with a curved side where the right line is movable (Figure 5-5). The area of the trapezoid $\Phi(x)$ changes with the position of x, and when x is given, the area $\Phi(x)$ is determined accordingly.

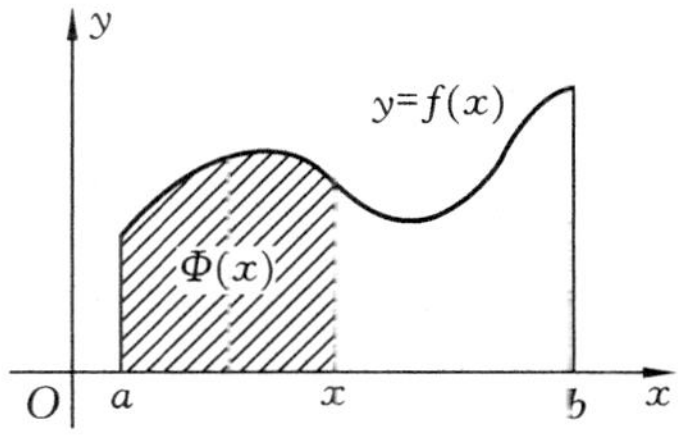

Figure 5-5

The following Theorem 5.2.1 provides an important property of the function of the integral upper limit $\Phi(x)$.

Theorem 5.2.1 If the function $f(x)$ is continuous on the interval $[a,b]$, then the function of the integral upper limit

$$\Phi(x) = \int_a^x f(t)\mathrm{d}t$$

is differentiable on $[a,b]$, and

$$\Phi'(x) = \frac{\mathrm{d}}{\mathrm{d}x}\int_a^x f(t)\mathrm{d}t = f(x) \quad (a \leqslant x \leqslant b).$$

Proof Let $x \in [a,b]$, $\Delta x \neq 0$, and $x+\Delta x \in [a,b]$. At this time, note that x takes the interval endpoint $x=a$, then $\Delta x > 0$, takes $x=b$, then $\Delta x < 0$. Since

$$\Delta\Phi = \Phi(x+\Delta x) - \Phi(x) = \int_a^{x+\Delta x} f(t)\mathrm{d}t - \int_a^x f(t)\mathrm{d}t$$
$$= \int_a^x f(t)\mathrm{d}t + \int_x^{x+\Delta x} f(t)\mathrm{d}t - \int_a^x f(t)\mathrm{d}t = \int_x^{x+\Delta x} f(t)\mathrm{d}t.$$

By the mean value theorem of integrals, it follows that $\Delta\Phi = f(\xi)\Delta x$, ξ is between x and $x+\Delta x$.

Since $f(x)$ is continuous at point x, and when $\Delta x \to 0$, $\xi \to x$, therefore

$$\Phi'(x) = \lim_{\Delta x \to 0} \frac{\Delta\Phi}{\Delta x} = \lim_{\xi \to x} f(\xi) = f(x),$$

that is

$$\frac{\mathrm{d}}{\mathrm{d}x}\int_a^x f(t)\mathrm{d}t = f(x) \quad (a \leqslant x \leqslant b).$$

Theorem 5.2.1 reveals the intrinsic connection between the concepts of differentiation (or derivative) and definite integral, which seem unrelated at first glance. It indicates that if a continuous function $f(x)$ is integrated with a variable upper limit x and then differentiated, the result reverts to $f(x)$ itself. Associating this with the definition of an antiderivative, it is known that $\Phi(x)$ is an antiderivative of $f(x)$. Thus, Theorem 5.2.1 also proves the fundamental conclusion that "a continuous function must have an antiderivative".

Theorem 5.2.2 If the function $f(x)$ is continuous on the interval $[a,b]$, then the function

$$\Phi(x) = \int_a^x f(t)\mathrm{d}t$$

is an antiderivative of $f(x)$ on $[a,b]$.

Using the chain rule for differentiation, we can further derive the following formulas:

(1) $\frac{\mathrm{d}}{\mathrm{d}x}\int_a^{\varphi(x)} f(t)\mathrm{d}t = f[\varphi(x)]\varphi'(x)$;

(2) $\frac{\mathrm{d}}{\mathrm{d}x}\int_{a(x)}^{b(x)} f(t)\mathrm{d}t = f[b(x)]b'(x) - f[a(x)]a'(x)$.

Example 5.2.1 Let $y = \int_0^x \cos^2 t\,\mathrm{d}t$, find $y'\left(\frac{\pi}{4}\right)$.

Solution $y' = \frac{\mathrm{d}}{\mathrm{d}x}\int_0^x \cos^2 t\,\mathrm{d}t = \cos^2 x$, hence

$$y'\left(\frac{\pi}{4}\right) = \cos^2\left(\frac{\pi}{4}\right) = \frac{1}{2}.$$

Example 5.2.2 Find $\frac{\mathrm{d}}{\mathrm{d}x}\int_1^{x^3} \mathrm{e}^{t^2}\,\mathrm{d}t$.

Solution Here $\int_1^{x^3} \mathrm{e}^{t^2}\,\mathrm{d}t$ is a function of x^3, and thus a composite function of x. Let

$x^3 = u$, then

$$\Phi(u) = \int_1^u e^{t^2} dt.$$

According to the chain rule for differentiation, we have

$$\frac{d}{dx}\int_1^{x^3} e^{t^2} dt = \frac{d}{du}\int_1^u e^{t^2} dt \frac{du}{dx} = e^{u^2} 3x^2 = 3x^2 e^{x^6}.$$

Example 5.2.3 Find $\lim\limits_{x \to 0} \dfrac{\int_{\cos x}^{1} e^{-t^2} dt}{x^2}$.

Solution The limit form is an indeterminate form of type "$\frac{0}{0}$", apply L'Hôpital's Rule to compute it. Since

$$\frac{d}{dx}\int_{\cos x}^{1} e^{-t^2} dt = -\frac{d}{dx}\int_1^{\cos x} e^{-t^2} dt = -e^{-\cos^2 x}(\cos x)' = \sin x\, e^{-\cos^2 x},$$

therefore,

$$\lim_{x \to 0} \frac{\int_{\cos x}^{1} e^{-t^2} dt}{x^2} = \lim_{x \to 0} \frac{\sin x\, e^{-\cos^2 x}}{2x} = \frac{1}{2e}.$$

Example 5.2.4 Let the function $y = y(x)$ be determined by the equation $\int_0^{y^2} e^{t^2} dt + \int_x^0 \sin t\, dt = 0$, find $\frac{dy}{dx}$.

Solution Differentiate both sides of the equation with respect to x, we get

$$\frac{d}{dx}\int_0^{y^2} e^{t^2} dt + \frac{d}{dx}\int_x^0 \sin t\, dt = 0,$$

thus,

$$\frac{d}{dy}\int_0^{y^2} e^{t^2} dt \cdot \frac{dy}{dx} + \frac{d}{dx}\int_x^0 \sin t\, dt = 0,$$

that is,

$$e^{y^4} \cdot 2y \cdot \frac{dy}{dx} - \sin x = 0.$$

Hence,

$$\frac{dy}{dx} = \frac{\sin x}{2y e^{y^4}}.$$

5.2.3 Newton-Leibniz Formula

Now, based on Theorem 5.2.2, we prove an important theorem that provides a formula for calculating definite integrals using the antiderivative.

Theorem 5.2.3 If the function $F(x)$ is an antiderivative of the continuous function $f(x)$ over the interval $[a, b]$, then

$$\int_a^b f(x)\mathrm{d}x = F(b) - F(a).$$

The equation is called the Newton-Leibniz formula, abbreviated as N-L formula.

Proof Given that the function $F(x)$ is an antiderivative of $f(x)$, and according to Theorem 5.2.2, we have

$$\Phi(x) = \int_a^x f(t)\mathrm{d}t .$$

$\Phi(x)$ is also an antiderivative of $f(x)$, so

$$F(x) - \Phi(x) = C\ (x \in [a, b]).$$

Let $x = a$ in the above equation, we get $F(a) - \Phi(a) = C$, and

$$\Phi(a) = \int_a^a f(t)\mathrm{d}t = 0.$$

Therefore $F(a) = C$, hence

$$\int_a^x f(t)\mathrm{d}t = F(x) - F(a).$$

Let $x = b$ in the above equation, then we get $\int_a^b f(x)\mathrm{d}x = F(b) - F(a)$. This equation is often denoted as

$$\int_a^b f(x)\mathrm{d}x = [F(x)]_a^b = F(x)\Big|_a^b = F(b) - F(a).$$

Since the antiderivative $F(x)$ of $f(x)$ can generally be obtained by finding the indefinite integral, the Newton-Leibniz formula links the calculation of definite integrals with indefinite integrals, transforming it into the problem of finding the increment of an antiderivative of the integrand over $[a, b]$.

The Newton-Leibniz formula is also known as the fundamental theorem of calculus.

Example 5.2.5 Calculate:

(1) $\int_0^{\ln a} a^x \mathrm{d}x\ (a > 0, a \neq 1)$; (2) $\int_{-4}^{-2} \frac{1}{x}\mathrm{d}x$.

Solution (1) Since $\frac{a^x}{\ln a}$ is an antiderivative of a^x, therefore

$$\int_0^{\ln a} a^x \mathrm{d}x = \left[\frac{a^x}{\ln a}\right]_0^{\ln a} = \frac{1}{\ln a}(a^{\ln a} - 1).$$

(2) When $x < 0$, an antiderivative of $\frac{1}{x}$ is $\ln|x|$, so

$$\int_{-4}^{-2} \frac{1}{x}\mathrm{d}x = [\ln|x|]_{-4}^{-2} = \ln 2 - \ln 4 = -\ln 2.$$

Example 5.2.6 Calculate $\int_{-\frac{\pi}{2}}^{\frac{\pi}{2}} f(x)\mathrm{d}x$, where $f(x) = \begin{cases} \cos x, & x \geq 0 \\ x, & x < 0 \end{cases}$.

Solution

$$\int_{-\frac{\pi}{2}}^{\frac{\pi}{2}} f(x)\mathrm{d}x = \int_{-\frac{\pi}{2}}^{0} x\,\mathrm{d}x + \int_0^{\frac{\pi}{2}} \cos x\,\mathrm{d}x = \frac{x^2}{2}\Big|_{-\frac{\pi}{2}}^{0} + \sin x\Big|_0^{\frac{\pi}{2}} = 1 - \frac{\pi^2}{8}.$$

Example 5.2.7 Calculate $\int_{-\frac{\pi}{2}}^{\frac{\pi}{3}} \sqrt{1-\cos^2 x}\, dx$.

Solution

$$\int_{-\frac{\pi}{2}}^{\frac{\pi}{3}} \sqrt{1-\cos^2 x}\, dx = \int_{-\frac{\pi}{2}}^{\frac{\pi}{3}} \sqrt{\sin^2 x} = \int_{-\frac{\pi}{2}}^{\frac{\pi}{3}} |\sin x|\, dx = -\int_{-\frac{\pi}{2}}^{0} \sin x\, dx + \int_{0}^{\frac{\pi}{3}} \sin x\, dx$$

$$= \cos x \Big|_{-\frac{\pi}{2}}^{0} - \cos x \Big|_{0}^{\frac{\pi}{3}} = \frac{3}{2}.$$

Example 5.2.8 A car is traveling at a speed of 36 km/h and needs to decelerate to a stop at a certain point. Suppose the car brakes with an acceleration of $a=-5t$ m/s^2. How far does the car travel from the start of braking to stopping?

Solution First, determine the velocity function of the car after braking and calculate the time taken from the start of braking to stopping. When $t=0$ s, the car's velocity

$$v_0 = 36 \text{ km/h} = \frac{36 \times 1000}{3600} \text{ m/s} = 10 \text{ m/s}.$$

The velocity of the car after braking

$$v(t) = \int a\, dt = \int (-5t)\, dt = -\frac{5}{2}t^2 + C.$$

Substituting $t=0$ s, $v=10$ m/s into the equation, we get $C=10$. Therefore, the velocity function of the car during deceleration is

$$v(t) = -\frac{5}{2}t^2 + 10.$$

When the car comes to a stop, $v(t)=0$, solving $-\frac{5}{2}t^2+10=0$ gives $t=2$ s. Thus, the distance traveled by the car during this time is

$$s = \int_0^2 v(t)\, dt = \int_0^2 \left(-\frac{5}{2}t^2 + 10\right) dt = \left[-\frac{5}{6}t^3 + 10t\right]_0^2 = \frac{40}{3} \text{ m}.$$

That is, after braking, the car needs to travel $\frac{40}{3}$ m to come to a stop.

Example 5.2.9 Suppose the function $f(x)$ is continuous on the closed interval $[a,b]$. Prove that there exists at least one point ξ in the open interval (a,b), such that

$$\int_a^b f(x)\, dx = f(\xi)(b-a) \quad (a<\xi<b).$$

Proof Since $f(x)$ is continuous, its primitive function exists, denoted as $F(x)$, i.e., assuming on $[a,b]$, $F'(x)=f(x)$. According to the Newton-Leibniz formula, we have

$$\int_a^b f(x)\, dx = F(b) - F(a).$$

Obviously, $F(x)$ satisfies the conditions of Lagrange's Mean Value Theorem on the interval $[a,b]$. Therefore, there is at least one point ξ in the open interval (a,b), such that

$$F(b) - F(a) = F'(\xi)(b-a) \quad (\xi \in (a,b)).$$

Hence

$$\int_a^b f(x)\mathrm{d}x = f(\xi)(b-a) \quad (\xi \in (a,b)).$$

The conclusion of this example is an improvement on the integral mean value theorem, and it is not difficult to see the connection between the integral mean value theorem and the differential mean value theorem from the proof.

Exercise 5.2

1. Calculate the following derivatives:

(1) $\dfrac{\mathrm{d}}{\mathrm{d}x}\displaystyle\int_0^x \sqrt{\sin t}\,\mathrm{d}t$; (2) $\dfrac{\mathrm{d}}{\mathrm{d}x}\displaystyle\int_0^x \mathrm{e}^{t^3}\,\mathrm{d}t$;

(3) $\dfrac{\mathrm{d}}{\mathrm{d}x}\displaystyle\int_x^0 \arctan t^2\,\mathrm{d}t$; (4) $\dfrac{\mathrm{d}}{\mathrm{d}x}\displaystyle\int_x^b \sqrt{1+\ln t}\,\mathrm{d}t$.

2. Find the following limits:

(1) $\lim\limits_{x\to 0}\dfrac{\int_0^x \cos t^2\,\mathrm{d}t}{x}$; (2) $\lim\limits_{x\to 0}\dfrac{\left(\int_0^x \mathrm{e}^{t^2}\,\mathrm{d}t\right)^2}{\int_0^x t\mathrm{e}^{2t^2}\,\mathrm{d}t}$.

3. Calculate the following definite integrals:

(1) $\displaystyle\int_{-2}^{-1}\frac{2}{x^2}\mathrm{d}x$; (2) $\displaystyle\int_0^{\pi}(1+\cos x)\,\mathrm{d}x$;

(3) $\displaystyle\int_0^{\frac{\pi}{3}} 2\sec^2 x\,\mathrm{d}x$; (4) $\displaystyle\int_0^{\frac{\pi}{2}}\frac{1+\cos 2t}{2}\mathrm{d}t$;

(5) $\displaystyle\int_{-1}^{4}|x|\,\mathrm{d}x$; (6) $\displaystyle\int_0^{\pi}\frac{1}{2}(\cos x+|\cos x|)\,\mathrm{d}x$;

(7) $\displaystyle\int_{-\frac{1}{\sqrt{3}}}^{\sqrt{3}}\frac{\mathrm{d}x}{1+x^2}$; (8) $\displaystyle\int_0^{\sqrt{3}a}\frac{\mathrm{d}x}{a^2+x^2}$;

(9) $\displaystyle\int_{-\frac{1}{2}}^{\frac{1}{2}}\frac{\mathrm{d}x}{\sqrt{1-x^2}}$; (10) $\displaystyle\int_0^{\frac{\pi}{4}}\tan^2 x\,\mathrm{d}x$;

(11) $\displaystyle\int_{-(\mathrm{e}+1)}^{-2}\frac{\mathrm{d}x}{1+x}$; (12) $\displaystyle\int_{-2}^{2}\max\{x,x^2\}\,\mathrm{d}x$;

(13) $\displaystyle\int_0^2 f(x)\mathrm{d}x$, where $f(x)=\begin{cases}x+1, & x\leqslant 1\\ \dfrac{1}{2}x^2, & x>1\end{cases}$.

4. Let k be a positive integer, prove:

(1) $\displaystyle\int_{-\pi}^{\pi}\cos kx\,\mathrm{d}x=0$; (2) $\displaystyle\int_{-\pi}^{\pi}\sin kx\,\mathrm{d}x=0$;

(3) $\displaystyle\int_{-\pi}^{\pi}\cos^2 kx\,\mathrm{d}x=\pi$; (4) $\displaystyle\int_{-\pi}^{\pi}\sin^2 kx\,\mathrm{d}x=\pi$.

5. Let k,l be a positive integer, $k\neq l$, prove (orthogonality of trigonometric functions):

(1) $\int_{-\pi}^{\pi} \cos kx \sin lx \, dx = 0$; (2) $\int_{-\pi}^{\pi} \cos kx \cos lx \, dx = 0$;

(3) $\int_{-\pi}^{\pi} \sin kx \sin lx \, dx = 0$.

6. Find the following derivatives:

(1) $\frac{d}{dx}\int_{0}^{x^2} \sqrt{1+t^2}\, dt$; (2) $\frac{d}{dx}\int_{x^2}^{x^3} \frac{dt}{\sqrt{1+t^4}}$;

(3) $\frac{d}{dx}\int_{\sin x}^{\cos x} e^{t^2}\, dt$; (4) $\frac{d}{dx}\int_{\sqrt{x}}^{0} \sin t^2 \, dt$.

7. Given the parametric equations $\begin{cases} x = \int_0^t \sin u^3 \, du \\ y = \int_0^{t^2} \cos(\sqrt{u}+1)\, du \end{cases}$, find $\frac{dy}{dx}$.

8. Given the implicit function $\int_0^y e^u \, du + \int_0^x \cos u^2 \, du = 0$, find $\frac{dy}{dx}$.

9. Given that $f(x)$ is continuous, find $\frac{d}{dx}\int_0^x t f(x^2 - t^2)\, dt$.

10. Given that $f(x)$ is continuous and $f(x) = \frac{1}{1+x^2} + 2\int_0^1 f(t)\, dt$, find $f(x)$.

11. Prove: If $f(x)$ is continuous on $[a, b]$ $(a < b)$, and $\int_a^b f(x)\, dx = 0$, then there exists at least one point ξ on $[a, b]$, such that $f(\xi) = 0$.

12. Given $f(x) = \begin{cases} x^2, & 0 \leqslant x < 1 \\ x, & 1 \leqslant x \leqslant 2 \end{cases}$, find the expression for $\Phi(x) = \int_0^x f(t)\, dt$ on $[0, 2]$, and discuss the continuity of $\Phi(x)$ within $(0, 2)$.

13. Given $f(x) = \begin{cases} \frac{1}{2}\sin x, & 0 \leqslant x \leqslant \pi \\ 0, & x < 0 \text{ or } x > \pi \end{cases}$, find the expression for $\Phi(x) = \int_0^x f(t)\, dt$ within $(-\infty, +\infty)$.

14. Given $F(x) = \int_0^x \frac{\sin t}{t}\, dt$, find $F'(0)$.

15. Given that $f(x)$ is continuous on $[a, b]$, differentiable within (a, b), and $f'(x) \leqslant 0$,

$$F(x) = \frac{1}{x-a}\int_a^x f(t)\, dt.$$

Prove: There exists $F'(x) \leqslant 0$ within (a, b).

5.3 Integration by Substitution and by Parts for Definite Integrals

Therefore, under certain conditions, definite integrals can be calculated using the method of substitution and integration by parts.

5.3.1 Integration by Substitution for Definite Integrals

To illustrate how to compute definite integrals using the substitution method, we first prove the following theorem:

Theorem 5.3.1 If the function $f(x)$ is continuous on the interval $[a,b]$, and the function $x=\varphi(t)$ satisfies the two following conditions:

(1) $\varphi(\alpha)=a$, $\varphi(\beta)=b$;

(2) $\varphi(t)$ has a continuous derivative on the interval $[\alpha,\beta]$ (or $[\beta,\alpha]$), and $a\leqslant\varphi(t)\leqslant b$.

Then the substitution formula for definite integrals holds.

$$\int_a^b f(x)\,\mathrm{d}x=\int_\alpha^\beta f[\varphi(t)]\varphi'(t)\,\mathrm{d}t.$$

Proof Since the integrands on both sides of it are continuous functions, their antiderivatives exist. Let $F(x)$ be an antiderivative of $f(x)$ on $[a,b]$, and $F[\varphi(t)]$ be the function composed by $F(x)$ and $x=\varphi(t)$. By the differentiation of composite functions, we have

$$\frac{\mathrm{d}}{\mathrm{d}t}F[\varphi(t)]=F'[\varphi(t)]\varphi'(t)=f[\varphi(t)]\varphi'(t).$$

It is evident that $F[\varphi(t)]$ is an antiderivative of $f[\varphi(t)]\varphi'(t)$. According to the Newton-Leibniz formula, the proof is complete.

$$\int_\alpha^\beta f[\varphi(t)]\varphi'(t)\,\mathrm{d}t=F[\varphi(\beta)]-F[\varphi(\alpha)]=F(b)-F(a)=\int_a^b f(x)\,\mathrm{d}x.$$

From the above proof, it is evident that when calculating a definite integral using the substitution method, once the primitive function in terms of the new variable is obtained, there is no need to revert the variable. Instead, one simply substitutes the new limits of integration and finds their difference. This is the distinction between the substitution method for definite integrals and that for indefinite integrals. The reason for this difference lies in the fact that indefinite integrals seek the primitive function of the integrand, which should retain the same independent variable as the original function. In contrast, the result of a definite integral is a specific number. If the definite integral on one side of the equation is calculated, the definite integral on the other side is naturally obtained.

Example 5.3.1 Calculate $\int_0^a \sqrt{a^2-x^2}\,\mathrm{d}x\ (a>0)$.

Solution Let $x=a\sin t$, then $\mathrm{d}x=a\cos t\,\mathrm{d}t$, and when $x=0$, $t=0$, when $x=a$, $t=\frac{\pi}{2}$. Therefore,

$$\int_0^a \sqrt{a^2-x^2}\,\mathrm{d}x=\int_0^{\frac{\pi}{2}} a\sqrt{1-\sin^2 t}\,a\cos t\,\mathrm{d}t=a^2\int_0^{\frac{\pi}{2}}\cos^2 t\,\mathrm{d}t$$
$$=\frac{a^2}{2}\int_0^{\frac{\pi}{2}}(1+\cos 2t)\,\mathrm{d}t=\frac{a^2}{2}\left(t+\frac{1}{2}\sin 2t\right)\Bigg|_0^{\frac{\pi}{2}}=\frac{\pi a^2}{4}.$$

Note: If the geometric meaning of the definite integral is used, this problem geometrically represents the area of the part of the circle with its center at the origin and radius a in the first quadrant. Thus, it is easy to obtain the calculation result directly.

The substitution formula for definite integrals can also be used in reverse, i.e., from right to left.

Example 5.3.2 Calculate $\int_0^{\frac{\pi}{2}} \cos^4 x \sin x \, dx$.

Solution Let $t = \cos x$, then $dt = -\sin x \, dx$, and when $x = 0, t = 1$, when $x = \frac{\pi}{2}, t = 0$. Therefore

$$\int_0^{\frac{\pi}{2}} \cos^4 x \sin x \, dx = -\int_1^0 t^4 \, dt = \int_0^1 t^4 \, dt = \frac{1}{5} t^5 \Big|_0^1 = \frac{1}{5}.$$

When using the substitution method for definite integrals, one can also avoid writing the new variable t and directly use the method of combining differentials. In this case, the upper and lower limits of the definite integral do not need to be changed. This example is recalculated as follows:

$$\int_0^{\frac{\pi}{2}} \cos^4 x \sin x \, dx = -\int_0^{\frac{\pi}{2}} \cos^4 x \, d(\cos x) = -\frac{1}{5}\cos^5 x \Big|_0^{\frac{\pi}{2}} = \frac{1}{5}.$$

Example 5.3.3 Calculate $\int_0^{\pi} \sqrt{\sin^3 x - \sin^5 x} \, dx$.

Solution Since

$$f(x) = \sqrt{\sin^3 x - \sin^5 x} = (\sin x)^{\frac{3}{2}} |\cos x|,$$

therefore

$$\int_0^{\pi} \sqrt{\sin^3 x - \sin^5 x} \, dx = \int_0^{\pi} (\sin x)^{\frac{3}{2}} |\cos x| \, dx = \int_0^{\frac{\pi}{2}} (\sin x)^{\frac{3}{2}} \cos x \, dx - \int_{\frac{\pi}{2}}^{\pi} (\sin x)^{\frac{3}{2}} \cos x \, dx$$

$$= \int_0^{\frac{\pi}{2}} (\sin x)^{\frac{3}{2}} d(\sin x) - \int_{\frac{\pi}{2}}^{\pi} (\sin x)^{\frac{3}{2}} d(\sin x)$$

$$= \frac{2}{5} (\sin x)^{\frac{5}{2}} \Big|_0^{\frac{\pi}{2}} - \frac{2}{5} (\sin x)^{\frac{5}{2}} \Big|_{\frac{\pi}{2}}^{\pi} = \frac{2}{5} - \left(-\frac{2}{5}\right) = \frac{4}{5}.$$

Example 5.3.4 Let $f(x)$ be continuous on $[-a, a]$, prove:

(1) If $f(x)$ is an even function, then $\int_{-a}^{a} f(x) dx = 2\int_0^a f(x) dx$;

(2) If $f(x)$ is an odd function, then $\int_{-a}^{a} f(x) dx = 0$.

Solution From a geometric perspective (Figure 5-6), the conclusion is quite obvious. Since

$$\int_{-a}^{a} f(x) dx = \int_{-a}^{0} f(x) dx + \int_0^a f(x) dx,$$

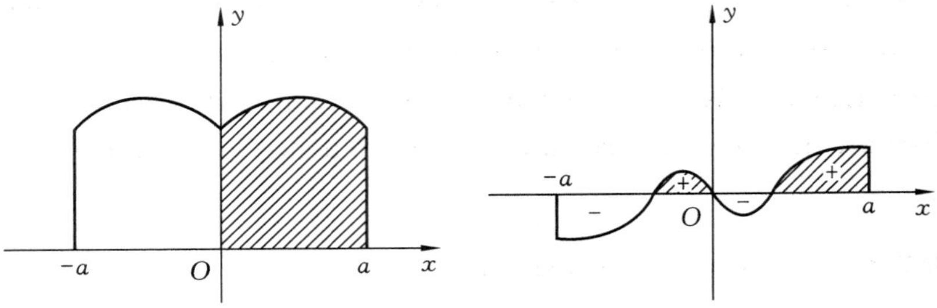

Figure 5-6

let $x=-t$ in the first term on the right-hand side of the equation, then

$$\int_{-a}^{0} f(x)\,\mathrm{d}x=-\int_{a}^{0} f(-t)\,\mathrm{d}t=\int_{0}^{a} f(-t)\,\mathrm{d}t=\int_{0}^{a} f(-x)\,\mathrm{d}x,$$

thus

$$\int_{-a}^{a} f(x)\,\mathrm{d}x=\int_{0}^{a} f(x)\,\mathrm{d}x+\int_{0}^{a} f(-x)\,\mathrm{d}x.$$

(1) If $f(x)$ is an even function, i.e., $f(-x)=f(x)$, then

$$\int_{-a}^{a} f(x)\,\mathrm{d}x=2\int_{0}^{a} f(x)\,\mathrm{d}x.$$

(2) If $f(x)$ is an odd function, i.e., $f(-x)=-f(x)$, then

$$\int_{-a}^{a} f(x)\,\mathrm{d}x=0.$$

Example 5.3.5 Calculate $\int_{-1}^{1}(|x|+\sin x)x^2\,\mathrm{d}x$.

Solution Since the integration interval is symmetric about the origin and $|x|\cdot x^2$ is an even function, $\sin x\cdot x^2$ is an odd function, therefore

$$\int_{-1}^{1}(|x|+\sin x)x^2\,\mathrm{d}x=\int_{-1}^{1}|x|x^2\,\mathrm{d}x=2\int_{0}^{1}x^3\,\mathrm{d}x=\frac{1}{2}.$$

Example 5.3.6 Calculate $\int_{-\frac{\pi}{2}}^{\frac{\pi}{2}}\frac{\mathrm{e}^x}{1+\mathrm{e}^x}\cos x\,\mathrm{d}x$.

Solution Because

$$\int_{-\frac{\pi}{2}}^{\frac{\pi}{2}}\frac{\mathrm{e}^x}{1+\mathrm{e}^x}\cos x\,\mathrm{d}x=\int_{-\frac{\pi}{2}}^{\frac{\pi}{2}}\left(1-\frac{1}{1+\mathrm{e}^x}\right)\cos x\,\mathrm{d}x=\int_{-\frac{\pi}{2}}^{\frac{\pi}{2}}\cos x\,\mathrm{d}x-\int_{-\frac{\pi}{2}}^{\frac{\pi}{2}}\frac{1}{1+\mathrm{e}^x}\cos x\,\mathrm{d}x.$$

Make a substitution in the second integral on the right-hand side: $x=-t$, then

$$\int_{-\frac{\pi}{2}}^{\frac{\pi}{2}}\frac{1}{1+\mathrm{e}^x}\cos x\,\mathrm{d}x=-\int_{\frac{\pi}{2}}^{-\frac{\pi}{2}}\frac{1}{1+\mathrm{e}^{-t}}\cos t\,\mathrm{d}t=\int_{-\frac{\pi}{2}}^{\frac{\pi}{2}}\frac{\mathrm{e}^t}{1+\mathrm{e}^t}\cos t\,\mathrm{d}t,$$

therefore

$$\int_{-\frac{\pi}{2}}^{\frac{\pi}{2}}\frac{\mathrm{e}^x}{1+\mathrm{e}^x}\cos x\,\mathrm{d}x=\int_{-\frac{\pi}{2}}^{\frac{\pi}{2}}\cos x\,\mathrm{d}x-\int_{-\frac{\pi}{2}}^{\frac{\pi}{2}}\frac{\mathrm{e}^t}{1+\mathrm{e}^t}\cos t\,\mathrm{d}t.$$

That is

$$\int_{-\frac{\pi}{2}}^{\frac{\pi}{2}}\frac{\mathrm{e}^x}{1+\mathrm{e}^x}\cos x\,\mathrm{d}x=\frac{1}{2}\int_{-\frac{\pi}{2}}^{\frac{\pi}{2}}\cos x\,\mathrm{d}x=\int_{0}^{\frac{\pi}{2}}\cos x\,\mathrm{d}x=1.$$

Example 5.3.7 Let $f(x)$ be continuous on $[0,1]$, prove that:

(1) $\int_0^{\frac{\pi}{2}} f(\sin x)\,dx = \int_0^{\frac{\pi}{2}} f(\cos x)\,dx$;

(2) $\int_0^{\pi} x f(\sin x)\,dx = \frac{\pi}{2}\int_0^{\pi} f(\sin x)\,dx$, hence calculate $\int_0^{\pi} \frac{x\sin x}{1+\cos^2 x}dx$.

Proof (1) Observing both sides of the equation, the transformation should turn $f(\sin x)$ into $f(\cos x)$. For this, let $x=\frac{\pi}{2}-t$, then $dx=-dt$, and when $x=0, t=\frac{\pi}{2}$, when $x=\frac{\pi}{2}, t=0$, so

$$\int_0^{\frac{\pi}{2}} f(\sin x)\,dx = -\int_{\frac{\pi}{2}}^{0} f\left[\sin\left(\frac{\pi}{2}-t\right)\right]dt = \int_0^{\frac{\pi}{2}} f(\cos t)\,dt = \int_0^{\frac{\pi}{2}} f(\cos x)\,dx.$$

(2) Multiply both sides of the equation by 2, obtaining $2\int_0^{\pi} x f(\sin x)\,dx = \pi\int_0^{\pi} f(\sin x)\,dx$, then rearrange to get

$$\int_0^{\pi} x f(\sin x)\,dx = \pi\int_0^{\pi} f(\sin x)\,dx - \int_0^{\pi} x f(\sin x)\,dx = \int_0^{\pi} (\pi - x) f(\sin x)\,dx.$$

Thus, let $x=\pi-t$, then $dx=-dt$, and when $x=0, t=\pi$, when $x=\pi, t=0$. Therefore,

$$\int_0^{\pi} x f(\sin x)\,dx = -\int_{\pi}^{0} (\pi - t) f[\sin(\pi - t)]\,dt = \int_0^{\pi} (\pi - t) f(\sin t)\,dt$$
$$= \pi\int_0^{\pi} f(\sin t)\,dt - \int_0^{\pi} t f(\sin t)\,dt = \pi\int_0^{\pi} f(\sin x)\,dx - \int_0^{\pi} x f(\sin x)\,dx.$$

Therefore,

$$\int_0^{\pi} x f(\sin x)\,dx = \frac{\pi}{2}\int_0^{\pi} f(\sin x)\,dx.$$

Using the above result, we calculate

$$\int_0^{\pi} \frac{x\sin x}{1+\cos^2 x}dx = \frac{\pi}{2}\int_0^{\pi} \frac{\sin x}{1+\cos^2 x}dx = -\frac{\pi}{2}\int_0^{\pi} \frac{d(\cos x)}{1+\cos^2 x}$$
$$= -\frac{\pi}{2}\arctan(\cos x)\Big|_0^{\pi} = -\frac{\pi}{2}\left(-\frac{\pi}{4}-\frac{\pi}{4}\right) = \frac{\pi^2}{4}.$$

Example 5.3.8 Calculate $\int_0^{a} \frac{dx}{x+\sqrt{a^2-x^2}}\ (a>0)$.

Solution Let $x=a\sin t$, then $dx=a\cos t\,dt$, and when $x=0, t=0$, when $x=a, t=\frac{\pi}{2}$. Therefore,

$$\int_0^{a} \frac{dx}{x+\sqrt{a^2-x^2}} = \int_0^{\frac{\pi}{2}} \frac{a\cos t\,dt}{a\sin t + a\cos t} = \int_0^{\frac{\pi}{2}} \frac{\cos x}{\sin x+\cos x}dx.$$

Using the conclusion from Example 5.3.7(1), we have

$$\int_0^{\frac{\pi}{2}} \frac{\cos x}{\sin x+\cos x}dx = \int_0^{\frac{\pi}{2}} \frac{\sin x}{\sin x+\cos x}dx.$$

Thus

$$\int_0^{\frac{\pi}{2}} \frac{\cos x}{\sin x + \cos x} dx = \frac{1}{2}\int_0^{\frac{\pi}{2}} \frac{\sin x + \cos x}{\sin x + \cos x} dx = \frac{1}{2}\int_0^{\frac{\pi}{2}} dx = \frac{\pi}{4}.$$

Therefore, there is

$$\int_0^a \frac{dx}{x + \sqrt{a^2 - x^2}} = \frac{\pi}{4}.$$

5.3.2 Integration by Parts in Definite Integrals

Based on the integration by parts method for indefinite integrals, the formula for integration by parts in definite integrals can be derived.

Since $d(uv) = u dv + v du$, by rearranging the terms, we obtain

$$u(x)d[v(x)] = d[u(x)v(x)] - v(x)d[u(x)],$$

thus,

$$\int_a^b u(x)d[v(x)] = \int_a^b d[u(x)v(x)] - \int_a^b v(x)d[u(x)] = [u(x)v(x)]_a^b - \int_a^b v(x)d[u(x)].$$

It is abbreviated as

$$\int_a^b u dv = [uv]_a^b - \int_a^b v du.$$

This is the formula for integration by parts in definite integrals. Unlike the formula for indefinite integrals, here the part of the antiderivative that has already been integrated uv can be substituted with the upper and lower limits first.

Example 5.3.9 Calculate $\int_1^e x^2 \ln x \, dx$.

Solution

$$\int_1^e x^2 \ln x \, dx = \frac{1}{3}\int_1^e \ln x \, dx^3 = \frac{1}{3}\left(x^3 \ln x \,\Big|_1^e - \int_1^e x^2 dx\right) = \frac{1}{3}\left(e^3 - \frac{1}{3}x^3 \Big|_1^e\right) = \frac{1}{9}(2e^3 + 1).$$

Example 5.3.10 Calculate $\int_0^{\frac{\pi}{4}} \frac{x \, dx}{1 + \cos 2x}$.

Solution

$$\int_0^{\frac{\pi}{4}} \frac{x \, dx}{1 + \cos 2x} = \int_0^{\frac{\pi}{4}} \frac{x \, dx}{2\cos^2 x} = \int_0^{\frac{\pi}{4}} \frac{x}{2} d(\tan x) = \frac{1}{2} x \tan x \Big|_0^{\frac{\pi}{4}} - \frac{1}{2}\int_0^{\frac{\pi}{4}} \tan x \, dx$$

$$= \frac{\pi}{8} + \frac{1}{2}[\ln|\cos x|]_0^{\frac{\pi}{4}} = \frac{\pi}{8} - \frac{\ln 2}{4}.$$

Example 5.3.11 Derive the recurrence formula for $I_n = \int_0^{\frac{\pi}{2}} \sin^n x \, dx$, and calculate $\int_0^{\frac{\pi}{2}} \sin^n x \, dx$ and $\int_0^{\frac{\pi}{2}} \cos^n x \, dx$ (n for non-negative integers).

Solution It is easy to see,

$$I_0=\int_0^{\frac{\pi}{2}} dx=\frac{\pi}{2}, I_1=\int_0^{\frac{\pi}{2}}\sin x\,dx=1.$$

When $n\geqslant 2$, using integration by parts, let $u=\sin^{n-1}x$, $dv=\sin x\,dx$, then

$$du=(n-1)\sin^{n-2}x\cos x\,dx, v=-\cos x.$$

Thus,

$$I_n=\left[-\sin^{n-1}x\cos x\right]_0^{\frac{\pi}{2}}+(n-1)\int_0^{\frac{\pi}{2}}\sin^{n-2}x\cos^2 x\,dx$$

$$=(n-1)\int_0^{\frac{\pi}{2}}\sin^{n-2}x\,dx-(n-1)\int_0^{\frac{\pi}{2}}\sin^n x\,dx=(n-1)I_{n-2}-(n-1)I_n.$$

Therefore, the recurrence formula is obtained: $I_n=\frac{n-1}{n}I_{n-2}$.

When n is even, let $n=2m$, then

$$I_{2m}=\frac{2m-1}{2m}\cdot\frac{2m-3}{2m-2}\cdot\frac{2m-5}{2m-4}\cdot\cdots\cdot\frac{5}{6}\cdot\frac{3}{4}\cdot\frac{1}{2}\cdot I_0$$

$$=\frac{2m-1}{2m}\cdot\frac{2m-3}{2m-2}\cdot\frac{2m-5}{2m-4}\cdot\cdots\cdot\frac{5}{6}\cdot\frac{3}{4}\cdot\frac{1}{2}\cdot\frac{\pi}{2}.$$

When n is odd, let $n=2m+1$, then

$$I_{2m+1}=\frac{2m}{2m+1}\cdot\frac{2m-2}{2m-1}\cdot\frac{2m-4}{2m-3}\cdot\cdots\cdot\frac{6}{7}\cdot\frac{4}{5}\cdot\frac{2}{3}\cdot I_1$$

$$=\frac{2m}{2m+1}\cdot\frac{2m-2}{2m-1}\cdot\frac{2m-4}{2m-3}\cdot\cdots\cdot\frac{6}{7}\cdot\frac{4}{5}\cdot\frac{2}{3}.$$

Also, from $\int_0^{\frac{\pi}{2}} f(\sin x)\,dx=\int_0^{\frac{\pi}{2}} f(\cos x)\,dx$, we get

$$\int_0^{\frac{\pi}{2}}\sin^n x\,dx=\int_0^{\frac{\pi}{2}}\cos^n x\,dx.$$

Summarizing the above, we obtain the integral formula

$$I_n=\int_0^{\frac{\pi}{2}}\sin^n x\,dx=\int_0^{\frac{\pi}{2}}\cos^n x\,dx$$

$$=\begin{cases}\frac{n-1}{n}\cdot\frac{n-3}{n-2}\cdot\cdots\cdot\frac{3}{4}\cdot\frac{1}{2}\cdot\frac{\pi}{2} & (n \text{ is a positive even number})\\ \frac{n-1}{n}\cdot\frac{n-3}{n-2}\cdot\cdots\cdot\frac{4}{5}\cdot\frac{2}{3} & (n \text{ is an odd number greater than } 1)\end{cases}.$$

Note: From these two equations, the famous Wallis Formula can be derived:

$$\frac{\pi}{2}=\lim_{m\to\infty}\left[\frac{(2m)!!}{(2m-1)!!}\right]^2\cdot\frac{1}{2m+1}.$$

In fact, from

$$\int_0^{\frac{\pi}{2}}\sin^{2m+1}x\,dx<\int_0^{\frac{\pi}{2}}\sin^{2m}x\,dx<\int_0^{\frac{\pi}{2}}\sin^{2m-1}x\,dx.$$

Substituting equations, we obtain

$$\frac{(2m)!!}{(2m+1)!!}<\frac{(2m-1)!!}{(2m)!!}\cdot\frac{\pi}{2}<\frac{(2m-2)!!}{(2m-1)!!}.$$

From this, we further obtain

$$A_m=\left[\frac{(2m)!!}{(2m-1)!!}\right]^2\frac{1}{2m+1}<\frac{\pi}{2}<\left[\frac{(2m)!!}{(2m-1)!!}\right]^2\frac{1}{2m}=B_m.$$

Because

$$0<B_m-A_m=\left[\frac{(2m)!!}{(2m-1)!!}\right]^2\frac{1}{2m(2m+1)}<\frac{1}{2m}\cdot\frac{\pi}{2}\to 0\ (m\to\infty),$$

therefore

$$\lim_{m\to\infty}(B_m-A_m)=0.$$

Also, since $0<\frac{\pi}{2}-A_m<B_m-A_m$, it follows that $\lim\limits_{m\to\infty}\left(\frac{\pi}{2}-A_m\right)=0$, that is,

$$\lim_{m\to\infty}A_m=\frac{\pi}{2}.$$

Wallis' Formula reveals an unusual relationship between π and integers.

Exercise 5.3

1. Calculate the following definite integrals:

(1) $\int_{-2}^{1}\frac{dx}{(11+5x)^3}$;

(2) $\int_{-\sqrt{2}}^{\sqrt{2}}\sqrt{8-2y^2}\,dy$;

(3) $\int_0^{\frac{\pi}{2}}\sin\varphi\cos^3\varphi\,d\varphi$;

(4) $\int_0^{\pi}(1-\cos^3\theta)\,d\theta$;

(5) $\int_0^{\sqrt{2}}\sqrt{2-x^2}\,dx$;

(6) $\int_{\frac{\pi^2}{6}}^{\frac{\pi}{2}}\cos^2 x\,dx$;

(7) $\int_0^{a}x^2\sqrt{a^2-x^2}\,dx$;

(8) $\int_{\frac{\sqrt{2}}{2}}^{1}\frac{\sqrt{1-x^2}}{x^2}dx$;

(9) $\int_1^{\sqrt{3}}\frac{dx}{x^2\sqrt{1+x^2}}$;

(10) $\int_{-1}^{1}\frac{dx}{\sqrt{5-4x}}$;

(11) $\int_1^4\frac{dx}{1+\sqrt{x}}$;

(12) $\int_{\frac{3}{4}}^{1}\frac{dx}{\sqrt{1-x}-1}$;

(13) $\int_0^1 te^{-\frac{t^2}{2}}dt$;

(14) $\int_0^{\sqrt{3a}}\frac{x\,dx}{\sqrt{3a^2-x^2}}$;

(15) $\int_1^{e^2}\frac{dx}{x\sqrt{1+\ln x}}$;

(16) $\int_{-2}^{0}\frac{dx}{x^2+2x+2}$;

(17) $\int_{-\frac{\pi}{2}}^{\frac{\pi}{2}}\sin x\cos 2x\,dx$;

(18) $\int_{-\frac{\pi}{2}}^{\frac{\pi}{2}}\sqrt{\cos x-\cos^3 x}\,dx$.

2. Calculate the following definite integrals:

(1) $\int_1^{e}x\ln x\,dx$;

(2) $\int_0^{2\pi}t\sin t\,dt$;

(3) $\int_{\frac{\pi}{4}}^{\frac{\pi}{3}} \frac{x}{\sin^2 x} dx$; (4) $\int_1^4 \frac{\ln x \, dx}{\sqrt{x}}$;

(5) $\int_{e^{-1}}^{e} |\ln x| dx$; (6) $\int_1^e \sin(\ln x) dx$;

(7) $\int_0^1 x \arctan x \, dx$; (8) $\int_0^{\frac{\pi}{2}} e^x \cos x \, dx$.

3. Given $f(x)=\begin{cases}1+x^2, & x\leqslant 0\\ e^{-x}, & x>0\end{cases}$, find $\int_1^3 f(x-2) dx$.

4. It is known that xe^x is an antiderivative of $f(x)$, find $\int_0^1 xf'(x)dx$.

5. Given that x^2 is a primitive function of $f(x)$, find $\int_0^{\frac{\pi}{2}} f(-\sin x)\cos x \, dx$.

6. Suppose $f(x)$ is continuous on $[-a, a]$, prove:

$$\int_{-a}^{a} f(x)dx = \int_{-a}^{a} f(-x) dx.$$

7. Prove: $\int_0^1 x^m (1-x)^n dx = \int_0^1 x^n (1-x)^m dx$.

8. Suppose $f(x)$ is a continuous function, prove:

$$\int_a^b f(x)dx = (b-a)\int_0^1 f[a+(b-a)x]dx.$$

9. If $f(x)$ is a continuous odd function, prove that $\int_0^x f(t)dt$ is an even function. If $f(x)$ is a continuous even function, prove that $\int_0^x f(t)dt$ is an odd function.

10. Suppose $f(x)$ is a continuous periodic function with period T, prove:

(1) $\int_{a+T}^{b+T} f(x)dx = \int_a^b f(x)dx$;

(2) $\int_a^{a+T} f(x)dx = \int_0^T f(x)dx$.

5.4 Improper Integrals

In discussing definite integrals, there are two fundamental restrictions: The finiteness of the integration interval and the boundedness of the integrand. However, in some practical problems, integrals with infinite intervals or integrands that are unbounded functions are often encountered, which are no longer within the scope of the previously discussed definite integrals (referred to as normal integrals). This section will extend the concept of definite integrals and introduce two types of improper integrals.

5.4.1 Improper Integrals over Infinite Intervals

Definition 5.4.1 Let the function $f(x)$ be continuous on the interval $[a, +\infty)$, take $u>a$, if the limit

$$\lim_{u \to +\infty} \int_a^u f(x)\,dx$$

exists, this limit is called the improper integral of the function $f(x)$ over the infinite interval $[a, +\infty)$, denoted as $\int_a^{+\infty} f(x)\,dx$.

That is,

$$\int_a^{+\infty} f(x)\,dx = \lim_{u \to +\infty} \int_a^u f(x)\,dx.$$

In this case, the improper integral $\int_a^{+\infty} f(x)\,dx$ is also said to converge. If the limit does not exist, the improper integral is said to diverge, and $\int_a^{+\infty} f(x)\,dx$ is merely a symbol and no longer represents a numerical value.

Similarly, the improper integral of $f(x)$ over $(-\infty, b]$ can be defined as

$$\int_{-\infty}^b f(x)\,dx = \lim_{u \to -\infty} \int_u^b f(x)\,dx.$$

For the improper integral of $f(x)$ over $(-\infty, +\infty)$, it is defined using the two types of improper integrals mentioned earlier:

If both improper integrals $\int_{-\infty}^0 f(x)\,dx$ and $\int_0^{+\infty} f(x)\,dx$ converge, the improper integral $\int_{-\infty}^{+\infty} f(x)\,dx$ is said to converge, otherwise, it is said to diverge. When it converges,

$$\int_{-\infty}^{+\infty} f(x)\,dx = \int_{-\infty}^0 f(x)\,dx + \int_0^{+\infty} f(x)\,dx.$$

The three forms of improper integrals defined above are collectively referred to as improper integrals over infinite intervals, simply called infinite integrals.

Let $F(x)$ be an antiderivative of $f(x)$ over $[a, +\infty)$. If $\lim\limits_{x \to +\infty} F(x)$ exists, then the improper integral

$$\int_a^{+\infty} f(x)\,dx = \lim_{u \to +\infty} \int_a^u f(x)\,dx = \lim_{u \to +\infty} [F(u) - F(a)] = \lim_{x \to +\infty} F(x) - F(a).$$

If $\lim\limits_{x \to +\infty} F(x)$ does not exist, the improper integral $\int_a^{+\infty} f(x)\,dx$ diverges.

If we denote $\lim\limits_{x \to +\infty} F(x) = F(+\infty)$, it can be abbreviated as

$$\int_a^{+\infty} f(x)\,dx = [F(x)]_a^{+\infty}.$$

Similarly, $\int_{-\infty}^b f(x)\,dx = [F(x)]_{-\infty}^b$. When $F(-\infty)$ does not exist, $\int_{-\infty}^b f(x)\,dx$ diverges, $\int_{-\infty}^{+\infty} f(x)\,dx = [F(x)]_{-\infty}^{+\infty}$, when either $F(-\infty)$ or $F(+\infty)$ does not exist, $\int_{-\infty}^{+\infty} f(x)\,dx$ diverges.

Example 5.4.1 Calculate the improper integral $\int_0^{+\infty} e^{-x}\,dx$.

Solution

$$\int_0^{+\infty} e^{-x}\,dx = -e^{-x}\Big|_0^{+\infty} = 0-(-1) = 1.$$

Example 5.4.2 Determine the convergence or divergence of the improper integral $\int_0^{+\infty} \sin x\,dx$.

Solution

$$\int_0^{+\infty} \sin x\,dx = -\cos x\Big|_0^{+\infty}.$$

Since $\lim\limits_{x\to+\infty}\cos x$ does not exist, the improper integral $\int_0^{+\infty}\sin x\,dx$ diverges.

Example 5.4.3 Calculate the improper integral $\int_{-\infty}^{+\infty} \frac{1}{1+x^2}dx$.

Solution

$$\int_{-\infty}^{+\infty} \frac{dx}{1+x^2} = \arctan x\Big|_{-\infty}^{+\infty} = \frac{\pi}{2} - \left(-\frac{\pi}{2}\right) = \pi.$$

The geometric meaning of this improper integral value is: When $a\to-\infty$, $b\to+\infty$, although the shaded area in Figure 5-7 extends infinitely to the left and right, its area has a finite limit value π. Simply put, it is the area of the figure located below the curve $y=\frac{1}{1+x^2}$ and above the x axis.

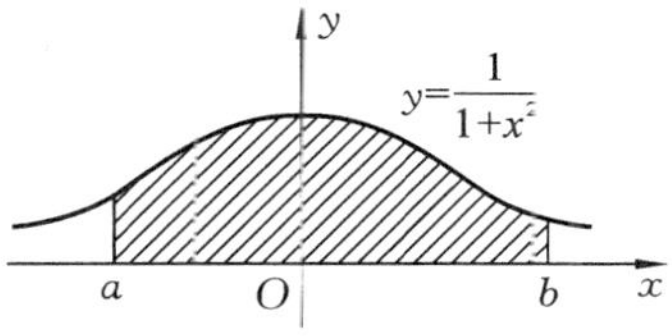

Figure 5-7

Example 5.4.4 Calculate the improper integral $\int_0^{+\infty} te^{-pt}\,dt$ (p is a constant and $p>0$).

Solution

$$\int_0^{+\infty} te^{-pt}\,dt = -\frac{1}{p}\int_0^{+\infty} t\,de^{-pt} = \left[-\frac{t}{p}e^{-pt} + \frac{1}{p}\int e^{-pt}\,dt\right]_0^{+\infty}$$

$$= \left[-\frac{t}{p}e^{-pt}\right]_0^{+\infty} - \left[\frac{1}{p^2}e^{-pt}\right]_0^{+\infty} = -\frac{1}{p}\lim_{t\to+\infty} te^{-pt} - 0 - \frac{1}{p^2}(0-1) = \frac{1}{p^2}.$$

The limit in the equation $\lim\limits_{t\to+\infty} te^{-pt}$ is an indeterminate form, which is solved using L'Hôpital's Rule.

Example 5.4.5 Discuss the convergence or divergence of the improper integral $\int_1^{+\infty}\frac{dx}{x^p}$.

Solution Since

$$\int_1^u \frac{1}{x^p}dx = \begin{cases} \frac{1}{1-p}(u^{1-p}-1), & p\neq 1 \\ \ln u, & p=1 \end{cases},$$

therefore

$$\lim_{u \to +\infty} \int_1^u \frac{1}{x^p} dx = \begin{cases} \frac{1}{p-1}, & p > 1 \\ +\infty, & p \leqslant 1 \end{cases}.$$

Thus, when $p>1$, the improper integral converges, and its value is $\frac{1}{p-1}$, when $p \leqslant 1$, the improper integral diverges.

5.4.2 Improper Integrals of Unbounded Functions

Another type of improper integral involves the integration of unbounded functions. If the function $f(x)$ is unbounded in any neighborhood of the point a, then the point a is called a singular point (or unbounded discontinuity) of the function $f(x)$.

Definition 5.4.2 Let the continuous function $f(x)$ be defined on the interval $(a, b]$, a is a singular point, and $u>a$, the limit

$$\lim_{u \to a^+} \int_u^b f(x) dx$$

exists, then this limit is called the improper integral of the unbounded function $f(x)$ over $(a, b]$, denoted as

$$\int_a^b f(x) dx = \lim_{u \to a^+} \int_u^b f(x) dx.$$

In this case, the improper integral $\int_a^b f(x) dx$ is said to converge. If the limit does not exist, the improper integral $\int_a^b f(x) dx$ is said to diverge. Improper integrals of unbounded functions are also known as singular integrals.

Similarly, a singular point can be defined as an improper integral of b.

$$\int_a^b f(x) dx = \lim_{u \to b^-} \int_a^u f(x) dx.$$

If $f(x)$ has a singular point at $c \in (a, b)$, the improper integral is defined as

$$\int_a^b f(x) dx = \int_a^c f(x) dx + \int_c^b f(x) dx = \lim_{u \to c^-} \int_a^u f(x) dx + \lim_{v \to c^+} \int_v^b f(x) dx.$$

The improper integral on the left side of equation converges if and only if both improper integrals on the right side converge.

The calculation of improper integrals can also be facilitated by Newton-Leibniz formula. Let $x=a$ be a singular point of $f(x)$ on $(a, b]$, and if the limit $\lim_{x \to a^+} F(x)$ exists, then the improper integral

$$\int_a^b f(x) dx = \lim_{u \to a^+} \int_u^b f(x) dx = \lim_{u \to a^+} [F(b) - F(u)] = F(b) - \lim_{x \to a^+} F(x) = F(b) - F(a^+).$$

If $\lim_{x \to a^+} F(x)$ does not exist, the improper integral $\int_a^b f(x) dx$ diverges.

Using the notation $[F(x)]_a^b$ to represent $F(b) - F(a^+)$, it is formally written as

$$\int_a^b f(x)\mathrm{d}x = [F(x)]_a^b.$$

For improper integrals with b as a singular point, a similar approach applies:

$$\int_a^b f(x)\mathrm{d}x = F(b^-) - F(a) = [F(x)]_a^b.$$

Example 5.4.6 Calculate the improper integral $\int_0^a \frac{\mathrm{d}x}{\sqrt{a^2 - x^2}} (a > 0)$.

Solution Because

$$\lim_{x \to a^-} \frac{1}{\sqrt{a^2 - x^2}} = +\infty,$$

therefore, the point a is a singular point, thus

$$\int_0^a \frac{\mathrm{d}x}{\sqrt{a^2 - x^2}} = \left[\arcsin\frac{x}{a}\right]_0^a = \lim_{x \to a^-} \arcsin\frac{x}{a} - 0 = \frac{\pi}{2}.$$

The geometric meaning of this improper integral: The area under the curve $y = \frac{1}{\sqrt{a^2 - x^2}}$, above the x axis, and between the lines $x = 0$ and $x = a$ (Figure 5-8).

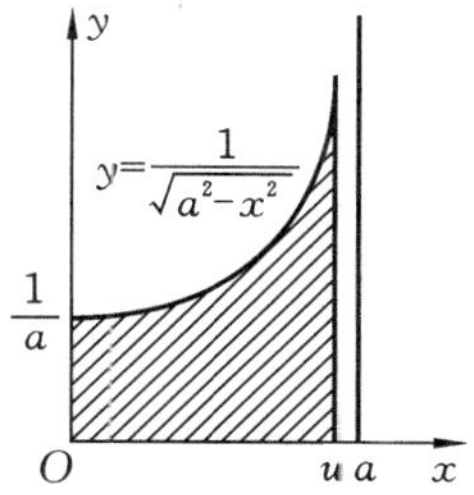

Figure 5-8

Example 5.4.7 Calculate the improper integral $\int_1^2 \frac{\mathrm{d}x}{x\ln x}$.

Solution Obviously, $x = 1$ is a point of discontinuity, thus

$$\int_1^2 \frac{\mathrm{d}x}{x\ln x} = \int_1^2 \frac{\mathrm{d}(\ln x)}{\ln x} = [\ln|\ln x|]_1^2 = +\infty.$$

Therefore, the improper integral diverges.

Example 5.4.8 Calculate the improper integral $\int_0^1 \frac{\arcsin\sqrt{x}}{\sqrt{x(1-x)}}\mathrm{d}x$.

Solution The integrand has two potential points of discontinuity: $x = 0$ and $x = 1$. Because

$$\lim_{x \to 0+} \frac{\arcsin\sqrt{x}}{\sqrt{x(1-x)}} = 1,$$

hence, $x = 1$ is the only point of discontinuity. Thus

$$\int_0^1 \frac{\arcsin\sqrt{x}}{\sqrt{x(1-x)}}\mathrm{d}x = 2\int_0^1 \arcsin\sqrt{x}\,\mathrm{d}(\arcsin\sqrt{x}) = (\arcsin\sqrt{x})^2\Big|_0^1 = \frac{\pi^2}{4}.$$

Example 5.4.9 Discuss the convergence of the improper integral $\int_0^1 \frac{1}{x^q}\mathrm{d}x \quad (q > 0)$.

Solution $x = 0$ is a point of discontinuity. Since

$$\int_u^1 \frac{\mathrm{d}x}{x^q} = \begin{cases} \frac{1}{1-q}(1 - u^{1-q}), & q \neq 1 \\ -\ln u, & q = 1 \end{cases},$$

therefore, when $0 < q < 1$, the improper integral converges, and

$$\int_0^1 \frac{dx}{x^q} = \lim_{u\to 0+}\int_u^1 \frac{dx}{x^q} = \frac{1}{1-q}.$$

While when $q \geqslant 1$, the improper integral diverges to $+\infty$.

If we connect Example 5.4.5 and Example 5.4.9, we examine the improper integral

$$\int_0^{+\infty} \frac{dx}{x^p}(p>0).$$

Denote

$$\int_0^{+\infty} \frac{dx}{x^p} = \int_0^1 \frac{dx}{x^p} + \int_1^{+\infty} \frac{dx}{x^p}.$$

From the results of Example 5.4.5 and Example 5.4.9, it is known that the two improper integrals on the right-hand side of the equation cannot converge simultaneously, hence the improper integral $\int_0^{+\infty} \frac{dx}{x^p}(p>0)$ diverges for any real number p.

Exercise 5.4

1. Determine whether the following improper integrals converge. If they converge, find their values.

(1) $\int_{-1}^{+\infty} \frac{dx}{x^2+5x+6}$; (2) $\int_0^{+\infty} \frac{dx}{\sqrt{x}(1+x)}$;

(3) $\int_1^{+\infty} \frac{dx}{x\sqrt{x^2-1}}$; (4) $\int_2^{+\infty} \frac{dt}{t^2-1}$;

(5) $\int_0^{+\infty} \frac{dx}{(1+x^2)(1+\arctan x)}$; (6) $\int_{-\infty}^{+\infty} e^{-|x|}dx$;

(7) $\int_{-\infty}^{+\infty} 2x e^{-x^2}dx$; (8) $\int_0^{+\infty} e^{-x}\sin x\, dx$;

(9) $\int_0^{\frac{\pi}{2}} \tan\theta\, d\theta$; (10) $\int_0^1 \frac{dx}{\sqrt{1-x^2}}$;

(11) $\int_0^2 \frac{t+1}{\sqrt{4-t^2}}dt$; (12) $\int_0^1 x\ln x\, dx$;

(13) $\int_{-1}^4 \frac{dx}{\sqrt{|x|}}$; (14) $\int_1^2 \frac{x\, dx}{\sqrt{x-1}}$.

2. Find the value of c, such that $\lim\limits_{x\to+\infty}\left(\frac{x+c}{x-c}\right)^x = \int_{-\infty}^c t e^{2t}dt$.

3. For what values of k does the improper integral $\int_2^{+\infty} \frac{dx}{x(\ln x)^k}$ converge? Diverge? Also, for what values of k does the improper integral achieve its minimum value?

4. Calculate the improper integral $I_n = \int_0^{+\infty} x^n e^{-x}dx$ using the recurrence formula.

5.5 Area of Plane

From this section onwards, we will apply the definite integral theory learned previously to analyze and solve problems in geometry and physics. The purpose is not only to master the formulas for calculating these practical problems but more importantly, to deeply understand the basic ideas and methods of solving practical problems using definite integrals—the method of elements, and to better grasp and continuously improve the application of mathematics.

5.5.1 The Element Method of Definite Integrals

All application problems of definite integrals are generally represented in the form of definite integrals through the steps of "partition, approximation, summation, and limit". However, in practice, the "element method" is often used for simplicity and practicality. Below, we introduce the basic ideas and methods of the element method. To this end, let us briefly review the problem of finding the area of a curved trapezoid discussed in Section 5.1. Suppose a curved trapezoid is bounded by a continuous curve $y = f(x)$ $(f(x) \geqslant 0)$, the x axis, and two straight lines $x = a$, $x = b$, and we seek to find its area A.

(1) Partition. Divide the interval $[a, b]$ into n subintervals of length Δx_i $(i = 1, 2, \cdots, n)$ using any set of partition points, correspondingly dividing the area of the curved trapezoid into n small curved trapezoid areas ΔA_i $(i = 1, 2, \cdots, n)$.

(2) Approximation. On the ith subinterval, take any point ξ_i and approximate the area of the ith small curved trapezoid ΔA_i with the area of a small rectangle $f(\xi_i)\Delta x_i$, i.e.,

$$\Delta A_i \approx f(\xi_i)\Delta x_i \quad (x_{i-1} \leqslant \xi_i \leqslant x_i).$$

(3) Summation. Obtain an approximate value for the area A:

$$A = \sum_{i=1}^{n} \Delta A_i \approx \sum_{i=1}^{n} f(\xi_i)\Delta x_i.$$

(4) Limit. Obtain the exact value of the area A:

$$A = \lim_{\lambda \to 0} \sum_{i=1}^{n} f(\xi_i)\Delta x_i = \int_a^b f(x)\mathrm{d}x,$$

where: $\lambda = \max\{\Delta x_1, \Delta x_2, \cdots, \Delta x_n\}$.

From the above process, it can be seen that when $[a, b]$ is divided into n small intervals, the area A (total quantity) to be found is also correspondingly divided into n small curved trapezoidal areas (partial quantities), and the total quantity is equal to the sum of the partial quantities $\left(\text{i.e.}, A = \sum_{i=1}^{n} \Delta A_i\right)$. This property is called additivity of the quantity with respect to the interval $[a, b]$.

For the sake of simplicity, the subscript i can be omitted in the above analysis process, and it is rewritten as follows:

(1) Partition. Arbitrarily divide $[a, b]$ into n small intervals, and take one of these small intervals $[x, x+\mathrm{d}x]$, ΔA to represent the area of the small curved trapezoid on $[x, x+\mathrm{d}x]$.

(2) Approximate. Take the left endpoint x as ξ, and use the function value $f(x)$ at point x as the height, $\mathrm{d}x$ as the base to form a small rectangle area $\mathrm{d}A = f(x)\mathrm{d}x$ as the approximate value of ΔA:

$$\Delta A \approx \mathrm{d}A = f(x)\mathrm{d}x.$$

(3) Summation. Obtain the approximate value of the area A:

$$A = \sum \Delta A \approx \sum f(x)\mathrm{d}x.$$

(4) Limit. Obtain the exact value of the area A:

$$A = \lim \sum f(x)\mathrm{d}x = \int_a^b f(x)\mathrm{d}x.$$

Among the four steps mentioned above, the second step is particularly important. Because as long as the approximate value $\mathrm{d}A = f(x)\mathrm{d}x$ of the area of the small curved trapezoid ΔA (the product of a certain continuous function $f(x)$ and $\mathrm{d}x$) is found, and then it is used as the integrand to perform a definite integral, the exact value of A can be obtained. Here, $\mathrm{d}A = f(x)\mathrm{d}x$ is called the area element or area differential.

This method of obtaining the total amount U by first finding its differential element $\mathrm{d}U$, and then using $\mathrm{d}U$ as the integrand to perform a definite integral to obtain the total amount U, is called the element method, also known as the differential element method.

The main steps of the element method (differential element method) are as follows:

(1) Obtain the differential element from the division. According to the specific problem, select an integral variable. For example, x is the integral variable, and determine its variation interval $[a, b]$. Take any small interval $[x, x+\mathrm{d}x] \subset [a, b]$, and find the approximate value $f(x)\mathrm{d}x$ of the partial quantity ΔU corresponding to this small interval, that is, obtain the differential element of the total amount U.

$$\mathrm{d}U = f(x)\mathrm{d}x.$$

(2) Write the integral from the differential element. According to $\mathrm{d}U = f(x)\mathrm{d}x$, write the definite integral representing the total amount U:

$$U = \int_a^b \mathrm{d}U = \int_a^b f(x)\mathrm{d}x.$$

When applying the element method to solve practical problems, the following two points should be noted:

(1) The total quantity sought U over the interval $[a, b]$ should be additive, i.e., if the interval $[a, b]$ is divided into many subintervals, then U is correspondingly divided into

many partial quantities, and U equals the sum of all partial quantities ΔU. This requirement is determined by the concept of the definite integral itself.

(2) The key to using the method of elements lies in correctly giving the approximate expression $f(x)dx$ for the partial quantity ΔU, i.e.,

$$f(x)dx = dU \approx \Delta U.$$

It is required here that $\Delta U - f(x)dx$ be a higher-order infinitesimal of dx.

In general, rigorously testing whether $\Delta U - f(x)dx$ is a higher-order infinitesimal of dx is often not an easy task, so in practical applications, the reasonableness of $dU = f(x)dx$ should be noted.

5.5.2 Area of Plane

1. Rectangular Coordinate Case

It is known that for a non-negative function $f(x)$, the definite integral $\int_a^b f(x)dx$ represents the area of the curvilinear trapezoid formed by the curve $y = f(x)$ and the straight lines $x=a$, $x=b$ and the x axis, where the integrand expression $f(x)dx$ is the area element, i.e., $dA = f(x)dx$.

If $f(x)$ is not non-negative, the area element should be

$$dA = |f(x)|dx.$$

The area of the enclosed figure (Figure 5-9) is

$$A = \int_a^b |f(x)|dx.$$

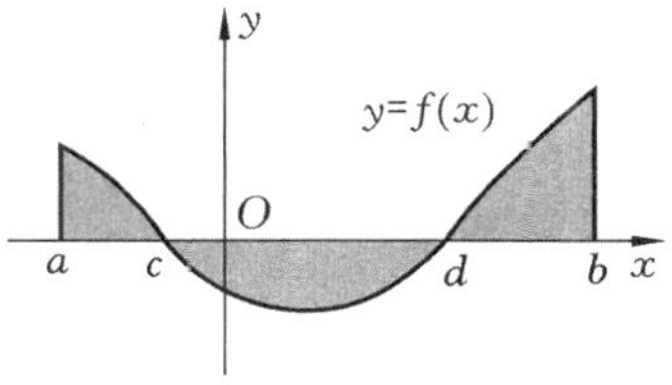

Figure 5-9

In general, the area of the figure enclosed by two curves $y=f(x)$, $y=g(x)$ $(f(x) \geqslant g(x))$ and the straight line $x=a$, $x=b$, as shown in Figure 5-10, is

$$A = \int_a^b f(x)dx - \int_a^b g(x)dx = \int_a^b [f(x) - g(x)]dx.$$

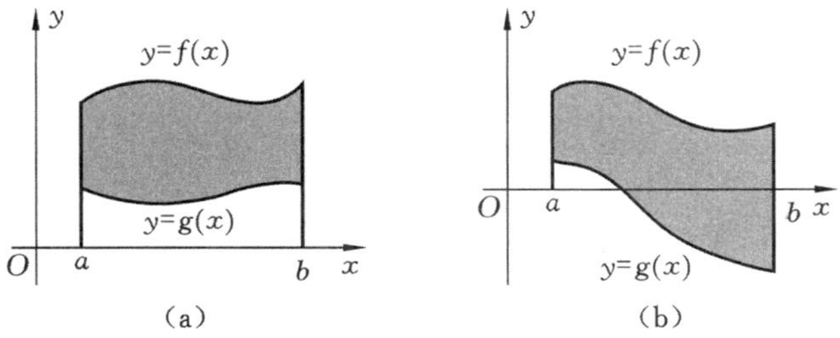

Figure 5-10

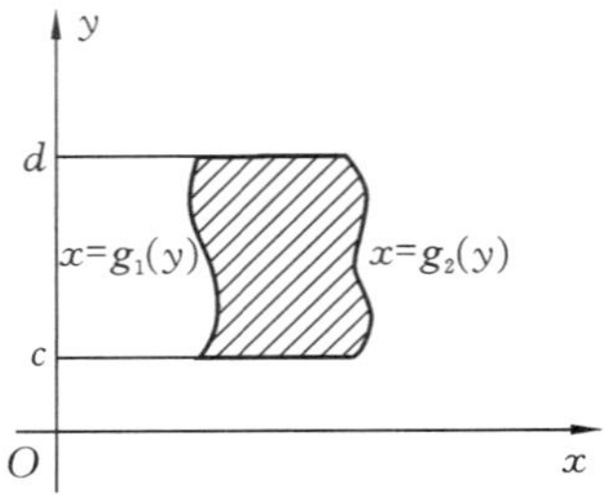

Figure 5-11

Similarly, by the method of definite integral elements, the area of the plane figure enclosed by the curve $x=g_1(y)$, $x=g_2(y)$ (where $g_1(y)\leqslant g_2(y)$) and $y=c$, $y=d$ (Figure 5-11), is

$$A=\int_c^d [g_2(y)-g_1(y)]\,\mathrm{d}y.$$

Example 5.5.1 Calculate the area of the figure enclosed by the two parabolas $y=x^2$ and $y^2=8x$.

Solution The figure enclosed by these two parabolas is shown in Figure 5-12.

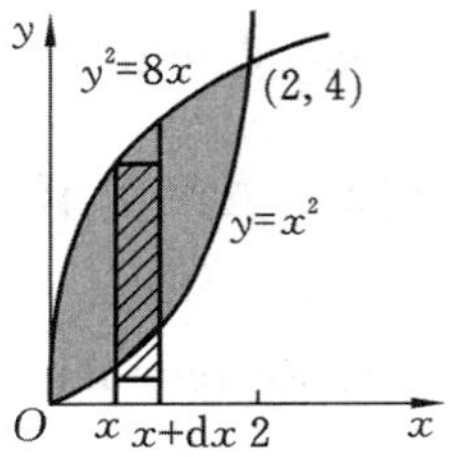

Figure 5-12

Solving the system of equations $\begin{cases} y=x^2 \\ y^2=8x \end{cases}$, we obtain their intersection points as (0,0), (2,4). Choose x as the integral variable, then the range of x is $[0,2]$. For any small interval $[x, x+\mathrm{d}x]\subset[0,2]$, the corresponding area element can be obtained as

$$\mathrm{d}A=(\sqrt{8x}-x^2)\,\mathrm{d}x.$$

Thus

$$A=\int_0^2(\sqrt{8x}-x^2)\,\mathrm{d}x=\left[\sqrt{8}\cdot\frac{2}{3}x^{\frac{3}{2}}-\frac{x^3}{3}\right]_0^2=\frac{8}{3}.$$

If y is chosen as the integration variable, then the range of y is $[0,4]$, and the area element can be obtained similarly, and the area A can be found. Please complete this by yourself.

Example 5.5.2 Calculate the area of the region bounded by the parabola $y^2=2x$ and the line $y=x-4$.

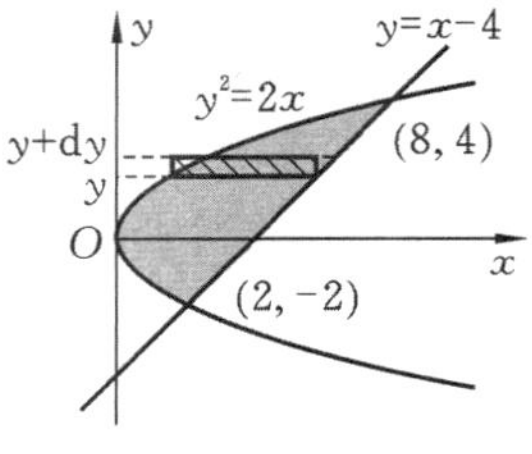

Figure 5-13

Solution Sketch the graph of the region, as shown in Figure 5-13.

The intersection points $(2,-2)$ and $(8,4)$ are obtained from the system of equations $\begin{cases} y^2=2x \\ y=x-4 \end{cases}$. Take y as the integration variable, then the range of y is $[-2,4]$. For any small interval $[y, y+\mathrm{d}y]\subset[-2,4]$, the area of the corresponding narrow strip is approximately equal to the area of a narrow rectangle with height $\mathrm{d}y$ and base $(y+4)-\frac{1}{2}y^2$, thus obtaining the area element,

$$\mathrm{d}A=\left(y+4-\frac{1}{2}y^2\right)\mathrm{d}y.$$

Therefore, the area sought is

$$A=\int_{-2}^4\mathrm{d}A=\int_{-2}^4\left(y+4-\frac{1}{2}y^2\right)\mathrm{d}y=\left[\frac{y^2}{2}+4y-\frac{y^3}{6}\right]_{-2}^4=18.$$

If x is chosen as the integration variable, then the range of x is $[0,8]$, but the area elements are different in the two intervals $[0,2]$ and $[2,8]$. When x varies in the interval $[0,2]$, the area element is

$$dA=\left[\sqrt{2x}-(-\sqrt{2x})\right]dx=2\sqrt{2x}\,dx.$$

When x varies in the interval $[2,8]$, the area element is

$$dA=\left[\sqrt{2x}-(x-4)\right]dx.$$

Thus, the area sought is

$$A=\int_0^2 2\sqrt{2x}\,dx+\int_2^8(\sqrt{2x}-x+4)\,dx.$$

The calculation of two definite integrals is required here, making the computation complex. It is evident that choosing an appropriate variable of integration can simplify the calculation.

Example 5.5.3 Find the area of the ellipse $\frac{x^2}{a^2}+\frac{y^2}{b^2}=1$.

Solution As shown in Figure 5-14, since the ellipse is symmetric about both coordinate axes, let A_1 be the area of the first quadrant part.

By the method of elements, $dA_1=y\,dx$, thus the area of the ellipse is

$$A=4A_1=4\int_0^a y\,dx.$$

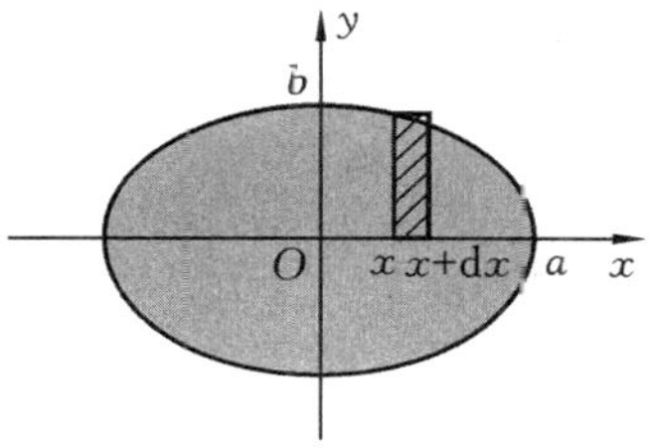

Figure 5-14

For convenience of calculation, use the parametric equations of the ellipse for substitution:

$$\begin{cases}x=a\cos t\\ y=b\sin t\end{cases}\quad\left(0\leqslant t\leqslant\frac{\pi}{2}\right).$$

When x varies from 0 to a, t varies from $\frac{\pi}{2}$ to 0, so

$$A=4\int_0^a y\,dx=4\int_{\frac{\pi}{2}}^0 b\sin t\,d(a\cos t)=4ab\int_0^{\frac{\pi}{2}}\sin^2 t\,dt=4ab\cdot\frac{1}{2}\cdot\frac{\pi}{2}=\pi ab.$$

When $a=b$, the ellipse becomes a circle, thus obtaining the formula for the area of a circle $A=\pi a^2$.

2. Polar Coordinates

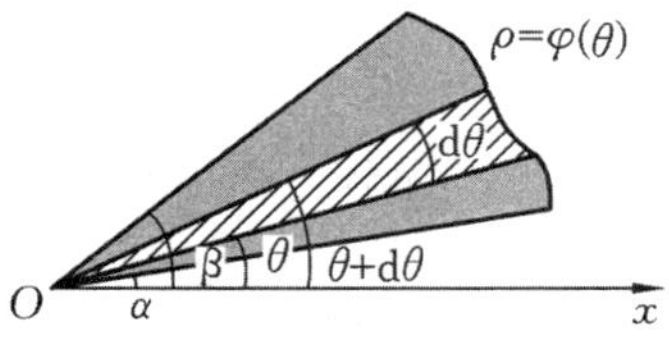

Figure 5-15

Some plane figures are more conveniently calculated using polar coordinates for their area.

The figure enclosed by the curve $\rho=\varphi(\theta)$ and the ray $\theta=\alpha$, $\theta=\beta$ is called a curved sector (Figure 5-15). Now, we need to calculate its area, where $\varphi(\theta)$ is continuous on

$[\alpha, \beta]$, and $\varphi(\theta) \geqslant 0$.

Since $\rho = \varphi(\theta)$ varies with θ, the area of the figure in question cannot be calculated using the formula for the area of a circular sector. Therefore, we consider solving it using the method of elements.

Choose the polar angle θ as the integral variable, then the range of θ is $[\alpha, \beta]$. For any small interval $[\theta, \theta + d\theta] \subset [\alpha, \beta]$, the area of the small curved sector corresponding to the interval $[\theta, \theta + d\theta]$ can be approximated by the area of a circular sector with radius $\rho = \varphi(\theta)$ and central angle $d\theta$, thus obtaining an approximate value for the area of this small curved sector, i.e., the area element of the curved sector.

$$dA = \frac{1}{2}[\varphi(\theta)]^2 d\theta.$$

Thus, the area of the curved sector is

$$A = \int_{\alpha}^{\beta} \frac{1}{2}[\varphi(\theta)]^2 d\theta.$$

Example 5.5.4 Calculate the area of the plane figure enclosed by the lemniscate $\rho^2 = a^2 \cos 2\theta$.

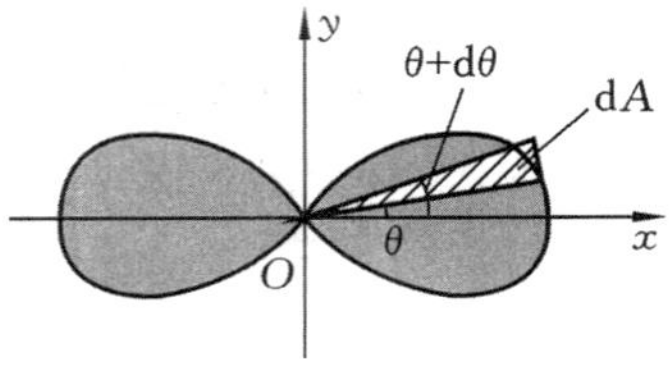

Figure 5-16

Solution The lemniscate is shown in Figure 5-16. Since $\rho^2 \geqslant 0$, the range of θ is $\left[-\frac{\pi}{4}, \frac{\pi}{4}\right]$ and $\left[\frac{3\pi}{4}, \frac{5\pi}{4}\right]$.

Due to the symmetry of the graph, it is sufficient to calculate the area of the portion in the first quadrant and then multiply it by 4. For the graph in the first quadrant, the range of θ is $\left[0, \frac{\pi}{4}\right]$, and for any small interval $[\theta, \theta + d\theta] \subset \left[0, \frac{\pi}{4}\right]$, the corresponding area element is obtained:

$$dA = \frac{1}{2} a^2 \cos 2\theta d\theta.$$

Thus, the area sought is

$$A = 4\int_0^{\frac{\pi}{4}} dA = 4\int_0^{\frac{\pi}{4}} \frac{1}{2} a^2 \cos 2\theta d\theta = a^2 \sin 2\theta \Big|_0^{\frac{\pi}{4}} = a^2.$$

Example 5.5.5 Calculate the area of the graph enclosed by the cardioid $\rho = a(1 + \cos\theta)$ $(a > 0)$.

Solution The graph of the cardioid is shown in Figure 5-17. The graph is symmetric about the polar axis, so the area A is twice the area of the portion above the polar axis.

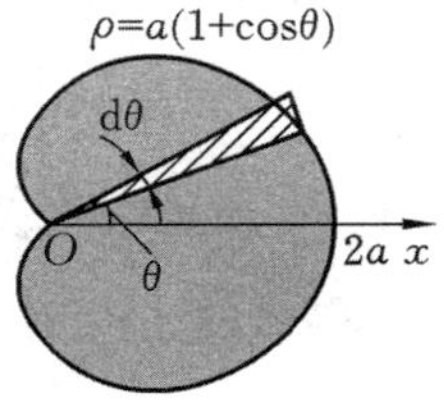

Figure 5-17

For the portion of the graph above the polar axis, θ ranges from $[0, \pi]$, and for any small interval $[\theta, \theta + d\theta] \subset [0, \pi]$, the corresponding area element is obtained:

$$\mathrm{d}A = \frac{1}{2}a^2(1+\cos\theta)^2\mathrm{d}\theta.$$

Thus, the area sought is

$$A = 2\int_0^{\pi}\mathrm{d}A = a^2\int_0^{\pi}(1+2\cos\theta+\cos^2\theta)\,\mathrm{d}6 = a^2\int_0^{\pi}\left(\frac{3}{2}+2\cos\theta-\frac{1}{2}\cos 2\theta\right)\mathrm{d}\theta$$

$$= a^2\left[\frac{3}{2}\theta + 2\sin\theta + \frac{1}{4}\sin 2\theta\right]_0^{\pi} = \frac{3}{2}\pi a^2.$$

Exercise 5.5

1. Find the area of the shaded region in the following figure.

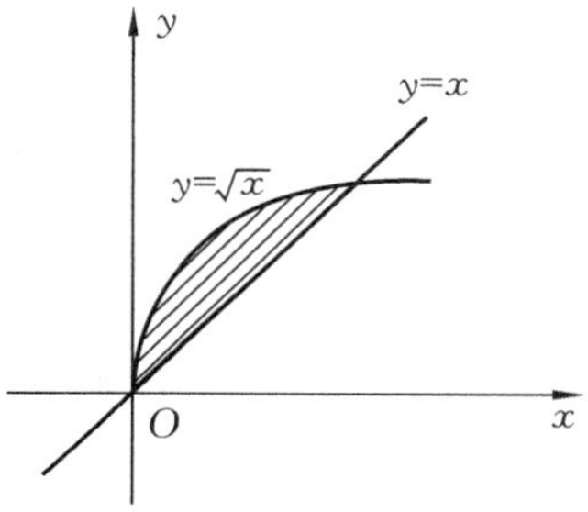

(1)

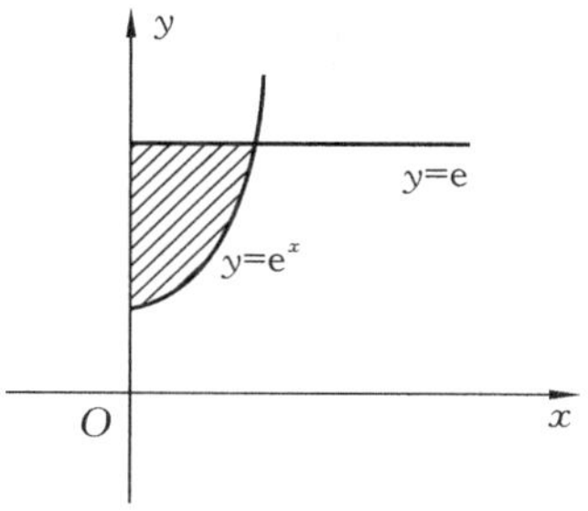

(2)

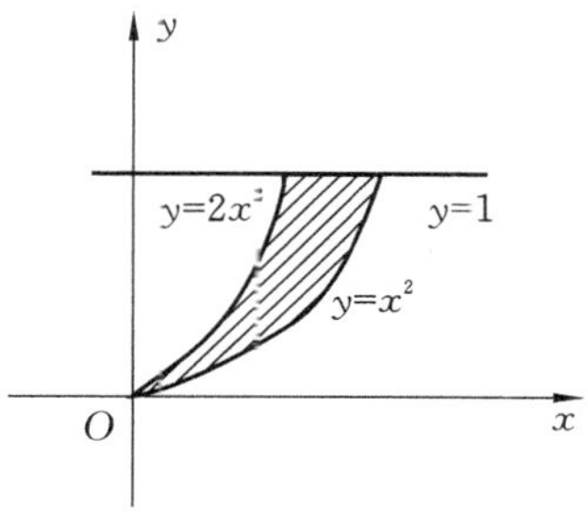

(3)

2. Find the area of the region enclosed by the following curves:

(1) $y=\frac{1}{2}x^2$ and $x^2+y^2=8$ (both parts need to be calculated);

(2) $y=\frac{1}{x}$ and the line $y=x$, $x=2$;

(3) the axis $y=\ln x$, y and the line $y=\ln a$, $y=\ln b$ $(b>a>0)$;

(4) $y=x^2$ and the line $y=x$, $y=2x$.

3. Find the area of the region enclosed by the parabola $y=-x^2+4x-3$ and its tangents at the points $(0,-3)$ and $(3,0)$.

4. Find the area of the region enclosed by the parabola $y^2=2px$ and its normal at the point $\left(\frac{p}{2}, p\right)$.

5. Find the area of the region enclosed by the curve $\rho=2a\cos\theta$.

6. Find the area of the region enclosed by the curve $x=a\cos^3 t$, $y=a\sin^3 t$.

7. Find the area of the region enclosed by the cycloid $x=a(t-\sin t)$, $y=a(1-\cos t)$ $(0\leqslant t\leqslant 2\pi)$ and the x axis.

8. Find the area of the common region enclosed by the curves $\rho=3\cos\theta$, $\rho=1-\cos\theta$.

9. Find the area of the common region enclosed by the curves $\rho=\sqrt{2}\sin\theta$, $\rho^2=\cos 2\theta$.

10. Find the area of the region bounded by the curve $y=\mathrm{e}^x$ below, the tangent line to the curve passing through the origin to the left, and above the x axis.

11. Find the parabola passing through the points $(0,0)$ and $(1,2)$ with the following

properties:

(1) Its axis of symmetry is parallel to the y axis and curves downward;

(2) It encloses the smallest area with the x axis.

12. Find a tangent line to the curve $y=\ln x$ within the interval $[2,6]$, such that the area of the region enclosed by this tangent line, the line $x=2$, $x=6$, and the curve $y=\ln x$ is minimized.

5.6 Volume of Solids

5.6.1 Volume of Solids of Revolution

A solid formed by rotating a plane figure around a line in the same plane is called a solid of revolution, and this line is called the axis of revolution.

For example, a cylinder can be viewed as a solid formed by rotating a rectangle around one of its sides. A cone can be viewed as a solid formed by rotating a right triangle around one of its legs. And a sphere can be viewed as a solid formed by rotating a semicircle around its diameter, and so on.

Generally, let the solid of revolution be formed by rotating a curvilinear trapezoid enclosed by the continuous curve $y=f(x)$, the straight lines $x=a$, $x=b$, and the x axis around the x axis. Now, we seek to find its volume V.

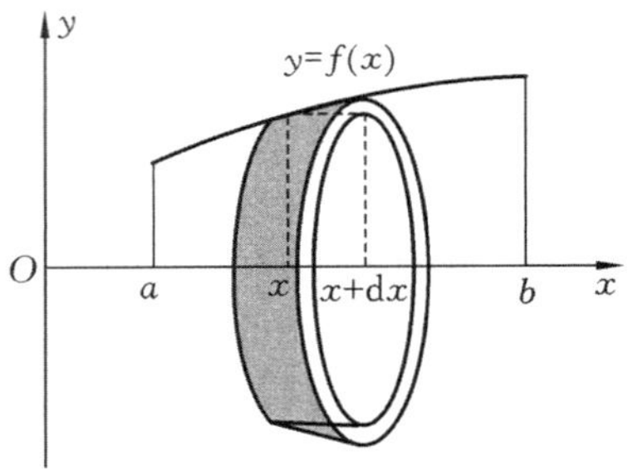

Figure 5-18

Take x as the variable of integration, with its range of variation being $[a,b]$. For any small interval $[x,x+\mathrm{d}x]\subset[a,b]$, the volume of the thin slice of the solid of revolution corresponding to the small interval $[x,x+\mathrm{d}x]$ can be approximated as the volume of a flat cylinder with a base radius of $|f(x)|$ and a height of $\mathrm{d}x$ (Figure 5-18), i.e., the volume element is

$$\mathrm{d}V=\pi\,[f(x)]^2\mathrm{d}x.$$

Thus, the volume of the solid of revolution is

$$V=\int_a^b \pi\,[f(x)]^2\mathrm{d}x.$$

Example 5.6.1 Find the volume of a right circular cone with height h and base radius r.

Solution The cone can be seen as a solid of revolution formed by rotating the plane figure enclosed by the line $y=\frac{r}{h}x$, $x=0$, $x=h$ around the x axis (Figure 5-19).

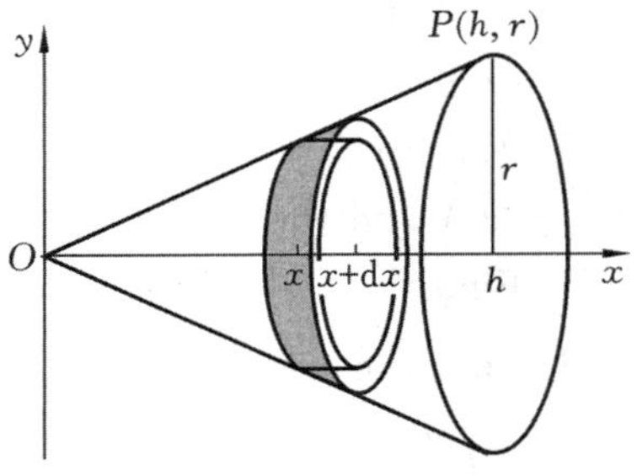

Figure 5-19

Take x as the variable of integration, with its range of variation being $[0,h]$. For any small interval $[x,x+\mathrm{d}x]\subset[0,h]$, the corresponding volume element is obtained as

$$\mathrm{d}V=\pi\left(\frac{r}{h}x\right)^2\mathrm{d}x.$$

Thus, the volume of the cone is found to be

$$V=\int_0^h\pi\left(\frac{r}{h}x\right)^2\mathrm{d}x=\frac{\pi r^2}{h^2}\left[\frac{x^3}{3}\right]_0^h=\frac{\pi r^2 h}{3}.$$

Example 5.6.2 Calculate the volume of the solid of revolution formed by rotating the plane region enclosed by the ellipse $\frac{x^2}{a^2}+\frac{y^2}{b^2}=1$ around the x axis.

Solution The solid of revolution can be considered as the solid formed by rotating the upper half of the ellipse $y=\frac{b}{a}\sqrt{a^2-x^2}$ and the x axis around the x axis (Figure 5-20).

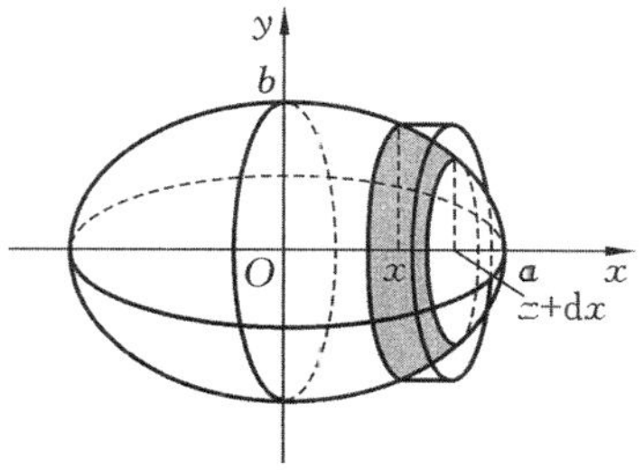

Figure 5-20

Take x as the variable of integration, with its range being $[-a,a]$. For any small interval $[x,x+\mathrm{d}x]\subset[-a,a]$, the volume of the thin slice corresponding to the small interval is approximately equal to the volume of a flat cylinder with a base radius of $\frac{b}{a}\sqrt{a^2-x^2}$ and a height of $\mathrm{d}x$, i.e., the volume element is

$$\mathrm{d}V=\pi\frac{b^2}{a^2}(a^2-x^2)\,\mathrm{d}x.$$

Thus, the volume of the ellipsoid of revolution is obtained as

$$V=\int_{-a}^a\mathrm{d}V=\int_{-a}^a\pi\frac{b^2}{a^2}(a^2-x^2)\,\mathrm{d}x=2\pi\frac{b^2}{a^2}\int_0^a(a^2-x^2)\,\mathrm{d}x=2\pi\frac{b^2}{a^2}\left(a^2x-\frac{x^3}{3}\right)\Bigg|_0^a=\frac{4}{3}\pi ab^2.$$

In particular, when $a=b=R$, the volume of a sphere with radius R can be obtained

$$V=\frac{4}{3}\pi R^3.$$

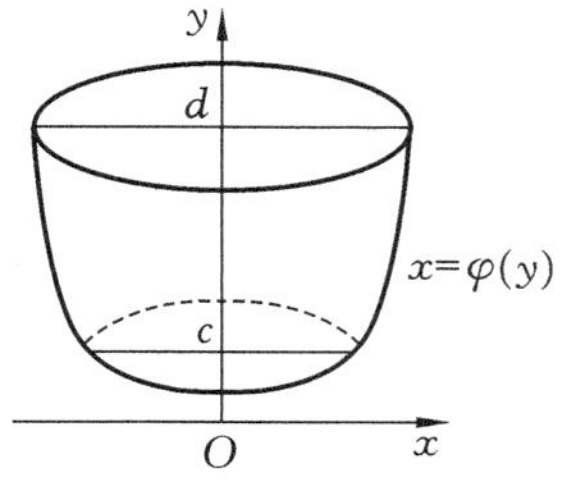

Figure 5-21

Using a similar method to the above, it can be deduced that the volume of the solid of revolution formed by rotating the curvilinear trapezoid enclosed by the continuous curve $x=\varphi(y)$, the straight lines $y=c$, $y=d$ $(c<d)$, and the y axis around the y axis (Figure 5-21) is

$$V=\int_c^d\pi\,[\varphi(y)]^2\mathrm{d}y.$$

Example 5.6.3 Find the volume of the solid of revolution formed by rotating the region bounded by the curve $xy=4$ and the line $y=1$, $x=0$ about the y axis (Figure 5-22).

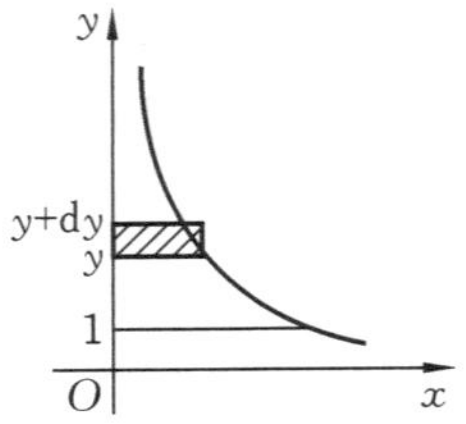

Figure 5-22

Solution　Take y as the variable of integration, with the range of y being $[1,+\infty)$. For any small interval $[y,y+dy]\subset[1,+\infty)$, it is easy to see that the volume element is

$$dV=\pi x^2 dy=\pi\frac{16}{y^2}dy.$$

Therefore, the volume sought is

$$V=\int_1^{+\infty}\pi\frac{16}{y^2}dy=\pi\left[-\frac{16}{y}\right]_1^{+\infty}=16\pi.$$

Example 5.6.4　Find the volume of the solid of revolution formed by rotating the region bounded by the curve $y=4-x^2$ and $y=0$ about the line $x=3$.

Solution　Sketch the plane figure (Figure 5-23).

Solve the system of equations $\begin{cases} y=4-x^2 \\ y=0 \end{cases}$, to find the intersection points $(-2,0)$, $(2,0)$.

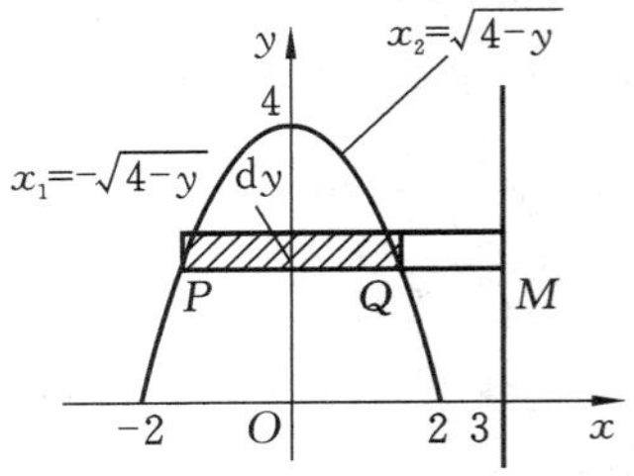

Figure 5-23

Take y as the variable of integration, with its range being $[0,4]$. For any small interval $[y,y+dy]$ within this range, the volume of the small slice corresponding to the interval is approximately equal to the volume of a flat torus cylinder with inner radius $\overline{QM}$, outer radius $\overline{PM}$, and height dy, i.e., the volume element is

$$dV=[\pi\,\overline{PM}^2-\pi\,\overline{QM}^2]dy=[\pi\,(3+\sqrt{4-y})^2-\pi\,(3-\sqrt{4-y})^2]dy=12\pi\sqrt{4-y}\,dy.$$

Thus, the volume of the solid of revolution is

$$V=12\pi\int_0^4\sqrt{4-y}\,dy=64\pi.$$

5.6.2　Volume of a Solid with Known Cross-Sectional Area

If a solid is not a rotational solid, but the cross-sectional areas perpendicular to a certain axis on the solid are known, then the volume of this solid can also be calculated using definite integrals.

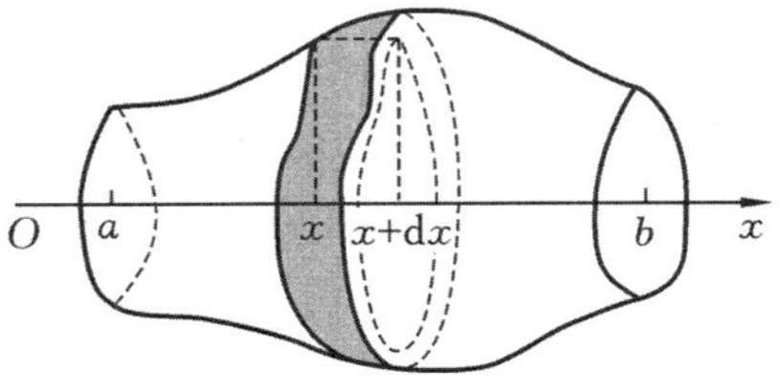

Figure 5-24

As shown in Figure 5-24, let the aforementioned fixed axis be the x axis, and the solid is located between two planes passing through point $x=a$, $x=b$, and perpendicular to the x axis.

$\forall x\in[a,b]$, passing through point x, a plane perpendicular to the x axis is made, and the cross-sectional area of the solid is $A(x)$, which is a known continuous function. That is, the cross-sectional area obtained by any plane perpendicular to the x axis is known, and now the volume of the solid is to be determined.

Take x as the integral variable, then the variation interval of x is $[a, b]$. On the corresponding small interval $[x, x+dx]$, the volume of the thin slice of the solid is approximately equal to the volume of a flat cylinder with a base area of $A(x)$ and a height of dx, that is, the volume element is

$$dV = A(x)dx.$$

Thus, the volume of the solid in question is

$$V = \int_a^b A(x)dx.$$

Example 5.6.5 Verify Zu's Theorem.

Zu Geng, a mathematician of the Qi and Liang Dynasties in China, was the son of Zu Chongzhi. He lived around the end of the 5th century to the beginning of the 6th century. Zu Geng's theorem states that if two solids have the same height and the same cross-sectional area at any given height, then these two solids have equal volume.

Solution As shown in Figure 5-25, the two solids Ω_A, Ω_B have the same height h, and their volumes are V_A, V_B. If their cross-sectional area functions $A(x)$ and $B(x)$ are continuous on $[0, h]$ and satisfy $A(x) = B(x)$, then it immediately follows from equation $V = \int_a^b A(x)dx$ that $V_A = V_B$.

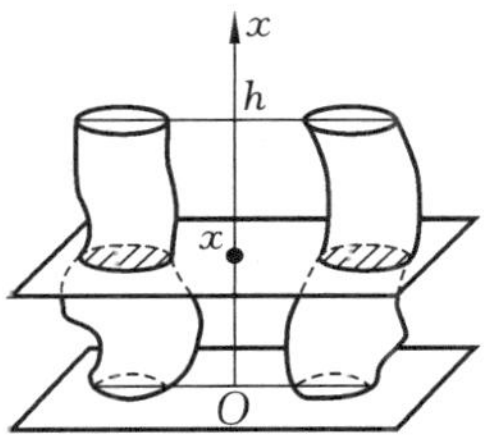

Figure 5-25

It was not until the 17th century that the Italian mathematician Cavalieri proposed a conclusion similar to Zu Geng's theorem, but more than a thousand years later than Zu Geng.

Example 5.6.6 Suppose the base radius of a cylinder is 3, and a plane intersects the cylinder at an angle α to the base, cutting off a portion of the cylinder to form a wedge. Find the volume of the wedge.

Solution Method 1: The shape of the solid is shown in Figure 5-26.

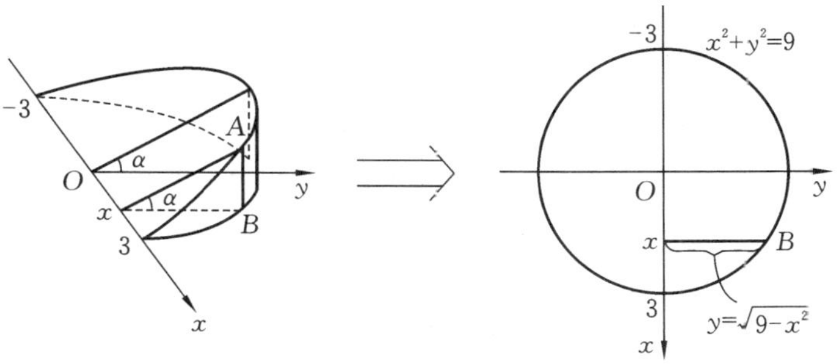

Figure 5-26

Take the intersection line of this plane with the bottom surface of the cylinder as the x axis, and the line on the bottom surface passing through the center and perpendicular to the x axis as the y axis. Then, the equation of the bottom circle is $x^2 + y^2 = 9$.

$\forall x \in (-3,3)$, passing through point x, make a plane perpendicular to the x axis to intercept the solid, the cross-section is a right angle ΔAxB, $\angle AxB=\alpha$, ΔAxB with an area of $A(x)=\frac{1}{2}|AB||Bx|$. Since

$$|AB|=|Bx|\tan\alpha,\ |Bx|=\sqrt{9-x},$$

therefore

$$A(x)=\frac{1}{2}|Bx|^2\tan\alpha=\frac{1}{2}(9-x^2)\tan\alpha.$$

Thus, the volume of the wedge-shaped solid is obtained as

$$V=\int_{-3}^{3}A(x)\mathrm{d}x=\frac{1}{2}\tan\alpha\int_{-3}^{3}(9-x^2)\,\mathrm{d}x=\tan\alpha\int_{0}^{3}(9-x^2)\,\mathrm{d}x=18\tan\alpha.$$

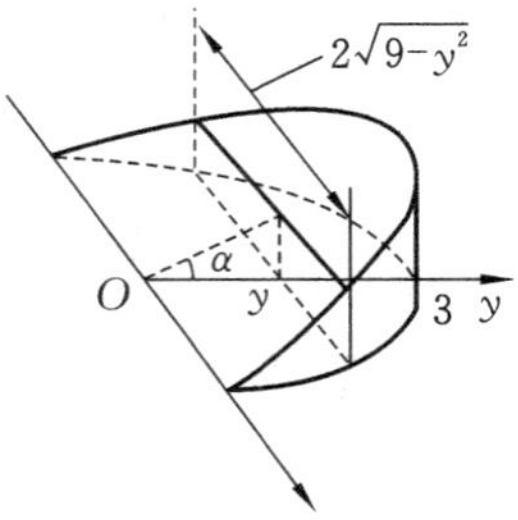

Figure 5-27

Method 2: If y is taken as the integral variable, then the range of y is $[0,3]$, $\forall y \in (0,3)$. Passing through point y, make a plane perpendicular to the y axis to intercept the solid, the cross-section is a rectangle (Figure 5-27), with an area of

$$B(y)=2\sqrt{9-y^2}\cdot y\tan\alpha.$$

Thus, the volume sought is

$$V=\int_{0}^{3}B(y)\mathrm{d}y=\tan\alpha\int_{0}^{3}\sqrt{9-y^2}\,2y\mathrm{d}y=\tan\alpha\int_{0}^{3}(9-y^2)^{\frac{1}{2}}\mathrm{d}y^2$$

$$=-\tan\alpha\left[\frac{2}{3}(9-y^2)^{\frac{3}{2}}\right]_0^3=18\tan\alpha.$$

Exercise 5.6

1. Rotate the region bounded by the parabola $y^2=4ax$ and the line $x=x_0\ (x_0>0)$ around the x axis, calculate the volume of the resulting rotating paraboloid.

2. For the region bounded by $y=x^3$, $x=2$, $y=0$, rotate it around the x axis and the y axis respectively, calculate the volumes of the two resulting solids of revolution.

3. Rotate the figure enclosed by the astroid $x^{2/3}+y^{2/3}=a^{2/3}$ around the x axis, and calculate the volume of the resulting solid of revolution.

4. Find the volume of the solid of revolution generated by rotating the figure enclosed by the given curves around the specified axis.

(1) $y=x^2$, $x=y^2$, around the y axis;

(2) $x^2+(y-5)^2=16$, around the x axis.

5. Find the volume of the solid of revolution formed by rotating the disk $x^2+y^2\leqslant a^2$ around $x=-b\ (b>a>0)$.

6. There is a solid whose base is the region enclosed by the curve $y^2=x$ and the line $y=x$ in the xOy plane, and each cross-section perpendicular to the x axis is a semicircle with its diameter on the base. Find the volume of this solid.

7. There is a solid whose base is a circle with radius R, and all cross-sections perpendicular to a fixed diameter of the base are equilateral triangles. Find the volume of this solid.

8. Prove that the volume of the solid of revolution formed by rotating the planar figure $0 \leqslant a \leqslant x \leqslant b, 0 \leqslant y \leqslant f(x)$ around the y axis is

$$V = 2\pi \int_a^b x f(x) \mathrm{d}x.$$

9. Using the conclusion from previous exercise, calculate the volume of the solid obtained by rotating the region enclosed by the curve $y = \sin x$ $(0 \leqslant x \leqslant \pi)$ and the x axis around the y axis.

10. Rotate the curve $y = \frac{\sqrt{x}}{1+x^2}$ around the x axis to form a solid of revolution.

(1) Find the volume V_∞ of this solid of revolution;

(2) Denote the volume of this solid between $x=0$ and $x=a$ as $V(a)$. For what value of a does the following hold:

$$V(a) = \frac{V_\infty}{2}.$$

11. (1) Find the volume $V(\xi)$ of the solid of revolution formed by rotating the plane region enclosed by the curve $y = \mathrm{e}^{-x}$ $(x>0)$, the line $x = \xi$ $(\xi>0)$, and the x and y axes around the x axis, and find the value of a that satisfies $\lim\limits_{\xi \to +\infty} V(\xi) = 2V(a)$.

(2) Find a point on this curve, such that the tangent at that point forms a plane region with the x and y axes that has the maximum area.

5.7 Applications of Definite Integrals in Physics

The applications of definite integrals are extensive and span various fields. The previous sections introduced some applications of definite integrals in geometry. This section further discusses their applications in physics.

5.7.1 Work

From physics, it is known that if a constant force F acts on an object, causing it to move a distance s in the direction of the force, the work done by the force F on the object is

$$W = FS.$$

If the force F acting on the object is not constant but varies in magnitude, this becomes a problem of work done by a variable force on the object. The following example illustrates this.

Example 5.7.1 Place a point charge with $+q$ charge at the origin O on the r axis, which generates an electric field. This electric field exerts a force on surrounding charges. According to physics, if a unit positive charge is placed in this electric field at a distance r from the origin O, the magnitude of the force exerted on it by the electric field is

$$F=k\frac{q}{r^2}\quad (k \text{ is a constant}).$$

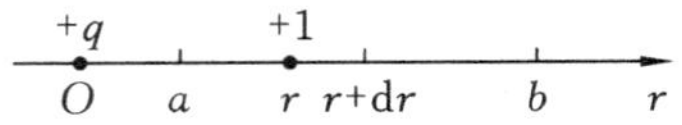

Figure 5-28

When this unit positive charge moves from $r=a$ along the r axis to $r=b$ $(a<b)$ in the electric field, calculate the work done by the electric force F on it (Figure 5-28).

Solution When a unit positive charge moves, its distance from the origin changes, and thus the magnitude of the electric field's force F on this unit positive charge varies. Taking r as the integral variable, its range of variation is $[a,b]$, and within this range, any small interval $[r,r+\mathrm{d}r]$ can be considered, where the electric field force can be approximated as a constant force. When the unit positive charge moves from r to $r+\mathrm{d}r$, the work done by the electric field force is approximately equal to $\frac{kq}{r^2}\mathrm{d}r$, thus obtaining the elemental work:

$$\mathrm{d}W=\frac{kq}{r^2}\mathrm{d}r.$$

Therefore, the work required is

$$W=\int_a^b \frac{kq}{r^2}\mathrm{d}r=kq\left[-\frac{1}{r}\right]_a^b=kq\left(\frac{1}{a}-\frac{1}{b}\right).$$

When calculating the electric potential at a certain point in the electrostatic field, one must consider the work done by the electric field force W in moving the unit positive charge from that point $(r=a)$ to infinity. At this point, we have

$$W=\int_a^{+\infty} \frac{kq}{r^2}\mathrm{d}r=\left[-\frac{kq}{r}\right]_a^{+\infty}=\frac{kq}{a}.$$

Example 5.7.2 A cylindrical water reservoir has a height of 10 m and a base radius of 6 m. The reservoir is filled with water. How much work is required to pump all the water out of the reservoir?

Solution As shown in Figure 5-29, an axis x is drawn, and the depth x is taken as the integral variable, with its range of variation being $[0,10]$. On $[0,10]$, any small interval $[x,x+\mathrm{d}x]$ is taken, corresponding to a thin layer of water with a height of $\mathrm{d}x$ (m) and a volume of $\mathrm{d}V=\pi\, 6^2\,\mathrm{d}x$ (m^3). Let the density of water be $\rho=1\ \mathrm{t/m^3}$, and the gravitational acceleration be $g=9.8\ \mathrm{m/s^2}$, then the weight of this thin layer of water is

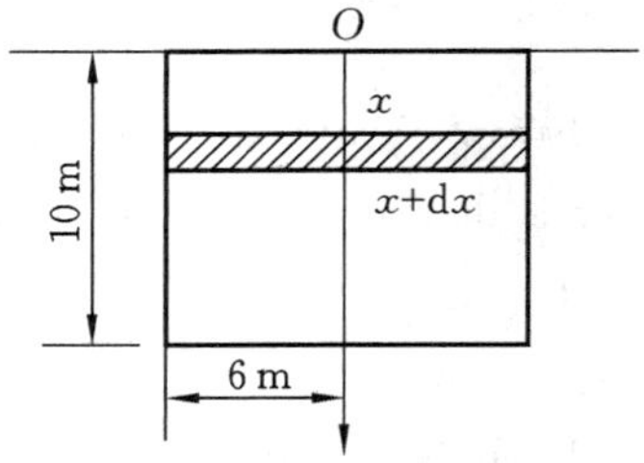

Figure 5-29

$$\rho g\,\mathrm{d}V = 1\times 9.8\times 36\pi\mathrm{d}x = 352.8\pi\mathrm{d}x\ (\mathrm{kN}).$$

Therefore, the work required to pump this thin layer of water out of the pool is approximately

$$\mathrm{d}W = 352.8\pi x\,\mathrm{d}x\ (\mathrm{kJ}).$$

This is the work element. Thus, the total work required is

$$W = \int_0^{10} 352.8\pi x\,\mathrm{d}x = 352.8\pi\cdot 50 \approx 55390\,(\mathrm{kJ}).$$

Note: In this problem, the force required to pump each thin layer of water out of the pool remains constant, so it is not a problem of variable force doing work. However, the distance moved by water at different depths varies, so a definite integral is also used to calculate it.

5.7.2 Pressure

Example 5.7.3 A horizontal cylindrical barrel is half-filled with water. Let the bottom radius of the barrel be R, and the density of the water be ρ. Calculate the pressure on one end of the barrel.

Analysis From physics, it is known that the pressure at a depth of h in water is $p=\rho gh$, where ρ is the density of the water and g is the gravitational acceleration. If a flat plate with an area of A is placed horizontally at a depth of h, the water pressure on one side of the plate is

$$P = pA.$$

If a flat plate is placed vertically in water, the pressure p at different depths is not equal, so the water pressure on one side of the plate cannot be calculated using the above method. In this problem, the end face of the bucket is also placed vertically. Below, we will solve it by using the method of elements.

Solution One end face of the bucket is a disc (Figure 5-30). Now, we need to calculate the water pressure on one side of a vertically placed semicircular disc when the water level passes through the center of the disc.

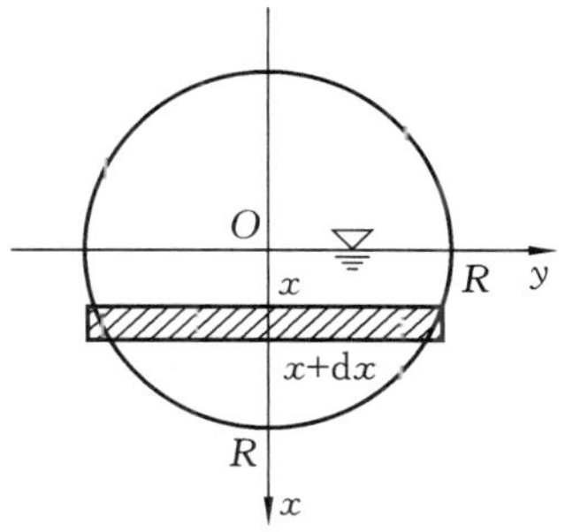

Figure 5-30

Take the straight line passing through the center and pointing vertically downward as the x axis, and the horizontal line passing through the center as the y axis. Then the equation of the semicircle is $x^2+y^2=R^2$ $(0\leqslant x\leqslant R)$. Take x as the integral variable, $x\in[0,R]$. On $[0,R]$, take any small interval $[x, x+\mathrm{d}x]$, the pressure at each point on the narrow strip corresponding to this interval is approximately ρgx, and the area of the narrow strip is approximately $2\sqrt{R^2-x^2}\,\mathrm{d}x$. Therefore, the approximate value of the water pressure on one side of the narrow strip, i.e., the pressure element, is

$$dP = 2\rho g x\sqrt{R^2 - x^2}\,dx.$$

Thus, the pressure sought is

$$P = \int_0^R 2\rho g x\sqrt{R^2 - x^2}\,dx = -\rho g\int_0^R (R^2 - x^2)^{\frac{1}{2}}d(R^2 - x^2)$$

$$= -\rho g\left[\frac{2}{3}(R^2 - x^2)^{\frac{3}{2}}\right]_0^R = \frac{2\rho g}{3}R^3.$$

Example 5.7.4 A circular plate with radius R, where the load on each point is $p = \ln(1+r)$ (the force per unit area, perpendicularly directed towards the plate), where r is the distance from any point on the plate to the center, find the total load on the plate.

Solution Since the load on each point on the same circumference with radius r is the same, the plate is divided into many thin rings by a series of concentric circles. Thus, the load on each point of a thin ring can be approximately considered constant.

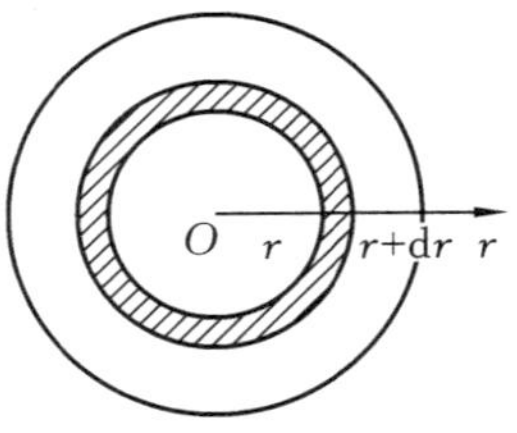

Figure 5-31

Construct the r axis as shown in Figure 5-31. Take r as the integral variable, $r \in [0, R]$. On $[0, R]$, take a small interval $[r, r+dr]$, the approximate load on each point of the corresponding thin ring is $p = \ln(1+r)$, and the area of the ring is approximately $dA = 2\pi r\,dr$. Therefore, the approximate load on the ring, i.e., the element of P, is

$$dP = \ln(1+r)\,dA = 2\pi r\ln(1+r)\,dr.$$

Thus, the total load on the plate is

$$P = \int_0^R 2\pi r\ln(1+r)\,dr = \int_0^R \pi\ln(1+r)\,dr^2 = \pi\left[r^2\ln(1+r)\Big|_0^R - \int_0^R \frac{r^2}{1+r}dr\right]$$

$$= \pi\left\{R^2\ln(1+R) - \left[\frac{r^2}{2} - r + \ln(1+r)\right]_0^R\right\} = \pi\left[(R^2 - 1)\ln(1+R) - \frac{R^2}{2} + R\right].$$

5.7.3 Force

Example 5.7.5 Suppose there is a uniform thin rod with length l and mass M. There is also a particle with mass m located on the same line as the rod, with a distance of a from the near end of the rod. Find the gravitational force between the rod and the particle.

Solution Denote the particle with mass m as A. If the rod is divided into many small segments, each segment can be approximated as a particle. Then, the gravitational force between each segment and the particle A can be calculated using the gravitational formula. Since the gravitational forces from all segments are in the same direction, they can be summed.

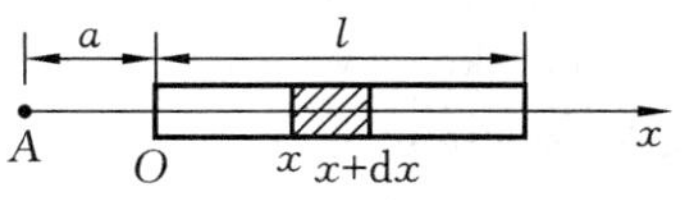

Figure 5-32

Construct the x axis as shown in Figure 5-32. Take x as the integration variable, $x \in [0, l]$. For any small interval $[x, x+dx]$ on the rod, the corresponding mass is $\frac{M}{l}dx$, and the distance between this segment and the particle A is approximately $x+a$. Thus, the approximate value of the gravitational force between this segment and the particle A, i.e., the element of the force F, is

$$\mathrm{d}F=G\frac{m\cdot\frac{M}{l}\mathrm{d}x}{(x+a)^2}=\frac{GmM}{l(x+a)^2}\mathrm{d}x.$$

Therefore, the gravitational force between the rod and the particle A is

$$F=\frac{GmM}{l}\int_0^l\frac{1}{(x+a)^2}\mathrm{d}x=\frac{GmM}{l}\left[-\frac{1}{x+a}\right]_0^l=\frac{GmM}{a(a+l)}.$$

Note: If the slender rod is placed vertically, the particle is located on the vertical bisector of the rod, and the distance from the rod is a. In this case, the gravitational forces of each small segment of the rod on the particle have different directions. In this situation, directly summing the gravitational forces is inconvenient, so they are decomposed into horizontal and vertical components, and then summed separately in the horizontal and vertical directions. How to find the horizontal and vertical components of the gravitational force of the rod on the particle is left for the reader to consider.

Example 5.7.6 Suppose there is a semicircular slender rod with a radius of R and a central angle of φ, and its linear density is a constant ρ. There is a particle with mass m at the center of the circle M. Find the gravitational force of this rod on the particle M.

Solution Establish the coordinate system as shown in Figure 5-33, with the mass point M as the origin, and the line connecting point M to the midpoint of the arc-shaped thin rod as the x axis. Take θ as the integral variable, $\theta\in\left[-\frac{\varphi}{2},\frac{\varphi}{2}\right]$, and on $\left[-\frac{\varphi}{2},\frac{\varphi}{2}\right]$, take a small interval $[\theta,\theta+\mathrm{d}\theta]$. The corresponding small segment of the arc-shaped thin rod in this interval can be regarded as a mass point, with its mass being $\rho R\,\mathrm{d}\theta$. Therefore, the gravitational force of this small segment on the mass point M is approximately

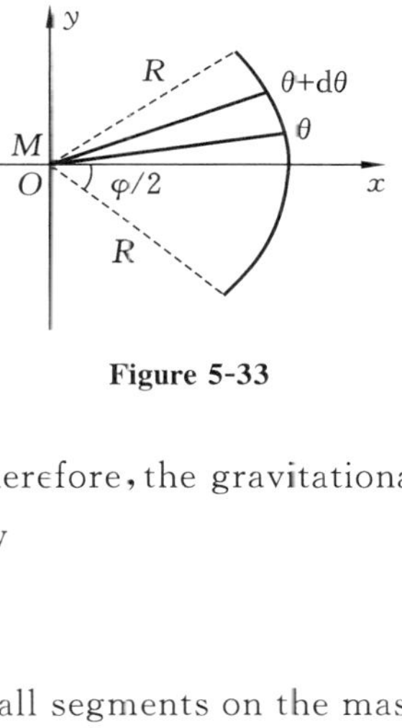

Figure 5-33

$$G\frac{m\rho R\,\mathrm{d}\theta}{R^2}=\frac{G\rho m}{R}\mathrm{d}\theta.$$

Since the directions of the gravitational forces of different small segments on the mass point M are different, they cannot be directly added together. As pointed out in the note of Example 5.7.5: They must be decomposed into horizontal and vertical components before they can be added together in the horizontal and vertical directions. The approximate value of the gravitational force of the small segment on the mass point M in the horizontal direction is

$$\frac{G\rho m}{R}\cos\theta\,\mathrm{d}\theta.$$

This is the element of the gravitational force of the arc-shaped thin rod on the mass point M in the horizontal direction F_x, that is,

$$dF_x = \frac{G\rho m}{R}\cos\theta\, d\theta.$$

Thus, the gravitational force in the horizontal direction is obtained as

$$F_x = \int_{-\frac{\varphi}{2}}^{\frac{\varphi}{2}} \frac{G\rho m}{R}\cos\theta\, d\theta = \frac{2G\rho m}{R}\sin\frac{\varphi}{2}.$$

By symmetry, the gravitational force in the vertical direction is $F_y = 0$.

Therefore, the gravitational force exerted by the slender rod on the particle M is $\frac{2G\rho m}{R}\sin\frac{\varphi}{2}$ in magnitude, and its direction is from M towards the midpoint of the arc.

Exercise 5.7

1. Experiments show that the force F (unit: N) required to stretch a spring is proportional to the elongation s (unit: cm), i.e.,

$$F = ks \quad (\text{where } k \text{ is the proportional constant}).$$

If the spring is stretched from its original length by 6 cm, calculate the work done.

2. A cylinder with a diameter of 20 cm and a height of 80 cm is filled with steam at a pressure of 10 N/cm^2. Assuming the temperature remains constant, how much work is required to reduce the volume of the steam by half?

3. A hammer is used to drive a nail into a wooden board. The resistance of the board to the nail is proportional to the depth the nail penetrates. During the first strike, the nail is driven into the board by 1 cm. If the work done by the hammer on each strike is equal, how much further will the nail penetrate during the second strike?

4. Consider a conical reservoir with a depth of 15 m and a diameter of 20 m, filled with water. How much work is required to pump all the water out using a pump?

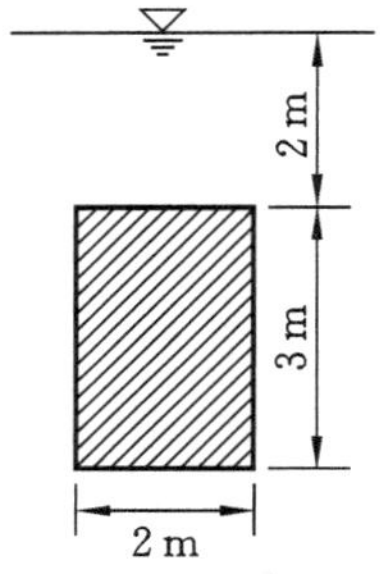

Figure 5-34

5. There is a gate with its shape and dimensions as shown in Figure 5-34. The water level is 2 meters above the top of the gate. Calculate the water pressure exerted on the gate.

6. An isosceles triangular thin plate is submerged vertically in water, with its base level with the water surface. The height of the plate is h and the base is a.

(1) Calculate the pressure on one side of the plate.

(2) If the plate is inverted, so that the vertex is level with the water surface and the base is parallel to the water surface. How much does the pressure on one side of the plate increase?

7. There is an isosceles trapezoidal gate with bottom sides of lengths 10 m and 6 m, and a height of 20 m. The longer bottom side is level with the water surface. Calculate the water pressure on one side of the gate.

8. Consider a uniform thin rod of length l and linear density ρ, with a mass point of mass m located at a perpendicular distance of a units from one end of the rod M. Find the gravitational force exerted by the rod on the mass point M.

Comprehensive Review 5

1. Calculate the following definite integrals:

(1) $\int_0^{\frac{\pi}{6}} \frac{\sin\theta}{\cos^3\theta} d\theta$; (2) $\int_0^{\frac{\pi}{4}} (\cos 2x + \sin 5x)\, dx$;

(3) $\int_1^4 \frac{dt}{\sqrt{t}\,(1+\sqrt{t})^3}$; (4) $\int_1^2 \frac{\left(1+\frac{1}{t}\right)^2}{t^2} dt$;

(5) $\int_{-\frac{\pi}{2}}^{\frac{\pi}{2}} \cos\theta \cos(\pi\sin\theta)\, d\theta$; (6) $\int_{-\frac{\pi}{4}}^{\frac{\pi}{4}} (|x| \sin^5 x + x^2 \tan x)\, dx$;

(7) $\int_0^{4\pi} |\sin 2x|\, dx$; (8) $\int_0^1 x^2 e^{-x}\, dx$;

(9) $\int_1^2 \frac{5^{\frac{1}{x}}}{x^2} dx$; (10) $\int_0^{8\pi} e^{-x} \sin x\, dx$.

2. Find the maximum and minimum values of the function $\int_0^x \frac{2t-1}{t^2-t+1} dt$ on the interval $[0,2]$.

3. Calculate the following improper integrals:

(1) $\int_{-\infty}^1 e^{3x}\, dx$; (2) $\int_1^{+\infty} x e^{-x}\, dx$;

(3) $\int_{\frac{\pi}{3}}^{\frac{\pi}{2}} \frac{\tan x\, dx}{[\ln(\cos x)]^2}$; (4) $\int_{-3}^3 \frac{x}{\sqrt{9-x^2}} dx$.

4. Fill in the blanks.

(1) If $f(x)$ is a continuous function, then $\int_{-1}^1 x f(x^2)\, dx =$ ________.

(2) $\lim\limits_{x\to 0+} \frac{\int_0^x \sqrt{t}\cos t\, dt}{x^2} =$ ________.

(3) The average value of $f(x) = \frac{x^2}{\sqrt{1-x^2}}$ over the interval $\left[\frac{1}{2}, \frac{\sqrt{3}}{2}\right]$ is ________.

(4) If $\lim\limits_{b\to+\infty} \int_a^b f(x) dx$ exists, it is called $\int_a^{+\infty} f(x) dx$ ________.

5. Find the area of the region enclosed by the following curves:

(1) Enclosed by the curve $y^2 = (4-x)^3$ and the vertical axis;

(2) A straight line $y=x$ divides the ellipse $x^2+3y^2=6y$ into two parts, with the smaller area being A and the larger area being B. Find A/B.

6. Find the volumes of the solids of revolution obtained by rotating the region enclosed by the curve $y = e^x$ $(x \leqslant 0)$, $x = 0$, $y = 0$ about the x axis and the y axis, respectively.

7. Establish a recurrence formula for $I_{2n} = \int_0^{\frac{\pi}{4}} \tan^{2n} x \, dx$ and calculate $\int_0^{\frac{\pi}{4}} \tan^6 x \, dx$.

8. Find $\lim\limits_{x \to 0} \dfrac{\int_0^{\sin^2 x} \ln(1+t)\, dt}{\sqrt{1+x^4} - 1}$.

9. Prove that the equation $\int_0^x \sqrt{1+t^4}\, dt + \int_{\cos x}^0 e^{-t^2}\, dt = 0$ has exactly one real root.

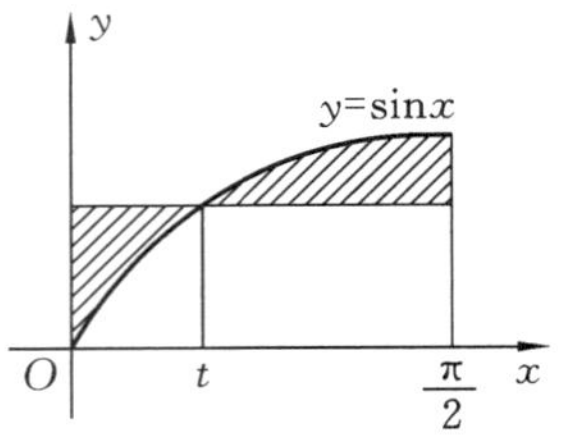

Figure 5-35

10. Given $f(x) = \int_x^1 e^{-y^2} dy$, calculate $I = \int_0^1 x^2 f(x) dx$.

11. Find the maximum and minimum areas of the shaded region in Figure 5-35.

12. Given $f(x) > 0$, and $f'(x)$ is continuous, $x \in \mathbf{R}$, $g(x) = \dfrac{\int_0^x t f(t) dt}{\int_0^x f(t) dt}$, when $x \neq 0$ and $g(0) = 0$.

(1) Find $g'(x)$;

(2) Prove that $g(x)$ is monotonically increasing.

13. If the area enclosed by the curve $y = \cos x$ $\left(0 \leqslant x \leqslant \dfrac{\pi}{2}\right)$ and the x axis, y axis is divided into three equal parts by $y = a\sin x$, $y = b\sin x$ $(a > b > 0)$, find a and b.

14. Suppose $f(x)$ is differentiable on $[0, +\infty)$, $f(0) = 0$, and its inverse function is $g(x)$. If $\int_0^{f(x)} g(t) dt = x^2 e^x$, find $f(x)$.

15. Given that $f(x)$ satisfies the equation $f(x) = 3x - \sqrt{1-x^2} \int_0^1 f^2(x) dx$, find $f(x)$.

16. Let the function $f(x)$ be continuous on $[a, b]$, and let $f(x)$ take the same value at points symmetric with respect to $x = \dfrac{a+b}{2}$. Prove:

$$\int_a^b f(x) dx = 2\int_a^{\frac{a+b}{2}} f(x) dx.$$

17. Let $\varphi(x)$ be the inverse function of the differentiable function $y = f(x)$, and let $f(1) = 0$. Prove:

$$\int_0^1 \left[\int_0^{f(x)} \varphi(t) dt\right] dx = 2\int_0^1 x f(x) dx.$$

18. Let $f(x)=\begin{cases}\dfrac{1}{1+x}, & x\geqslant 0\\ \dfrac{1}{1+e^x}, & x<0\end{cases}$, find $\int_0^2 f(x-1)\,dx$.

19. Let $f(x)$ be continuous on the interval $[a,b]$, and let $g(x)$ be continuous and non-changing in sign on the interval $[a,b]$. Prove that: there exists at least one point $\xi\in[a,b]$, such that

$$\int_a^b f(x)g(x)\,dx=f(\xi)\int_a^b g(x)\,dx$$

holds.

20. If the function $f(x)$ has a continuous derivative on the closed interval $[2,4]$, and $f(2)=f(4)=0$. Prove:

$$\left|\int_2^4 f(x)\,dx\right|\leqslant\max_{2\leqslant x\leqslant 4}|f'(x)|.$$

21. Let the parabola $y=ax^2+bx+c$ pass through the point $(0,0)$, and let $y\geqslant 0$ when $x\in[0,1]$. Determine the value of a,b,c, such that the area enclosed by the parabola $y=ax^2+bx+c$ and the line $x=1, y=0$ is $\dfrac{4}{9}$, and the volume of the solid of revolution formed by rotating this area around the x axis is minimized.

22. A rectangular thin plate with sides of length a and b is submerged at an angle of α to the liquid surface, with the longer side parallel to the surface and located at a depth of h. Given that $a>b$, the density of the liquid is ρ. Calculate the pressure on each side of the plate.

23. Suppose the linear density at each point on the astroid $x=a\cos^3 t, y=a\sin^3 t$ is equal to the cube of the distance from that point to the origin. There is a unit mass at the origin O. Find the gravitational force exerted by the arc segment of the astroid in the first quadrant on this mass.

Chapter 6 Differential Equations

(1) The Dao of Yin-Yang Transformation (Background of Differential Equations)

The *I Ching* emphasizes the mutual transformation and balance between Yin and Yang, proposing that all things in the world exist in a state of perpetual change, always tending toward a dynamic equilibrium of Yin and Yang. This transformation is not merely static balance but manifests in dynamic processes. Differential equations describe the dynamic relationships between independent and dependent variables, often involving derivatives of the dependent variable with respect to the independent variable. These derivatives can be interpreted as the instantaneous rate of change—akin to the momentary equilibrium between opposing forces (e.g., growth and decay of a function). Differential equations formalize how this equilibrium evolves with the independent variable, embodying the "Dao of Yin-Yang Transformation".

(2) The Art of Dynamic Equilibrium (Core of Differential Equations)

The concept of "dynamic equilibrium" in the *I Ching* reflects how systems achieve new states of balance through continuous adjustments between Yin and Yang. This equilibrium is not static but adapts to evolving conditions. At the heart of differential equations lies the quest to solve functions that satisfy specific dynamic equilibrium conditions. These conditions, defined by the equations themselves, describe the relationship between a function's instantaneous rate of change (derivative) and its value. Solving a differential equation reveals the function that adheres to this dynamic equilibrium, mirroring how systems evolve toward balance. For example, consider a swinging pendulum: The differential equation governs how its velocity (derivative of position) relates to its position at each moment. Solving it unveils the pendulum's position over time—its dynamic equilibrium state.

In the process of solving practical problems in production and science, it is necessary to establish mathematical models to identify the relationships between relevant variables. For mathematical models related to the laws of continuous quantity changes, it is often necessary to analyze the problem and establish a relationship that satisfies an unknown function and its derivative (or differential). This relationship is known as a differential equation. After establishing the differential equation, studying it to find the unknown function is known as solving the differential equation.

Differential equations have developed alongside calculus and serve as an important bridge for connecting mathematics with practical applications. They are a powerful tool for scientific research across various disciplines. Differential equations are not only applied in geometry, mechanics, and physics, but their applications are also expanding and penetrating into various fields such as machinery, telecommunications, chemical engineering, biology, economics, and other social sciences. This chapter primarily introduces some basic concepts of differential equations and several common methods for solving them.

6.1 Basic Concepts of Differential Equations

Equations are quite familiar to us, and they must satisfy two conditions: unknowns and equality relationships. In elementary mathematics, there are various types of equations, such as linear equations, quadratic equations, exponential equations, logarithmic equations, trigonometric equations, and systems of equations. These equations aim to identify the relationships between known and unknown quantities in the problems being studied, listing one or more equations involving one or several unknowns, and then solving these equations. For example,

$$x^2+3x+2=0 \quad \text{(quadratic equation in one variable)},$$

$$\sin x+\cos x=1 \quad \text{(trigonometric function equation)},$$

$$\frac{1}{x+1}-\frac{4x}{x^2-1}=2+x \quad \text{(rational equation)},$$

$$2^{x-1}+2^{x-2}+x^{x-7}=14 \quad \text{(exponential equation)}.$$

Their unknowns are all x, the above equation can be written as:

$$F(x)=0.$$

For example, an equation where the unknown is a function: $\forall x, y\in \mathbf{R}$, if

$$f(x+y)=f(x)f(y).$$

Find the function $f(x)$. It is evident that $f(x)=a^x$ $(a>0, a\neq 1)$ is a solution to the equation.

However, in practical work, problems often arise that are completely different in nature from the above equations.

Example 6.1.1 A curve passes through the point $(1,2)$ and the slope of the tangent line at any point $M(x,y)$ on the curve is $3x^2$. Find the equation of this curve.

Solution Let the equation of the desired curve be $y=y(x)$. According to the geometric meaning of the derivative, it is known that the unknown function $y=y(x)$ should satisfy the relationship

$$\frac{dy}{dx}=3x^2.$$

The above equation is an equation containing the derivative of the unknown function. In addition, the unknown function $y=y(x)$ should also satisfy the following conditions:

When $x=1, y=2$, integrate both sides of equation,

$$y=\int 3x^2\,dx,$$

to obtain

$$y=x^3+C,$$

where: C is an arbitrary constant.

Substitute the condition "$x=1$ when $y=2$" into equation, we get $2=1^3+C$, from which $C=1$ is determined. Substituting $C=1$ into equation, we obtain the desired curve equation (referred to as the solution satisfying the condition $y|_{x=1}=2$ of the differential equation):

$$y=x^3+1.$$

Example 6.1.2 An object is thrown vertically upward from a height of s_0 with an initial velocity of v_0. Assuming that the object's motion is only affected by gravity, find the functional relationship between the object's path s and time t.

Solution Let the required functional relationship be $s=s(t)$. According to the problem statement and the mechanical significance of the second derivative, $s=s(t)$ should satisfy the following relationship:

$$\frac{\mathrm{d}^2 s}{\mathrm{d}t^2}=-g.$$

The above equation is an equation containing the second derivative of the unknown function. Additionally, $s=s(t)$ also satisfies the conditions

$$s|_{t=0}=s_0,$$

$$\left.\frac{\mathrm{d}s}{\mathrm{d}t}\right|_{t=0}=v_0.$$

Integrating both sides of equation, we get

$$v=\frac{\mathrm{d}s}{\mathrm{d}t}=-gt+C_1.$$

Integrating once more, we get

$$s=-\frac{1}{2}gt^2+C_1t+C_2,$$

where: C_1, C_2 is an arbitrary constant.

Substituting conditions into these equations, we get $C_2=s_0$ and $C_1=v_0$. Thus, the functional relationship between the path s and time t is obtained as

$$s=-\frac{1}{2}gt^2+v_0t+s_0.$$

Based on the analysis of the two examples above, we provide the definition of a differential equation.

Definition 6.1.1 An equation containing the independent variable, the unknown function of the independent variable, and the derivative or differential of the unknown function is called a differential equation.

It must be noted that in differential equations, the independent variable and the unknown function may not appear, but the derivative or differential of the unknown function must appear, and it can also be of higher order. A differential equation where the unknown function is a single-variable function is called an ordinary differential equation. A differential equation where the unknown function is a multi-variable function is called a partial differential equation. This chapter only discusses ordinary differential equations. For convenience, ordinary differential equations mentioned below will be referred to simply as differential equations (or equations).

For example, $y'=2x$, $(y-2xy)\,dx+x^3\,dy=0$, $\frac{d^2y}{dx^2}=xy'''+3y'+4y=0$, are all differential equations, and $\frac{\partial z}{\partial x}=xy$ is a partial differential equation.

The order of the highest derivative of the unknown function that appears in the differential equation is called the order of the differential equation. Differential equations of second order and higher are called higher-order differential equations.

The general form of a first-order differential equation is

$$F(x,y,y')=0 \quad \text{or} \quad y'=f(x,y).$$

The general form of an n-order differential equation is

$$F(x,y,y',\cdots,y^{(n)})=0 \quad \text{or} \quad y^{(n)}=f(x,y,y',\cdots,y^{(n-1)}),$$

where: x is the independent variable, y is the unknown function, F is a function of $n+2$ variables, and it must contain the variable $y^{(n)}$, while the variable x, y, y', $\cdots$, $y^{(n-1)}$ may not appear.

A function that satisfies the differential equation (substituting the function into the differential equation makes it an identity) is called a solution to the differential equation. Specifically, let the function $y=\varphi(x)$ have n continuous derivatives on the interval I, and if on the interval I,

$$F[x,\varphi(x),\varphi'(x),\cdots,\varphi^{(n)}(x)]\equiv 0,$$

then the function $y=\varphi(x)$ is called a solution to the differential equation $F(x,y,y',\cdots,y^{(n)})=0$ on the interval I.

It is easy to verify that the functions:

$$y=x^2+1,\ y=x^2+2,\ y=x^2+6,\ y=x^2+C$$

are all solutions to equation $y'=2x$.

If the solution to a differential equation contains the same number of independent arbitrary constants as the order of the equation, it is called the general solution (or complete solution) of the differential equation. The function $y=x^2+C$ is the general solution to equation $y'=2x$.

The conditions that determine the values of the arbitrary constants in the general solution of a differential equation are called initial conditions. For example, when

$$x=x_0,\ y=y_0,\ y'=y_0',$$

are generally written as

$$y\big|_{x=x_0}=y_0, y'\big|_{x=x_0}=y'_0.$$

A solution obtained after the arbitrary constants in the general solution are determined is called a particular solution to the equation.

The problem of finding a particular solution of the differential equation $y'=f(x,y)$ that satisfies the initial condition $y\big|_{x=x_0}=y_0$, is called the initial value problem for a first-order differential equation. It is denoted as

$$\begin{cases} y'=f(x,y) \\ y\big|_{x=x_0}=y_0 \end{cases}.$$

Solving an initial value problem is to find the particular solution of the equation.

The initial value problem for a second-order differential equation can be denoted as

$$\begin{cases} f(x,y,y',y'')=0 \\ y\big|_{x=x_0}=y_0 \\ y'\big|_{x=x_0}=y'_0 \end{cases}.$$

In general, the graph of a particular solution of a differential equation is a curve, which is called the integral curve of the differential equation. The general solution is a family of integral curves. The initial problem $\begin{cases} y'=f(x,y) \\ y\big|_{x=x_0}=y_0 \end{cases}$, is the integral curve of the differential equation passing through the point (x_0, y_0).

Example 6.1.3 Verifies the equation

$$xy'=2y.$$

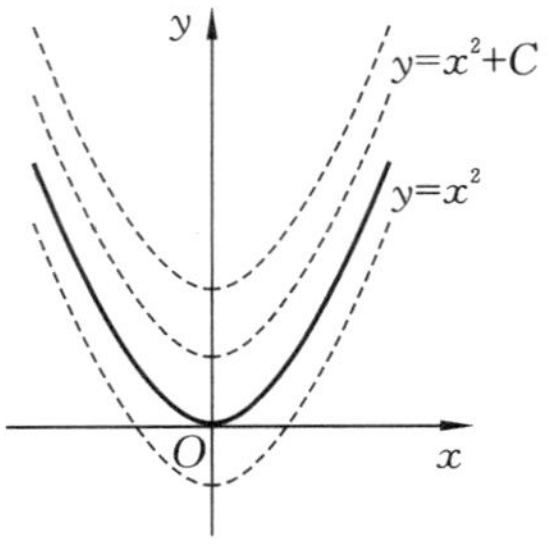

Figure 6-1

Solution Differentiating equation gives $y'=2Cx$, substituting y and y' into both sides of the original equation, we get

$$xy'=2Cx^2, 2y=2Cx^2.$$

Therefore, the function is the general solution of equation. Substituting the initial condition into equation, we get $C=1$, hence the particular solution is $y=x^2$, and its graph is shown in Figure 6-1.

Exercise 6.1

1. Indicate whether the function in each of the following questions is a solution to the given differential equation:

(1) $y=e^{-x^2}, \dfrac{dy}{dx}=-2xy$;

(2) $y=\arctan(x+y)+C, y'=\dfrac{1}{(x+y)^2}$;

(3) $y=xe^x, y''-2y'+y=0$;

(4) $\int_1^y e^{-\frac{t^2}{2}}dt+x+1=0, y'+e^{\frac{1}{2}y^2}=0$.

2. State the order of each of the following equations:

(1) $y=x(y')^2-2yy'+x=0$; (2) $(y'')^3+5(y')^4-y^5+x^7=0$;

(3) $xy'''+2y''+x^2y=0$; (4) $(x^2-y^2)dx+(x^2+y^2)dy=0$;

(5) $(7x-6y)dx+(x+y)dy=0$; (6) $(y''')^2-y^4=e^x$.

3. From the family of integral curves, find the curve that satisfies the given initial conditions:

(1) $y-x^3=C, y(0)=1$;

(2) $y=C_1e^x+C_2e^{-x}, y(0)=1, y'(0)=0$;

(3) $y=C_1\sin(x+C_2), y(\pi)=1, y'(\pi)=0$.

4. Find the particular solution of the following differential equations that satisfies the given initial conditions:

(1) $\frac{dy}{dx}=\sin x, y|_{x=0}=1$;

(2) $\frac{d^2y}{dx^2}=6x, y|_{x=0}=0, y'|_{x=0}=2$.

5. Write the differential equation and initial conditions satisfied by the curve determined by the following conditions:

(1) The slope of the tangent to the curve at any point is twice the abscissa of that point, and the curve passes through the point $(1,4)$;

(2) The curve passes through the point $(-1,1)$ and the abscissa of the point where the tangent at any point on the curve intersects the x axis is equal to the square of the abscissa of the tangent point.

6. Express a physical proposition with a differential equation: The rate of change of gas pressure P with respect to temperature T is proportional to the pressure and inversely proportional to the square of the temperature.

6.2 Separable Differential Equations and Homogeneous Equations

The most basic type of differential equation is the first-order differential equation, which generally takes the form

$$F(x,y,y')=0.$$

If y' can be solved from equation $F(x,y,y')=0$, the first-order differential equation can be expressed as

$$y'=f(x,y).$$

First-order differential equations can sometimes be written in the following symmetric form:

$$P(x,y)dx+Q(x,y)dy=0.$$

In this equation, the variables x and y are symmetric.

If x is regarded as the independent variable and y as the unknown function, then when $Q(x,y)\neq 0$, there is

$$\frac{dy}{dx}=-\frac{P(x,y)}{Q(x,y)}.$$

If y is regarded as the independent variable and x as the unknown function, then when $P(x,y)\neq 0$, there is

$$\frac{dx}{dy}=-\frac{Q(x,y)}{P(x,y)}.$$

There is no unified elementary method for solving general first-order differential equations. This section and the following sections of this chapter will introduce several types of solvable differential equations and their corresponding solution methods—elementary methods, which transform the problem of solving differential equations into an integration problem.

6.2.1 Separable Differential Equations

If the first-order differential equation is

$$\frac{dy}{dx}=f(x) \text{ or } dy=f(x)dx.$$

Then it suffices to integrate both sides of the equation to obtain

$$y=\int f(x)dx+C.$$

However, not all first-order differential equations can be solved in this manner, for example,

$$\frac{dy}{dx}=x^3y.$$

The method described above cannot be used because the right-hand side of the equation contains an unknown function, and the integral $\int x^3y\,dx$ cannot be evaluated. To address this difficulty, multiply both sides of the equation by $\frac{dx}{y}$, transforming the equation into $\frac{dy}{y}=x^3dx$. This way, the variables y and x are separated on either side of the equation, and integrating both sides yields

$$\int\frac{dy}{y}=\int x^3dx+C_1,$$

$$\ln|y|=\frac{1}{4}x^4+C_1,$$

$$y=Ce^{\frac{1}{4}x^4},$$

where: C is an arbitrary constant.

The reader can verify that the function is indeed a solution to the original differential equation.

In general, if a first-order differential equation can be written as

$$y'=f(x)g(y)$$

or

$$\psi(y)\mathrm{d}y=\varphi(x)\mathrm{d}x.$$

In the form, that is, if the differential equation can be written such that one side contains only functions of y and $\mathrm{d}y$, and the other side contains only functions of x and $\mathrm{d}x$, then the original equation is called a separable differential equation.

For solving equation $y'=f(x)g(y)$, first separate the variables by moving $f(x)$, $\mathrm{d}x$ and $g(y)$, $\mathrm{d}y$ to opposite sides of the equation:

$$\frac{\mathrm{d}y}{g(y)}=f(x)\mathrm{d}x.$$

Then integrate both sides to obtain

$$\int\frac{\mathrm{d}y}{g(y)}=\int f(x)\mathrm{d}x,$$

which gives the general solution of the differential equation:

$$G(y)=F(x)+C.$$

The implicit function determined by equation $G(y)=F(x)+C$ is the implicit general solution to the original equation.

Note: It is only possible to move terms when $g(y)\neq 0$, if $g(y)=0$, then we may assume $y=y_0$ is a zero point of $g(y)=0$, i.e., $g(y_0)=0$, substituting into the original equation shows that the constant function $y=y_0$ is obviously a particular solution to equation.

Example 6.2.1 Find the general solution to the differential equation $\frac{\mathrm{d}y}{\mathrm{d}x}=y\sin x$.

Solution Separate the variables in the equation to obtain

$$\frac{\mathrm{d}y}{y}=\sin x\,\mathrm{d}x,$$

integrate both sides to get

$$\int\frac{\mathrm{d}y}{y}=\int\sin x\,\mathrm{d}x.$$

Thus, we have

$$\ln|y|=\cos x+C_1 \quad \text{or} \quad |y|=\mathrm{e}^{\cos x+C_1}.$$

Therefore, $y=\pm\mathrm{e}^{C_1}\mathrm{e}^{\cos x}$, i.e., $y=C\mathrm{e}^{\cos x}$ (let $C=\pm\mathrm{e}^{C_1}$).

Hence, the general solution to the equation is $y=C\mathrm{e}^{\cos x}$ (where C is an arbitrary constant).

Note: In solving this differential equation, $y\neq 0$ or $y=0$ was not specified. Typically, no discussion is added, and it is assumed to be solved under meaningful conditions. In fact, $y=0$

is also a solution to the equation. Similar treatments can be applied to similar cases in the future.

Example 6.2.2 Find the solution to the equation $\frac{dy}{dx}=\frac{x-1}{y+1}$.

Solution After separating the variables, we get

$$(y+1)dy=(x-1)dx,$$

integrate both sides to obtain

$$\frac{1}{2}(y+1)^2=\frac{1}{2}(x-1)^2-\frac{1}{2}C.$$

Therefore, the solution is

$$(x-1)^2-(y+1)^2=C \quad (C \text{ is any constant number }).$$

Example 6.2.3 The rate of decay of uranium is proportional to the amount of undecayed atoms M at that time. It is known that when $t=0$, the amount of uranium is M_0. Find the law of variation of uranium content $M(t)$ with time t during the decay process.

Solution The rate of decay of uranium is the derivative of $M(t)$ with respect to time t, denoted as $\frac{dM}{dt}$. Since the rate of decay is proportional to its content, we obtain the differential equation

$$\frac{dM}{dt}=-\lambda M,$$

where: λ ($\lambda>0$) is a constant, the negative sign before λ indicates that M decreases monotonically as t increases. That is, $\frac{dM}{dt}<0$. According to the problem, the initial condition is $M|_{t=0}=M_0$.

Separating the variables in the equation gives

$$\frac{dM}{M}=-\lambda dt,$$

integrating both sides yields

$$\int\frac{dM}{M}=\int(-\lambda)dt.$$

We get $\ln M=-\lambda t+\ln C$, that is, $M=Ce^{-\lambda t}$. Using the initial condition, we get $M_0=Ce^0=C$, so the law of variation of uranium content $M(t)$ with time t is

$$M=M_0e^{-\lambda t}.$$

Example 6.2.4 (Population Problem) The total population N of a certain region is a function of time t, i.e., $N=N(t)$. If the birth rate of this region is n and the death rate is m, examine the total population $N(t)$ at any moment t.

Solution Treat $N(t)$ as a continuous and differentiable function (since the total population is large, it can be approximately treated this way, which is a discretization of discrete variables), within the time period $[t, t+\Delta t]$, the change in population is $\Delta N=$

$N(t+\Delta t)-N(t)$, which should be equal to the difference between the number of births and deaths during this period, i.e.,

$$\Delta N=N(t+\Delta t)-N(t)=nN(t)\Delta t-mN(t)\Delta t.$$

Divide both sides by Δt, and let $\Delta t \to 0$, we get

$$\frac{dN}{dt}=(n-m)N,$$

and assume the total population at time $t=0$ is N_0, thus

$$\begin{cases}\dfrac{dN}{dt}=(n-m)N \\ N|_{t=0}=N_0\end{cases}.$$

Separate the variables to get

$$\frac{dN}{N}=(n-m)dt.$$

Integrate both sides to get

$$\ln N=(n-m)t+\ln C.$$

That is $N=Ce^{(n-m)t}$. Substituting the initial condition into the above equation yields

$$N=N_0e^{(n-m)t}.$$

This equation indicates that when $n>m$, the population grows exponentially over time without bound. This is the famous Malthusian (Thomas Robert Malthus, 1766-1834, British) model. This model accurately reflected the total world population between 1700-1961. However, using the example of the U. S. population, when the results of the Malthusian model were compared with population data, significant discrepancies were found, especially when predicting the total future Earth population far into the future. According to this model, by 2670, there would be 36000 billion people on Earth. If the Earth's surface were entirely land (in fact, 80% of the Earth's surface is covered by water), we would have to stand shoulder to shoulder in two layers, which is absurd. In reality, constrained by factors such as food, energy, disease, and living space, population cannot grow indefinitely, and birth and death rates are not fixed constants. Based on statistical data, it is reasonable to assume that n and m are both linear functions of N (within a certain range of N), i.e.,

$$n=a-bN,\ m=p+qN,$$

where: a, b, p, q are all positive constants. Thus, the differential equation becomes

$$\frac{dN}{dt}=(a-p)N-(b+q)N^2=(b+q)N\left(\frac{a-p}{b+q}-N\right).$$

Let $k=b+q$, $C=\dfrac{a-p}{b+q}$, and we obtain a differential equation that more closely aligns with the actual population growth law:

$$\frac{dN}{dt}=kN(C-N).$$

Separating variables yields

$$\frac{dN}{N(C-N)}=k\,dt.$$

Integrating both sides and substituting the initial conditions gives

$$\frac{1}{C}\ln\frac{N(C-N_0)}{N_0(C-N)}=kt$$

or

$$N=\frac{CN_0}{N_0+(C-N_0)e^{-kt}}.$$

Equation is the logistic function commonly used in biological sciences. When used to calculate population growth in a region, it often remarkably matches actual data. Based on the total population statistics for 1949 and 1957, using this equation, it was predicted that the total population of China in 1978 would be 958 million, and in 2000, 1.247 billion.

Example 6.2.5 A certain object is placed in air, and at time $t=0$, its temperature is measured as $u_0=150$ ℃. After 10 minutes, the temperature is measured as $u_1=100$ ℃. Determine the relationship between the object's temperature and time, and calculate the object's temperature after 20 minutes. Assume the air temperature remains constant at $u_a=24$ ℃.

Solution Let the temperature of the object at time t be $u=u(t)$. According to Newton's law of cooling, we have

$$\begin{cases}\dfrac{du}{dt}=-k(u-u_a)\\ u\big|_{t=0}=u_0\end{cases},$$

where: k is a constant related to the object's medium. Separating variables gives

$$\frac{du}{u-u_a}=-k\,dt,$$

integrating both sides gives

$$\ln(u-u_a)=-kt+\ln C,$$

C is an arbitrary constant, i.e.,

$$u=u_a+Ce^{-kt}.$$

According to the initial conditions, when $t=0$, $u=u_0$, we obtain the constant $C=u_0-u_a$. Thus,

$$u=u_a+(u_0-u_a)e^{-kt}.$$

According to the condition, when $t=10$ minutes, $u=u_1$, we obtain

$$u_1=u_a+(u_0-u_a)e^{-10k},$$

$$k=\frac{1}{10}\ln\frac{u_0-u_a}{u_1-u_a}.$$

Substituting $u_0=150$, $u_1=100$, $u_a=24$ into the above equation, we get

$$k=\frac{1}{10}\ln\frac{150-24}{100-24}=\frac{1}{10}\ln 1.66\approx 0.051,$$

hence,

$$u=24+126e^{-0.051t}.$$

From the above equation, it is known that when $t=20$ minutes, the temperature of the object is $u_2 \approx 70$ ℃, and when $t \to +\infty$, $u \to 24$ ℃. This can explain a common knowledge: After a period of time, the temperature of the object and the air temperature will be almost the same. In fact, after 2 hours, the temperature of the object has become 24 ℃, which is quite close to the air temperature.

6.2.2 Homogeneous Equations

If a first-order differential equation

$$\frac{dy}{dx}=f(x,y),$$

the function $f(x,y)$ in the equation can be written as a function of $\frac{y}{x}$, i.e., $f(x,y)=\varphi\left(\frac{y}{x}\right)$, then the equation is called a homogeneous equation.

For example, $xy'-y-\sqrt{y^2-x^2}=0$ is a homogeneous equation, because

$$\frac{dy}{dx}=\frac{y+\sqrt{y^2-x^2}}{x}=\frac{y}{x}+\sqrt{\left(\frac{y}{x}\right)^2-1}.$$

Some equations are not separable variable equations, but can be transformed into separable variable equations after appropriate variable substitutions. Homogeneous equations are one such type.

For homogeneous equations

$$\frac{dy}{dx}=\varphi\left(\frac{y}{x}\right),$$

let $u=\frac{y}{x}$, that is, $y=ux$, we have

$$\frac{dy}{dx}=u+x\frac{du}{dx},$$

$$u+x\frac{du}{dx}=\varphi(u).$$

Separate the variables to get

$$\frac{du}{\varphi(u)-u}=\frac{dx}{x}.$$

Integrate both sides to get

$$\int\frac{du}{\varphi(u)-u}=\int\frac{dx}{x}.$$

Denote $G(u)$ as an antiderivative of $\frac{1}{\varphi(u)-u}$, and substitute $u=\frac{y}{x}$ back in, then the general solution of the homogeneous equation can be obtained as

$$G\left(\frac{y}{x}\right)=\ln|x|+C \quad (C \text{ is any constant number }).$$

Example 6.2.6 Find the general solution of the differential equation $y'=\frac{y}{x}+\tan\frac{y}{x}$.

Solution Let $\frac{y}{x}=u$, then $y=ux$, $\frac{dy}{dx}=u+x\frac{du}{dx}$, substitute into the equation to get

$$x\frac{du}{dx}=\tan u.$$

Separate the variables to get

$$\cot u\,du=\frac{dx}{x}.$$

Integrate to get $\ln(\sin u)=\ln x+\ln C$, $\sin u=Cx$. Substitute back to the original variables to obtain the general solution

$$\sin\frac{y}{x}=Cx.$$

Example 6.2.7 Solve the differential equation

$$\begin{cases} y^2 dx+(x^2-xy)dy=0 \\ y|_{x=1}=1 \end{cases}.$$

Solution The original equation is

$$\frac{dy}{dx}=\frac{y^2}{xy-x^2}.$$

Dividing both the numerator and the denominator by x^2 yields

$$\frac{dy}{dx}=\frac{\left(\frac{y}{x}\right)^2}{\frac{y}{x}-1}.$$

Let $u=\frac{y}{x}$, then $y=ux$, $\frac{dy}{dx}=u+x\frac{du}{dx}$, substituting into the equation gives

$$x\frac{du}{dx}=\frac{u}{u-1}.$$

Separating the variables yields

$$\frac{u-1}{u}du=\frac{1}{x}dx.$$

Integrating both sides yields $u-\ln u=\ln x+\ln C$, i.e., $u=\ln(Cux)$. Therefore, $\frac{y}{x}=\ln(Cy)$. Thus, the general solution is

$$y=Ce^{\frac{y}{x}}.$$

Substituting the initial condition $x=1, y=1$ into the above equation, we get $C=e^{-1}$, so the particular solution of the equation is

$$y=e^{\frac{y}{x}-1}.$$

The process of solving a homogeneous equation essentially involves transforming the equation into a separable one through variable substitution. Variable substitution plays a special role in solving differential equations. Generally, there is no conventional choice for the substitution. It often depends on the characteristics of the differential equation being considered.

Example 6.2.8 Find the general solution of the differential equation $\frac{dy}{dx}=x^2+2xy+y^2$.

Solution Since $\frac{dy}{dx}=x^2+2xy+y^2=(x+y)^2$, let $u=x+y$, then

$$y=u-x, \frac{dy}{dx}=\frac{du}{dx}-1.$$

The original equation becomes

$$\frac{du}{dx}-1=u^2,$$

that is

$$\frac{du}{u^2+1}=dx.$$

Integrating both sides, we get $\arctan u=x+C$, substituting u back with $x+y$, we obtain the general solution of the original equation as

$$x+y=\tan(x+C).$$

Example 6.2.9 Solve the differential equation $\frac{dy}{dx}=\frac{y}{x-\sqrt{x^2+y^2}}$ $(y\neq 0)$.

Solution The equation is transformed into

$$\frac{dx}{dy}=\frac{x-\sqrt{x^2+y^2}}{y}=\frac{x}{y}\pm\sqrt{1+\left(\frac{x}{y}\right)^2}.$$

Let $u=\frac{x}{y}$, we get $\frac{dx}{dy}=u+y\frac{du}{dy}$, thus

$$u+y\frac{du}{dy}=u\pm\sqrt{1+u^2}.$$

Separation of variables

$$\frac{du}{\sqrt{1+u^2}}=\pm\frac{dy}{y}.$$

Integrate both sides to get

$$\int\frac{du}{\sqrt{1+u^2}}=\pm\int\frac{dy}{y}.$$

That is

$$\ln\left(u+\sqrt{1+u^2}\right)=\pm\ln y+\ln C,$$

thus

$$u+\sqrt{1+u^2}=\frac{C}{y} \text{ or } u+\sqrt{1+u^2}=Cy,$$

substituting $u=\frac{x}{y}$ into the equation, the general solution of the original equation is

$$x+\sqrt{x^2+y^2}=C \text{ or } x+\sqrt{x^2+y^2}=Cy^2.$$

Exercise 6.2

1. Find the general solution of the following separable variable differential equations:

(1) $y\,dy=x\,dx$;

(2) $\frac{dy}{dx}=y\ln y$;

(3) $\frac{dy}{dx}=e^{x-y}$;

(4) $\tan y\,dx-\cot x\,dy=0$;

(5) $\sec^2 x\tan y\,dx+\sec^2 y\tan x\,dy=0$;

(6) $\frac{dy}{dx}=10^{x+y}$;

(7) $(e^{x+y}-e^x)\,dx+(e^{x+y}+e^y)\,dy=0$;

(8) $\cos x\sin y\,dx+\sin x\cos y\,dy=0$;

(9) $(y+1)^2\frac{dy}{dx}+x^3=0$;

(10) $y\,dx+(x^2-4x)\,dy=0$.

2. Find the solution of the following equations satisfying the given initial value conditions:

(1) $\frac{dy}{dx}=y(y-1)$, $y|_{x=0}=1$;

(2) $(x^2-1)y'+2xy^2=0$, $y|_{x=0}=1$;

(3) $y'\sin x=y\ln y$, $y|_{x=\frac{\pi}{2}}=e$;

(4) $\cos y\,dx+(1+e^{-x})\sin y\,dy=0$, $y|_{x=0}=\frac{\pi}{4}$;

(5) $x\,dy+2y\,dx=0$, $y|_{x=2}=1$.

3. Solve the following equations:

(1) $xy'-y-\sqrt{y^2-x^2}=0$;

(2) $(y^2-2xy)\,dx+x^2\,dy=0$;

(3) $(x^2+y^2)\,dx-xy\,dy=0$;

(4) $xy'-y=x\tan\frac{y}{x}$;

(5) $xy'-y=(x+y)\ln\frac{x+y}{x}$;

(6) $(1+2e^{\frac{x}{y}})\,dx+2e^{\frac{x}{y}}\left(1-\frac{x}{y}\right)dy=0$.

4. Find the particular solution of the following homogeneous equations that satisfies the given initial conditions:

(1) $(y^2-3x^2)\,dy+2xy\,dx=0$, $y|_{x=0}=1$;

(2) $y'=\frac{x}{y}+\frac{y}{x}$, $y|_{x=1}=2$;

(3) $(x^2+2xy-y^2)\,dx+(y^2+2xy-x^2)\,dy=0$, $y|_{x=1}=1$.

5. Find a curve such that the tangent at each point of the curve, the radius vector from

the origin to the point of tangency, and the x axis forms an isosceles triangle (with the x axis as the base), and passes through the point $(1,2)$.

6. Consider a concave-upward curve arc $\overset{\frown}{OA}$ connecting points $O(0,0)$ and $A(1,1)$, for any point on $\overset{\frown}{OA}$ $P(x,y)$, the area of the figure enclosed by the curve arc $\overset{\frown}{OP}$ and the straight line segment $\overline{OP}$ is x^2, find the equation of the curve arc $\overset{\frown}{OA}$.

6.3 First-Order Linear Differential Equations

In the form of

$$\frac{dy}{dx}+p(x)y=Q(x),$$

the equation is called a first-order linear differential equation, where $p(x), Q(x)$ is a known continuous function of x. When $Q(x)\equiv 0$, this equation becomes

$$\frac{dy}{dx}+p(x)y=0.$$

This equation is called a first-order linear homogeneous differential equation. If $Q(x)\neq 0$, the equation $\frac{dy}{dx}+p(x)y=Q(x)$ is called a first-order linear nonhomogeneous differential equation, and the equation $\frac{dy}{dx}+p(x)y=0$, is called the corresponding linear homogeneous differential equation to $\frac{dy}{dx}+p(x)y=Q(x)$.

6.3.1 Solution Method for First-Order Linear Equations

The homogeneous linear equation $\frac{dy}{dx}+p(x)y=0$, is a separable equation. Separating the variables yields

$$\frac{dy}{y}=-P(x)dx.$$

Integrating both sides gives

$$\ln|y|=-\int P(x)dx+C_1,$$

or

$$y=Ce^{-\int P(x)dx}\ (C=\pm e^{C_1}).$$

This is the general solution of the homogeneous linear equation (no arbitrary constant is added in the integration).

Example 6.3.1 Find the general solution of the equation $(x-2)\frac{dy}{dx}=y$.

Solution This is a homogeneous linear equation. Separating the variables gives

$$\frac{dy}{y}=\frac{dx}{x-2}.$$

Integrating both sides yields

$$\ln|y|=\ln|x-2|+\ln C,$$

the general solution of the equation is

$$y=C(x-2).$$

The equation can also be solved using the formula method. Transforming the differential equation into $\frac{dy}{dx}-\frac{y}{x-2}=0$, $P(x)=-\frac{1}{x-2}$ and substituting into equation, we get

$$y=Ce^{-\int P(x)dx}=Ce^{\int\frac{1}{x-2}dx}=Ce^{\ln(x-2)}=C(x-2).$$

Next, we discuss the solution method for the first-order nonhomogeneous linear equation

$$\frac{dy}{dx}+p(x)y=Q(x).$$

First, note that equation $\frac{dy}{dx}+p(x)y=Q(x)$ differs from equation $\frac{dy}{dx}+p(x)y=0$ only by a nonhomogeneous term $Q(x)$, so there should be some connection between them. Second, when C is a constant, substituting the function $Ce^{-\int p(x)dx}$ into equation $\frac{dy}{dx}+p(x)y=0$ always yields

$$[Ce^{-\int P(x)dx}]'+p(x)Ce^{-\int P(x)dx}\equiv 0.$$

Therefore, it is only when C is not a constant, i.e., when $C=C(x)$, that it may be possible to satisfy

$$[C(x)e^{-\int P(x)dx}]'+p(x)C(x)e^{-\int P(x)dx}\neq 0.$$

It then follows that, is it possible to determine $C(x)$, such that

$$[C(x)e^{-\int P(x)dx}]'+p(x)C(x)e^{-\int P(x)dx}=Q(x),$$

that is,

$$C'(x)e^{-\int P(x)dx}=Q(x),$$

$$C'(x)=Q(x)e^{\int P(x)dx}.$$

Integrating both sides yields

$$C(x)=\int Q(x)e^{\int P(x)dx}dx+C.$$

Thus, we obtain

$$y=C(x)e^{-\int P(x)dx}=\left[\int Q(x)e^{\int P(x)dx}dx+C\right]e^{-\int P(x)dx}$$

or

$$y=Ce^{-\int P(x)dx}+e^{-\int P(x)dx}\int Q(x)e^{\int P(x)dx}dx.$$

This equation is the general solution to the nonhomogeneous linear equation. The general solution to a nonhomogeneous linear equation is the sum of the general solution to the corresponding homogeneous linear equation and a particular solution to the nonhomogeneous linear equation.

The above solution method can be summarized as follows:

Step 1: Find the general solution to the homogeneous equation $\frac{dy}{dx}+p(x)y=0$,

$$y=Ce^{-\int P(x)dx}.$$

Step 2: Let $C=C(x)$, that is,

$$y=C(x)e^{-\int P(x)dx}.$$

Step 3: Substitute equation $y=C(x)e^{-\int P(x)dx}$ into the nonhomogeneous linear equation $\frac{dy}{dx}+p(x)y=Q(x)$ to obtain a differential equation for $C(x)$, and solve for $C(x)$.

Step 4: Substitute $C(x)$ into equation $y=C(x)e^{-\int P(x)dx}$ to obtain the general solution to the nonhomogeneous linear equation $\frac{dy}{dx}+p(x)y=Q(x)$.

Since Step 2, this method is called the method of variation of parameters.

Example 6.3.2 Find the general solution to the equation $y'+2xy=xe^{-x^2}$.

Solution Solve using the method of variation of parameters. The general solution to the corresponding homogeneous equation $y'+2xy=0$ is $y=Ce^{-x^2}$. Assume the general solution to the original equation is $y=C(x)e^{-x^2}$, and differentiate to obtain

$$y'=C'(x)e^{-x^2}-2xC(x)e^{-x^2}.$$

Substituting y and y' into the original equation, we obtain

$$C'(x)e^{-x^2}-2xC(x)e^{-x^2}+2xC(x)e^{-x^2}=xe^{-x^2},$$

that is $C'(x)=x$, integrating, we obtain

$$C(x)=\int x\,dx=\frac{1}{2}x^2+C.$$

Substituting the above equation into $y=C(x)e^{-x^2}$, we obtain the general solution of the original equation as

$$y=\left(\frac{1}{2}x^2+C\right)e^{-x^2}.$$

The solution can also be obtained using the formula method.

Substituting $P(x)=2x$, $Q(x)=xe^{-x^2}$ into the general solution formula, we obtain

$$y=e^{-\int 2x dx}\left(\int xe^{-x^2}e^{\int 2x dx}dx+C\right)=e^{-x^2}\left(\int xe^{-x^2}e^{x^2}dx+C\right)=e^{-x^2}\left(\frac{1}{2}x^2+C\right).$$

This is the general solution of the original equation.

Example 6.3.3 Solve the initial value problem $\begin{cases} y' - y\cot x = 2x\sin x \\ y\big|_{x=\frac{\pi}{2}} = \dfrac{\pi^2}{4} \end{cases}$.

Solution Substituting $P(x) = -\cot x$, $Q(x) = 2x\sin x$ into the general solution formula, we obtain the general solution of the original equation

$$y = e^{\int \cot x\,dx}\left(\int 2x\sin x\, e^{-\int \cot x\,dx}\,dx + C\right) = e^{\ln(\sin x)}\left(\int 2x\sin x\, e^{-\ln(\sin x)}\,dx + C\right)$$

$$= (\sin x)\left(\int 2x\sin x\,\frac{1}{\sin x}\,dx + C\right) = (x^2 + C)\sin x.$$

Substituting the initial condition $y\big|_{x=\frac{\pi}{2}} = \dfrac{\pi^2}{4}$ into the above equation, we obtain $C=0$. Therefore, the solution to the original problem is $y = x^2\sin x$.

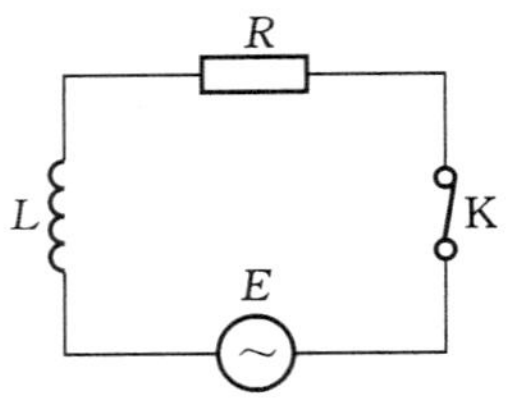

Figure 6-2

Example 6.3.4 There is a circuit as shown in Figure 6-2, where the power supply electromotive force is $E = E_m\sin(\omega t)$ (E_m, ω are both constants), the resistance R and inductance L are constants. Find the current $i(t)$.

Solution From electrical theory, it is known that when the current changes, an induced electromotive force $-L\dfrac{di}{dt}$ is generated on L. This is derived from the loop voltage law.

$$E - L\frac{di}{dt} - iR = 0,$$

that is

$$\frac{di}{dt} + \frac{R}{L}i = \frac{E}{L}.$$

Substituting $E = E_m\sin\omega t$ into the above equation, we get

$$\frac{di}{dt} + \frac{R}{L}i = \frac{E_m}{L}\sin(\omega t).$$

The initial condition is $i\big|_{t=0} = 0$.

The equation $\dfrac{di}{dt} + \dfrac{R}{L}i = \dfrac{E_m}{L}\sin(\omega t)$ is a nonhomogeneous linear equation, where

$$P(t) = \frac{R}{L},\ Q(t) = \frac{E_m}{L}\sin(\omega t).$$

From the general solution formula, we get

$$i(t) = e^{-\int P(t)dt}\left[\int Q(t)e^{\int P(t)dt}dt + C\right] = e^{-\int \frac{R}{L}dt}\left[\int \frac{E_m}{L}\sin(\omega t)e^{\int \frac{R}{L}dt}dt + C\right]$$

$$= \frac{E_m}{L}e^{-\frac{R}{L}t}\left[\int \sin(\omega t)e^{\frac{R}{L}t}dt + C\right] = \frac{E_m}{R^2 + \omega^2L^2}\left[R\sin(\omega t) - \omega L\cos(\omega t)\right] + Ce^{-\frac{R}{L}t}.$$

Where: C is an arbitrary constant.

Substituting the initial condition $i|_{t=0}=0$ into the general solution, we get $C=\frac{\omega LE_m}{R^2+\omega^2L^2}$, thus, the required function $i(t)$ is

$$i(t)=\frac{\omega LE_m}{R^2+\omega^2L^2}e^{-\frac{R}{L}t}+\frac{E_m}{R^2+\omega^2L^2}[R\sin(\omega t)-\omega L\cos(\omega t)].$$

Example 6.3.5 Find the general solution of the equation $y\mathrm{d}x+(y-x)\mathrm{d}y=0$.

Solution Method 1: Transform the equation into

$$\frac{\mathrm{d}y}{\mathrm{d}x}=\frac{y}{x-y}=\frac{\frac{y}{x}}{1-\frac{y}{x}}.$$

Let $u=\frac{y}{x}$ be substituted into the above equation, simplifying to

$$\frac{(1-u)\mathrm{d}u}{u^2}=\frac{\mathrm{d}x}{x}.$$

Integrating both sides yields

$$-\frac{1}{u}-\ln|u|=\ln|x|-C,$$

that is

$$\frac{x}{y}+\ln|y|=C.$$

Method 2: If we treat x as the dependent variable and y as the independent variable, the original equation can be transformed into a first-order linear differential equation for x. By rearranging the equation to $\frac{\mathrm{d}x}{\mathrm{d}y}-\frac{1}{y}x=-1$, and substituting into the formula, we get

$$x=e^{\int\frac{1}{y}\mathrm{d}y}\left[\int(-1)e^{-\int\frac{1}{y}\mathrm{d}y}\mathrm{d}y+C\right]=y(C-\ln|y|),$$

that is

$$\frac{x}{y}+\ln|y|=C.$$

6.3.2 Bernoulli Equation

The equation

$$\frac{\mathrm{d}y}{\mathrm{d}x}+P(x)y=Q(x)y^n \quad (n\neq 0,1)$$

is known as the Bernoulli (Jacob Bernoulli, 1654-1705, Swiss mathematician) equation. It is not a linear equation, but through appropriate variable substitutions, it can be transformed into a linear equation. In fact, dividing both sides of equation by y^n, we get

$$y^{-n}\frac{\mathrm{d}y}{\mathrm{d}x}+P(x)y^{1-n}=Q(x).$$

Let $z=y^{1-n}$, we obtain a linear equation

$$\frac{dz}{dx}+(1-n)P(x)z=(1-n)Q(x).$$

After obtaining the general solution using the formula for solving linear differential equations, substituting back the original variable z yields the general solution for the Bernoulli equation as

$$y^{1-n}=e^{-\int(1-n)P(x)dx}\left[\int(1-n)Q(x)e^{\int(1-n)P(x)dx}dx+C\right].$$

Example 6.3.6 Solve the equation $\frac{dy}{dx}-y=xy^5$.

Solution The original equation can be rearranged as

$$\frac{1}{y^5}\frac{dy}{dx}-\frac{1}{y^4}=x.$$

Let $z=\frac{1}{y^4}$, then

$$\frac{dz}{dx}+4z=-4x.$$

This is a first-order linear differential equation, and its general solution is

$$z=e^{-\int 4dx}\left[\int(-4x)e^{\int 4dx}dx+C\right]=e^{-4}\left(-4\int xe^{4x}dx+C\right)=-x+\frac{1}{4}+Ce^{-4x}.$$

Therefore, the general solution of the original equation is $\frac{1}{y^4}=-x+\frac{1}{4}+Ce^{-4x}$.

Through variable substitution, certain equations can be transformed into separable variable equations, or into equations whose solution methods are already known.

Example 6.3.7 Solve the equation $\frac{dy}{dx}=\frac{1}{x-y}+1$.

Solution Let $u=x-y$, then the original equation becomes

$$1-\frac{du}{dx}=\frac{1}{u}+1,$$

that is, $dx=-u\,du$. Integrating both sides gives

$$x=-\frac{1}{2}u^2+C_1.$$

Substituting $u=x+y$ into the above equation yields the general solution of the original equation

$$x=-\frac{1}{2}(x-y)^2+C_1,$$

that is,

$$(x-y)^2=-2x+C\quad(C=2C_1).$$

Exercise 6.3

1. Find the general solution of the following differential equations:

(1) $\frac{dy}{dx}=y+\sin x$; (2) $\frac{dx}{dt}+3x=e^{2t}$;

(3) $\frac{ds}{dt}=-s\cos t+\frac{1}{2}\sin 2t$;

(4) $\frac{dy}{dx}-\frac{x}{n}y=e^{x}x^{n}$ (n is a constant);

(5) $\frac{dy}{dx}+\frac{1-2x}{x^{2}}y-1=0$;

(6) $\frac{d\rho}{d\theta}+3\rho=2$;

(7) $(x-2xy-y^{2})dy+y^{2}dx=0$;

(8) $x(1+x^{2})dy=(y+x^{2}y-x^{2})dx$;

(9) $(x-2)\frac{dy}{dx}=y+2(x-2)^{3}$;

(10) $(y^{2}-6x)\frac{dy}{dx}+2y=0$.

2. Find the particular solution of the following differential equations that satisfy the given initial conditions:

(1) $\frac{dy}{dx}-y\tan x=\sec x, y|_{x=0}=0$;

(2) $\frac{dy}{dx}+\frac{y}{x}=\frac{\sin x}{x}, y|_{x=\pi}=1$;

(3) $\frac{dy}{dx}+y\cot x=5e^{\cos x}, y|_{x=\frac{\pi}{2}}=-4$;

(4) $\frac{dy}{dx}+3y=8, y|_{x=0}=2$;

(5) $\frac{dy}{dx}+\frac{2-3x^{2}}{x^{3}}y=1, y|_{x=1}=0$;

(6) $(1-x^{2})y'+xy=1, y|_{x=0}=1$.

3. Establish and solve the differential equations for curves with the following properties:

(1) The vertical intercept of the tangent line at any point on the curve is equal to the square of the abscissa of the point of tangency;

(2) The vertical intercept of the tangent line at any point on the curve is the arithmetic mean of the abscissa and ordinate of the point of tangency.

4. Find the general solution of the following Bernoulli equations:

(1) $\frac{dy}{dx}+y=y^{2}(\cos x-\sin x)$;

(2) $\frac{dy}{dx}-3xy=xy^{2}$;

(3) $y'+\frac{y}{x+1}+y^{2}=0$;

(4) $\frac{dy}{dx}=\frac{y}{2x}+\frac{x^{2}}{2y}$.

5. Find the general or particular solution of the following equations through appropriate transformations:

(1) $\frac{dy}{dx}=\frac{1}{x^{2}+y^{2}+2xy}$;

(2) $\frac{dy}{dx}=(x-y)^{2}+1$;

(3) $xy'+y=y(\ln x+\ln y)$;

(4) $x\frac{dy}{dx}+x+\sin(x+y)=0, y\left(\frac{\pi}{2}\right)=0$.

6. There is a smooth, upward-convex curve connecting points $A(0,1)$ and $B(1,0)$, with point $P(x,y)$ being any point on the curve. It is known that the area between the curve and the chord AP is x^{3}, find the equation of the curve.

7. Suppose the curve $\int_L yf(x)\mathrm{d}x + [2xf(x) - x^2]\mathrm{d}y$ is path-independent in the right half-plane ($x>0$), where $f(x)$ is differentiable, and $f(1)=1$, find $f(x)$.

8. Consider a particle of mass m moving in a straight line, starting from rest, with a force acting on it in the direction of motion that is proportional to time (with a proportionality constant of k_1), and additionally subjected to a resistance proportional to velocity (with a proportionality constant of k_2). Find the functional relationship between the particle's velocity and time.

6.4 Total Differential Equations

A first-order differential equation $\frac{\mathrm{d}y}{\mathrm{d}x}=f(x,y)$ can be written in the following symmetric form:

$$P(x,y)\mathrm{d}x+Q(x,y)\mathrm{d}y=0.$$

Suppose $P(x,y)$, $Q(x,y)$ is a continuous function of x, y in a region D, and has continuous first-order partial derivatives. If there exists a differentiable function $u(x,y)$, such that

$$\mathrm{d}u(x,y)=P(x,y)\mathrm{d}x+Q(x,y)\mathrm{d}y,$$

then equation $P(x,y)\mathrm{d}x+Q(x,y)\mathrm{d}y=0$ is called a total differential equation. Here

$$\frac{\partial u}{\partial x}=P(x,y),\frac{\partial u}{\partial y}=Q(x,y),$$

and the equation $P(x,y)\mathrm{d}x+Q(x,y)\mathrm{d}y=0$ can be written as

$$\mathrm{d}u(x,y)=0.$$

Thus, $u(x,y) \equiv C$ is the implicit general solution of equation $P(x,y)\mathrm{d}x + Q(x,y)\mathrm{d}y=0$, where C is an arbitrary constant that makes the function meaningful.

Although the left-hand side of equation has the same structure as the total differential of a function, it is not necessarily the total differential of some function. So under what conditions is the expression $P(x,y)\mathrm{d}x+Q(x,y)\mathrm{d}y$ the total differential of some binary function $u(x,y)$? When such a binary function exists, how can we find this binary function $u(x,y)$?

In the process of finding the total differential of a binary function, the following conclusion is drawn: In a simply connected domain D of the plane, if the functions $P(x,y)$ and $Q(x,y)$ have continuous first-order partial derivatives within D, then $P(x,y)\mathrm{d}x+Q(x,y)\mathrm{d}y$ is the total differential of some function $u(x,y)$ if and only if the equation

$$\frac{\partial P}{\partial y}=\frac{\partial Q}{\partial x}$$

holds identically in D.

Therefore, condition $\frac{\partial P}{\partial y}=\frac{\partial Q}{\partial x}$ is the necessary and sufficient condition for equation to be a total differential equation, and the binary function can be found using the second type of line integral:

$$u(x,y)=\int_{x_0}^{x} P(x,y)\mathrm{d}x+\int_{y_0}^{y} Q(x_0,y)\,\mathrm{d}x \quad ((x_0,y_0)\in D).$$

The binary function $u(x,y)$ can also be found using indefinite integrals. Because

$$\frac{\partial u(x,y)}{\partial x}=P(x,y),\frac{\partial u(x,y)}{\partial y}=Q(x,y),$$

then

$$u(x,y)=\int \frac{\partial u(x,y)}{\partial x}\mathrm{d}x=\int P(x,y)\mathrm{d}x+C(y),$$

differentiating the above equation gives

$$\frac{\partial u}{\partial y}=Q=\frac{\partial}{\partial y}\int P(x,y)\mathrm{d}x+C'(y),$$

by comparison, we get

$$C'(y)=Q-\frac{\partial}{\partial y}\int P(x,y)\mathrm{d}x,$$

integrating y gives

$$C(y)=\int Q(x,y)\mathrm{d}y-\int\left[\frac{\partial}{\partial y}\int P(x,y)\mathrm{d}x\right]\mathrm{d}y,$$

thus

$$u(x,y)=\int P(x,y)\mathrm{d}x+\int Q(x,y)\mathrm{d}y-\int\left[\frac{\partial}{\partial y}\int P(x,y)\mathrm{d}x\right]\mathrm{d}y.$$

Example 6.4.1 Solve $(2x+y)\,\mathrm{d}x+(4y+x)\,\mathrm{d}y=0$.

Solution Method 1: Solving $P=2x+y$, $Q=x+4y$, and since $\frac{\partial P}{\partial y}=1$, $\frac{\partial Q}{\partial x}=1$, i.e., $\frac{\partial Q}{\partial x}=\frac{\partial P}{\partial y}=1$, this equation is a total differential equation.

$$\begin{aligned}u(x,y)&=\int_{(0,0)}^{(x,y)}(2x+y)\,\mathrm{d}x+(4y+x)\,\mathrm{d}y=\int_{(0,0)}^{(x,0)}(2x+y)\,\mathrm{d}x+\int_{(x,0)}^{(x,y)}(4y+x)\,\mathrm{d}y\\&=\int_0^x 2x\,\mathrm{d}x+\int_0^y(4y+x)\,\mathrm{d}y=x^2+2y^2+xy.\end{aligned}$$

Therefore, the general solution of the equation is $x^2+2y^2+xy=C$.

Method 2: Since $\frac{\partial u}{\partial x}=P=2x+y$, $\frac{\partial u}{\partial y}=Q=4y+x$, then

$$u=\int\frac{\partial u}{\partial x}\mathrm{d}x=\int(2x+y)\,\mathrm{d}x=x^2+xy+C(y).$$

Differentiating the above equation with respect to y yields

$$\frac{\partial u}{\partial y}=x+C'(y).$$

However, since $\frac{\partial u}{\partial y}=Q=4y+x$, by comparison we get $C'(y)=4y$, hence $C(y)=2y^2+C$, thus the general solution of the equation is

$$x^2+xy+2y^2+C=0.$$

Method 3: The equation $(2x+y)\mathrm{d}x+(4y+x)\mathrm{d}y=0$ can be written as

$$2x\,\mathrm{d}x+4y\,\mathrm{d}y+(y\,\mathrm{d}x+x\,\mathrm{d}y)=0,$$

then

$$\mathrm{d}x^2+\mathrm{d}(2y^2)+\mathrm{d}(xy)=0,$$

that is

$$\mathrm{d}(x^2+2y^2+xy)=0,$$

that is also

$$x^2+2y^2+xy=C.$$

If $\frac{\partial P}{\partial y}\neq\frac{\partial Q}{\partial x}$, then this equation is not an exact differential equation, but there exists a function $\mu=\mu(x,y)$ $(\mu(x,y)\neq 0)$.

Such that the equation

$$\mu(x,y)P(x,y)\mathrm{d}x+\mu(x,y)Q(x,y)\mathrm{d}y=0$$

is an exact differential equation, then the function $\mu(x,y)$ is the integrating factor.

Example 6.4.2 Find the integrating factor of the equation by observation and find its general solution:

(1) $y\mathrm{d}x-x\mathrm{d}y=0$; (2) $(1+xy)y\mathrm{d}x+(1-xy)x\mathrm{d}y=0$.

Solution (1) The equation $y\mathrm{d}x-x\mathrm{d}y=0$ is not an exact differential equation. Because

$$\mathrm{d}\left(\frac{x}{y}\right)=\frac{y\mathrm{d}x-x\mathrm{d}y}{y^2},$$

therefore, $\frac{1}{y^2}$ is an integrating factor for the equation $y\mathrm{d}x-x\mathrm{d}y=0$, and thus $\frac{y\mathrm{d}x-x\mathrm{d}y}{y^2}=0$ becomes an exact differential equation. The general solution to the given equation is $\frac{x}{y}=C$.

(2) The equation $(1+xy)y\mathrm{d}x+(1-xy)x\mathrm{d}y=0$ is not an exact differential equation. Rearranging the terms of the equation, we get

$$(y\mathrm{d}x+x\mathrm{d}y)+xy(y\mathrm{d}x-x\mathrm{d}y)=0,$$

rewriting it as

$$\mathrm{d}(xy)+x^2y^2\left(\frac{\mathrm{d}x}{x}-\frac{\mathrm{d}y}{y}\right)=0.$$

It is easy to see that $\frac{1}{(xy)^2}$ is an integrating factor at this point. Multiplying the equation by this integrating factor, it becomes

$$\frac{d(xy)}{(xy)^2}+\frac{dx}{x}-\frac{dy}{y}=0,$$

integrating, we obtain the general solution

$$-\frac{1}{xy}+\ln\left|\frac{x}{y}\right|=\ln C,$$

that is

$$\frac{x}{y}=Ce^{\frac{1}{xy}}.$$

One can also solve a first-order linear equation using the method of integrating factors.

$$y'+P(x)y=Q(x).$$

It can be verified that $\mu(x)=e^{\int P(x)dx}$ is an integrating factor for the first-order linear equation $y'+P(x)y=Q(x)$. Multiplying both sides of the first-order linear equation by $\mu(x)=e^{\int P(x)dx}$, we get

$$y'e^{\int P(x)dx}+yP(x)e^{\int P(x)dx}=Q(x)e^{\int P(x)dx},$$

that is

$$y'e^{\int P(x)dx}+y\ [e^{\int P(x)dx}]'=Q(x)e^{\int P(x)dx},$$

that is

$$[ye^{\int P(x)dx}]'=Q(x)e^{\int P(x)dx}.$$

Integrating both sides yields the general solution:

$$ye^{\int P(x)dx}=\int Q(x)e^{\int P(x)dx}dx+C \text{ or } y=e^{-\int P(x)dx}\left[\int Q(x)e^{\int P(x)dx}dx+C\right].$$

Example 6.4.3 Find the general solution of $\frac{dy}{dx}+2xy=4x$ using an integrating factor.

Solution The integrating factor for solving the equation is

$$\mu(x)=e^{\int 2x\,dx}=e^{x^2}.$$

Multiplying both sides of the equation by e^{x^2} gives

$$y'e^{x^2}+2xe^{x^2}y=4xe^{x^2},$$

that is,

$$(e^{x^2}y)'=4xe^{x^2},$$

thus,

$$e^{x^2}y=\int 4xe^{x^2}dx=2e^{x^2}+C.$$

Therefore, the general solution of the original equation is

$$y=\int 4xe^{x^2}dx=2+Ce^{-x^2}.$$

Exercise 6.4

1. Verify that the following equations are exact differential equations and find their general solutions:

(1) $2xy\,dx + (x^2 - y^2)\,dy = 0$;

(2) $e^{-y}\,dx - (2y + xe^{-y})\,dy = 0$;

(3) $(5x^4 + 3xy^2 - y^3)\,dx + (3x^2y - 3xy^2 + y^2)\,dy = 0$;

(4) $(a^2 - 2xy - y^2)\,dx - (x + y)^2\,dy = 0$;

(5) $e^y\,dx + (xe^y - 2y)\,dy = 0$;

(6) $(x^2 + y)\,dx + (x - 2y)\,dy = 0$;

(7) $(y - 3x^2)\,dx - (4y - x)\,dy = 0$;

(8) $\left[\dfrac{y^2}{(x-y)^2} - \dfrac{1}{x}\right]dx + \left[\dfrac{1}{y} - \dfrac{x^2}{(x-y)^2}\right]dy = 0$;

(9) $2(3xy^2 + 2x^3)\,dx + 3(2x^2y + y^2)\,dy = 0$;

(10) $\left(\dfrac{1}{y}\sin\dfrac{x}{y} - \dfrac{y}{x^2}\cos\dfrac{y}{x} + 1\right)dx + \left(\dfrac{1}{x}\cos\dfrac{y}{x} - \dfrac{x}{y^2}\sin\dfrac{x}{y} + \dfrac{1}{y^2}\right)dy = 0$.

2. Find the integrating factors and integrals for the following equations:

(1) $(x + y)(dx - dy) = dx + dy$;

(2) $y\,dx - x\,dy + y^2x\,dx = 0$;

(3) $y^2(x - 3y)\,dx + (1 - 3y^2x)\,dy = 0$;

(4) $(x^2 + y^2 + x)\,dx + xy\,dy = 0$;

(5) $(2xy^4e^y + 2xy^3 + y)\,dx + (x^2y^4e^y - x^2y^2 - 3x)\,dy = 0$;

(6) $(x^4 + y^4)\,dx - xy^3\,dy = 0$.

3. Prove that $\dfrac{1}{xy[f(xy) - g(xy)]}$ is an integrating factor for the differential equation $yf(xy)\,dx + xg(xy)\,dy = 0$, and find the general solutions for the following equations:

(1) $y(x^2y^2 + 2)\,dx + x(2 - 2x^2y^2)\,dy = 0$;

(2) $y(2xy + 1)\,dx + x(1 + 2xy - x^3y^3)\,dy = 0$.

Comprehensive Review 6

1. Fill in the blanks.

(1) $xy''' + 2x^2y'^2 + x^3y = x^4 + 1$ is ________.

(2) $y_1(x)$, $y_2(x)$ are two particular solutions of the equation $y' + p(x)y - q(x) = 0\ (q(x) \neq 0)$, if $\alpha y_1 + \beta y_2$ is also a solution of this equation, then α, β should satisfy ________.

(3) The initial value problem of the differential equation equivalent to the integral

equation $y=\int_{x_0}^{x} f(x,y)\mathrm{d}x$ is ________.

(4) The differential equation with the general solution $y=C_1\mathrm{e}^x+C_2\mathrm{e}^{-2x}$ is ________.

2. Multiple choice questions.

(1) The differentiable function $f(x)$ that satisfies the equation $\int_0^1 f(tx)\,\mathrm{d}t=nf(x)$ (where n is a natural number greater than 1) is ( ).

A. $Cx^{\frac{1-n}{n}}$ (C is a constant) B. C C. $C\sin(nx)$ D. $C\cos(nx)$

(2) The differential equation $y'-xy'=2(y^2+y')$ is ( ).

A. homogeneous equation B. Bernoulli equation

C. linear nonhomogeneous equations D. separable variable equations

(3) Given that $y_1=x$ is a solution to $y''+y=x$, and $y^2=\frac{1}{2}\mathrm{e}^x$ is a solution to $y''+y=\mathrm{e}^x$, the general solution to the differential equation $y''+y=x+\mathrm{e}^x$ is $y=$( ).

A. $x+\frac{1}{2}\mathrm{e}^x$ B. $C_1\cos x+C_2\sin x+\frac{1}{2}\mathrm{e}^x+x$

C. $C_1\cos x+C_2\sin x+x$ D. $C_1\cos x+C_2\sin x$

(4) Given that $xy''+y'=4x$ has a particular solution x^2, and the corresponding homogeneous equation $xy''+y'=0$ has a particular solution $\ln x$, the general solution to the original equation is $y=$( ).

A. $C_1\ln x+C_2+x^2$ B. $C_1\ln x+C_2x+x^2$

C. $C_1\ln x+C_2\mathrm{e}^x+x^2$ D. $C_1\ln x+C_2\mathrm{e}^{-x}+x^2$

3. Find the general solution to the following differential equations:

(1) $xy'+(1-x)y=\mathrm{e}^{2x}$;

(2) $x\mathrm{d}y+(2xy^2-y)\mathrm{d}x=0$;

(3) $\left(\cos x+\frac{1}{y}\right)\mathrm{d}x+\left(\frac{1}{y}-\frac{x}{y^2}\right)\mathrm{d}y=0$;

(4) $(x^2-y^2-2y)\mathrm{d}x+(x^2+2x-y^2)\mathrm{d}y=0$;

(5) $(\sqrt{x^2+y^2})+\frac{x}{\sqrt{x^2+y^2}}\mathrm{d}x+\frac{y}{\sqrt{x^2+y^2}}\mathrm{d}y=0$;

(6) $\sqrt{1+x^2}\sin 2y\frac{\mathrm{d}y}{\mathrm{d}x}=2x\sin^2 y+\mathrm{e}^{2\sqrt{1+x^2}}$;

(7) $y''+2y'+y=x\mathrm{e}^x$;

(8) $y''+y=x+\cos x$.

4. Find the particular solution to the following differential equations:

(1) $(2\cos 2x)y'\cdot y''+2\sin 2x\cdot y'^2=\sin 2x$, $y|_{x=0}=1$, $y'|_{x=0}=0$;

(2) $xy\frac{\mathrm{d}y}{\mathrm{d}x}=x^2+y^2$, $y|_{x=\mathrm{e}}=2\mathrm{e}$;

(3) $yy''=2(x'^2-y'),y|_{x=0}=1,y'|_{x=0}=2$;

(4) $(x+y)^2\dfrac{dy}{dx}=k^2$ (k is a constant), $y|_{x=0}=0$.

5. Given that $y_1=x^2, y_2=x+x^2, y_3=e^x+x^2$ are all solutions to the equation $(x-1)y''-xy'+y=-x^2+2x-2$, find the general solution to this equation.

6. Suppose a particular solution to the second-order linear differential equation with constant coefficients $y''+\alpha y'+\beta y=\gamma e^x$ is $y=e^{2x}+(1+x)e^x$. Determine the constants α, β, γ and find the general solution to the equation.

7. Let $y=e^x$ be a solution to the differential equation $xy'+P(x)y=x$. Find the particular solution of this equation that satisfies the condition $y|_{x=\ln 2}=0$.

8. Let $f(x)$ have continuous second derivatives, $f(0)=f'(0)=0$, and the line integral

$$\int_L (e^x+\sin x)y\,dx+[f'(x)-f(x)]\,dy$$

is independent of the path of integration. Find $f(x)$.

9. Let $y_1(x), y_2(x)$ be two solutions to the second-order homogeneous linear equation $y''+p(x)y'+q(x)y=0$. Let

$$W(x)=\begin{vmatrix} y_1(x) & y_2(x) \\ y'_1(x) & y'_2(x) \end{vmatrix}=y_1(x)y'_2(x)-y'_1(x)y_2(x).$$

Prove:

(1) $W(x)$ satisfies the equation $W'+p(x)W=0$;

(2) $W(x)=W(x_0)e^{-\int_{x_0}^{x}p(t)dt}$.

10. A rabbit is located 100 m directly west of a wolf. Assuming both the rabbit and the wolf spot each other simultaneously and start running at the same time, the rabbit runs directly north towards its den 60 m, while the wolf chases the rabbit. Given that both the rabbit and the wolf run at constant speeds and the wolf's speed is twice that of the rabbit, find the equation of the path of the wolf's motion and determine whether the rabbit can safely reach its den.

Chapter 7 Infinite Series

The *I Ching* emphasizes the mutual transformation and harmony between Yin and Yang, and holds that all things in the world are in a state of balance between Yin and Yang. The harmony of Yin and Yang is not only reflected in the static balance, but also in the dynamic change process. Through the superposition of Yin and Yang, things gradually develop into a new state or form. Infinite series can be seen as a quantitative description of the "cascading effect" of a function or sequence of numbers. This cascading effect reflects the harmonization and balance of the two forces (such as the positive and negative terms of the series) in the process of change. The sum of infinite series is the ultimate embodiment of this Yin and Yang harmony and layering.

The thought of "infinity" is embodied in the infinite exploration and understanding of the process of change. According to the *I Ching*, everything in the world is in constant change, and this change is infinite. The calculation process of infinite series is to approximate a limit value by the accumulation of finite terms. This "art of infinite approximation" embodies the fine grasp of the change of a function or sequence of numbers in infinite terms, and approximates the real sum value by increasing the number of terms. Imagine an infinite mountain, and the infinite series is like dividing the mountain into countless tiny stones, then calculating the weight of each stone (that is, each term in the series), and finally summing the weight of all the stones to get the total weight of the mountain. Although it is practically impossible to compute the sum of infinite numbers, through the "art of infinite approximation", we can get an approximation with arbitrary precision.

Chinese tradition emphasizes "flexibility" and "application", and holds that only when theory is applied to practice can it exert its value. As an important part of mathematical analysis, infinite series is widely used. Infinite series is widely used in physics, engineering, computer science and other fields. It can be used not only to calculate the value of a function, but also to solve differential equations and perform numerical analysis. This wide application is precisely the value of infinite series as a mathematical tool.

The concept and theory of infinite series are developments and applications of calculus theory. Historically, people understood infinite series as a generalization of addition and achieved many results in numerical calculation and function research. For example, Newton extended the binomial theorem from finite terms to infinite terms, resulting in an infinite polynomial.

$$\frac{1}{1+x}=(1+x)^{-1}=1-x+x^2-x^3+\cdots,$$

this represents the function as a power series. Taylor transformed a formula used by Newton for approximate calculation into

$$f(a+x)=f(a)+f'(a)\frac{x}{1!}+f''(a)\frac{x^2}{2!}+\cdots,$$

thus obtaining the so-called Taylor series. At the same time, they also obtained puzzling results. For example, in the equation $\frac{1}{1+x}=(1+x)^{-1}=1-x+x^2-x^3+\cdots$, letting $x=1$, one obtains

$$\frac{1}{2}=1-1+1-1+\cdots.$$

According to the associative law of addition, we have

$$\frac{1}{2}=(1-1)+(1-1)+\cdots=0,$$

and

$$\frac{1}{2}=1-(1-1)-(1-1)-\cdots=1.$$

This contradiction prompts people to clarify the essential characteristics of infinite series and the differences between "infinite sums" and "finite sums" operations.

Infinite series are significant in analysis, with important applications in numerical computation, function approximation, and differential equations.

This chapter will introduce the basic theory of numerical series. For function series, we will discuss power series.

7.1 Concepts and Properties of Constant Term Series

People's understanding of the quantitative characteristics of things often involves a process, from approximation to precision, during which the problem of adding an infinite number of quantities arises.

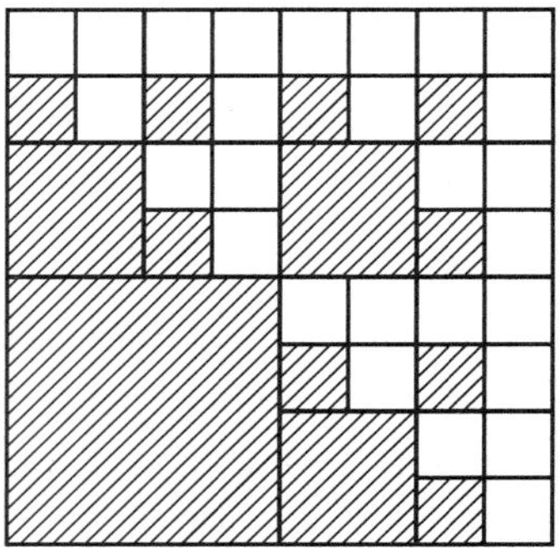

Figure 7-1

7.1.1 Constant Term Series

As shown in Figure 7-1, divide a unit square into four smaller squares, shade the bottom-left square, then divide the other three squares into four smaller squares each, and shade the bottom-left square of each, and so on. Find the area of the shaded parts.

Let the area of the shaded small square in the lower left corner of the unit square divided into four equal parts be a_1, then $a_1=\left(\frac{1}{2}\right)^2$. Let the sum of the areas of the shaded small squares in the lower left corners of the other three small squares be a_2, then $a_2=3\cdot\left(\frac{1}{4}\right)^2$. By continuing this process, the area of the shaded parts in Figure 7-1 is the sum of an infinite number of quantities: $a_1+a_2+a_3+\cdots+a_n+\cdots$, that is,

$$\left(\frac{1}{2}\right)^2+3\cdot\left(\frac{1}{4}\right)^2+3^2\cdot\left(\frac{1}{8}\right)^2+\cdots+3^{n-1}\left(\frac{1}{2^n}\right)^2+\cdots.$$

Generally, let $\{u_n\}$ be a sequence, and the expression

$$\sum_{n=1}^{\infty}u_n=u_1+u_2+u_3+\cdots+u_n+\cdots$$

is called an infinite series of constant terms, simply referred to as a constant term series, where the nth term u_n is called the general term of the series.

How should we understand the addition of an infinite number of quantities in series? Obviously, as in the previous examples, calculating the "infinite sum" of the infinite series $1-1+1-1+\cdots$ using the method of "finite sum" leads to contradictions. In connection with the introductory example, the problem of finding the area of the shaded parts in Figure 7-1 can be approached by starting with the sum of a finite number of terms, observing their changing trends, and thereby understanding the meaning of adding an infinite number of quantities.

Take the sum of the first n terms of series:

$$s_n=u_1+u_2+\cdots+u_n=\sum_{i=1}^{n}u_i,$$

s_n is the partial sum of the series, and when n takes $1,2,3,\cdots$ successively, they form a new sequence:

$$s_1=u_1, s_2=u_1+u_2, s_3=u_1+u_2+u_3, \cdots, s_n=u_1+u_2+\cdots+u_n, \cdots.$$

If this sequence has a limit, this limit value is taken as the "infinite sum" of the series.

Definition 7.1.1 If the sequence of partial sums $\{s_n\}$ of the series has a limit s, i.e., $\lim\limits_{n\to\infty}s_n=s$, then the series is said to converge, and s is called the sum of the series, i.e.,

$$s=u_1+u_2+\cdots+u_n+\cdots.$$

If $\{s_n\}$ does not have a limit, the series is said to diverge. In particular, when $\lim\limits_{n\to\infty}s_n$ is $+\infty$ or $-\infty$, the series is said to diverge to $+\infty$ or $-\infty$. When the series converges, its partial sum s_n is an approximation of the sum s, and the difference is denoted as

$$r_n=s-s_n=u_{n+1}+u_{n+2}+\cdots,$$

and it is called the remainder of the series, while the error generated by substituting s_n for s is denoted as $|r_n|=|s-s_n|$.

It is evident from Definition 7.1.1 that there is a close connection between series and the limits of sequences.

Given a series $\sum_{n=1}^{\infty} u_n$,the sequence of partial sums $\{s_n\}$ is uniquely determined by the series,i.e., $\left\{s_n = \sum_{i=1}^{n} u_i\right\}$. Conversely,if the sequence of partial sums $\{s_n\}$ is known,the corresponding series is uniquely determined.In fact,at this point,we have

$$s_1 + (s_2 - s_1) + \cdots + (s_n - s_{n-1}) + \cdots = s_1 + \sum_{n=2}^{\infty} (s_n - s_{n-1}) = \sum_{n=1}^{\infty} u_n,$$

where: $u_1 = s_1, u_n = s_n - s_{n-1}\ (n \geqslant 2)$ .According to Definition 7.1.1,it is easy to see that the series $\sum_{n=1}^{\infty} u_n$ and the sequence $\{s_n\}$ converge or diverge together,and when they converge, we have $\sum_{n=1}^{\infty} u_n = \lim_{n\to\infty} s_n$,i.e., $\sum_{n=1}^{\infty} u_n = \lim_{n\to\infty} \sum_{i=1}^{n} u_i$.

With Definition 7.1.1 in hand,we can now find the area of the shaded region in Figure 7-1 in this section.

Let the partial sum be s_n ,then

$$s_n = \left(\frac{1}{2}\right)^2 + 3 \cdot \left(\frac{1}{4}\right)^2 + 3^2 \cdot \left(\frac{1}{8}\right)^2 + \cdots + 3^{n-1}\left(\frac{1}{2^n}\right)^2 = \frac{1}{2^2} + \frac{3}{2^4} + \frac{3^2}{2^6} + \cdots + \frac{3^{n-1}}{2^{2n}}$$

$$= \frac{1}{2^2}\left(1 + \frac{3}{2^2} + \frac{3^2}{2^4} + \cdots + \frac{3^{n-1}}{2^{2(n-1)}}\right) = \frac{1}{4}\left[1 + \frac{3}{2^2} + \left(\frac{3}{2^2}\right)^2 + \cdots + \left(\frac{3}{2^2}\right)^{n-1}\right]$$

$$= \frac{1}{4} \cdot \frac{1 - \left(\frac{3}{4}\right)^n}{1 - \frac{3}{4}} = 1 - \left(\frac{3}{4}\right)^n,$$

thus,the area sought is $\lim_{n\to\infty} s_n = \lim_{n\to\infty}\left[1 - \left(\frac{3}{4}\right)^n\right] = 1.$

Example 7.1.1 Discuss the convergence of the geometric series

$$a + aq + aq^2 + \cdots + aq^{n-1} + \cdots = \sum_{n=1}^{\infty} aq^{n-1} \quad (a \neq 0).$$

Solution If $q \neq 1$,then the partial sum

$$s_n = a + aq + aq^2 + \cdots + aq^{n-1} = a \cdot \frac{1 - q^n}{1 - q} \quad (q \neq 1).$$

When $|q| < 1$,we have $\lim_{n\to\infty} q^n = 0$,thus $\lim_{n\to\infty} s_n = \lim_{n\to\infty} a\frac{1-q^n}{1-q} = \frac{a}{1-q}$. When $|q| > 1$,we have $\lim_{n\to\infty} q^n = \infty$,thus $\lim_{n\to\infty} s_n = \lim_{n\to\infty} a\frac{1-q^n}{1-q} = \infty$. When $q = -1$,the geometric series becomes
$a - a + a - a + \cdots + (-1)^{n-1}a + \cdots$,

$$s_n = \begin{cases} 0 & (n \text{ is even}) \\ a & (n \text{ is odd}) \end{cases}.$$

Thus,when $\lim_{n\to\infty} s_n$ does not exist,the series diverges. When $q = 1$,the series is $a + a + \cdots + a + \cdots$, $s_n = na$,thus $\lim_{n\to\infty} s_n = \infty$,the series diverges.

In summary, when $|q|<1$, the geometric series converges; when $|q|\geqslant 1$, the geometric series diverges.

Example 7.1.2 Determine the convergence of the infinite series

$$\frac{1}{1\cdot 2}+\frac{1}{2\cdot 3}+\cdots+\frac{1}{n(n+1)}+\cdots.$$

Solution Since $\frac{1}{n(n+1)}=\frac{1}{n}-\frac{1}{n+1}$, therefore

$$s_n=\frac{1}{1\cdot 2}+\frac{1}{2\cdot 3}+\cdots+\frac{1}{n(n+1)}=\left(1-\frac{1}{2}\right)+\left(\frac{1}{2}-\frac{1}{3}\right)+\cdots+\left(\frac{1}{n}-\frac{1}{n+1}\right)=1-\frac{1}{n+1}.$$

Thus $\lim\limits_{n\to\infty}s_n=\lim\limits_{n\to\infty}\left(1-\frac{1}{n+1}\right)=1$, the original series converges.

Example 7.1.3 Prove that the series $\sum\limits_{n=1}^{\infty}\ln\left(1+\frac{1}{n}\right)$ diverges.

Proof

$$s_n=\sum_{i=1}^{n}\ln\left(1+\frac{1}{i}\right)=\sum_{i=1}^{n}[\ln(i+1)-\ln i]=\ln(n+1).$$

thus $\lim\limits_{n\to\infty}s_n=\lim\limits_{n\to\infty}\ln(n+1)=\infty$, the original series diverges.

From the above examples, it is known that using Definition 7.1.1 to determine the convergence of a series requires finding the sum of the first n terms s_n. However, s_n can only be found in extremely special cases, and finding the sum of the series is even more difficult. However, if it is known that an infinite series converges, its sum can be approximated by the sum of its first n terms. Therefore, it is necessary to find convenient and effective methods to determine the convergence of infinite series.

7.1.2 Basic Properties of Convergent Series

Property 7.1.1 (Necessity) If the infinite series $\sum\limits_{n=1}^{\infty}u_n$ converges, then $\lim\limits_{n\to\infty}u_n=0$.

Proof Let $s_n=\sum\limits_{i=1}^{n}u_i$, since the series $\sum\limits_{n=1}^{\infty}u_n$ converges, let the sum of the convergent series be s, then from $u_n=s_n-s_{n-1}$, we have

$$\lim_{n\to\infty}u_n=\lim_{n\to\infty}(s_n-s_{n-1})=\lim_{n\to\infty}s_n-\lim_{n\to\infty}s_{n-1}=s-s=0,$$

that completes the proof.

From the necessity of series convergence, when the general term of the series u_n does not tend to zero, the series $\sum\limits_{n=1}^{\infty}u_n$ must diverge. For example, the series $\sum\limits_{n=1}^{\infty}(\sqrt{n^2+n}-n)$, the general term

$$u_n=\sqrt{n^2+n}-n=\frac{n}{\sqrt{n^2+n}+n}\to\frac{1}{2}\neq 0\quad(n\to\infty).$$

Thus, the series diverges.

It should be noted that the converse of Property 7.1.1 is not true, i.e., when $\lim_{n\to\infty} u_n = 0$, $\sum_{n=1}^{\infty} u_n$ does not necessarily converge.

The harmonic series $\sum_{n=1}^{\infty} \frac{1}{n}$ is an example. Here $u_n = \frac{1}{n} \to 0\ (n \to \infty)$, but it diverges. The following proof by contradiction is used: Suppose the harmonic series converges to s, s_n as the sum of its first n terms, then $s_n \to s$, $s_{2n} \to s\ (n\to\infty)$, thus $s_{2n} - s_n \to s - s = 0\ (n\to\infty)$, but

$$s_{2n} - s_n = \frac{1}{n+1} + \frac{1}{n+2} + \cdots + \frac{1}{2n} > \frac{1}{2n} + \frac{1}{2n} + \cdots + \frac{1}{2n} = \frac{1}{2}.$$

Hence, $s_{2n} - s_n$ does not tend to zero, which contradicts the previous conclusion. This contradiction is clearly due to the assumption that the series converges.

Example 7.1.4 Let the series $\sum_{n=1}^{\infty} (u_n + v_n)$ converge, and $v_n \to 0\ (n \to \infty)$, and let the partial sum of the first odd terms of the series $\sum_{n=1}^{\infty} u_n$ be $s_{2k+1} \to s\ (k \to \infty)$. Prove that the series $\sum_{n=1}^{\infty} u_n = s$.

Proof By the necessary condition for convergence, $u_n + v_n \to 0\ (n \to \infty)$, and by $v_n \to 0$, it can be deduced that $u_n \to 0\ (n \to \infty)$, thus $u_{2k+1} \to 0\ (k \to \infty)$. Since $s_{2k} = s_{2k+1} - u_{2k+1} \to s\ (k \to \infty)$, then for any positive integer n, there is $s_n \to s\ (n \to \infty)$, i.e., $\sum_{n=1}^{\infty} u_n$ converges to s.

Property 7.1.2 Let $k \neq 0$, then the series $\sum_{n=1}^{\infty} u_n$ and $\sum_{n=1}^{\infty} ku_n$ have the same convergence property.

Proof Let s_n, σ_n be the partial sums of $\sum_{n=1}^{\infty} u_n$, $\sum_{n=1}^{\infty} ku_n$, since $\sigma_n = ks_n$ and $k \neq 0$, s_n should have the same convergence property as σ_n, i.e., the series $\sum_{n=1}^{\infty} u_n$, $\sum_{n=1}^{\infty} ku_n$ has the same convergence property.

Property 7.1.3 Let the series $\sum_{n=1}^{\infty} u_n$, $\sum_{n=1}^{\infty} v_n$ converge to s and σ respectively, then for any constant k_1, k_2, the series $\sum_{n=1}^{\infty} (k_1 u_n + k_2 v_n)$ also converges, with its sum being $k_1 s + k_2 \sigma$.

Proof Let s_n, σ_n be the first n terms sum of $\sum_{n=1}^{\infty} u_n$ and $\sum_{n=1}^{\infty} v_n$ respectively, then the first n terms sum of the series $\sum_{n=1}^{\infty}(k_1u_n+k_2v_n)$,

$$\Delta_n=\sum_{i=1}^{n}(k_1u_i+k_2v_i)=k_1\sum_{i=1}^{n}u_i+k_2\sum_{i=1}^{n}v_i=k_1s_n+k_2\sigma_n.$$

Since $s_n\to s,\sigma_n\to\sigma\ (n\to\infty)$, thus $\lim\limits_{n\to\infty}\Delta_n=k_1s+k_2\sigma$.

In particular, if the series $\sum_{n=1}^{\infty}u_n$, $\sum_{n=1}^{\infty}v_n$ converge to s and σ respectively, then the series

$$\sum_{n=1}^{\infty}(u_n\pm v_n)=s\pm\sigma=\sum_{n=1}^{\infty}u_n\pm\sum_{n=1}^{\infty}v_n.$$

Note:

(1) If the series $\sum_{n=1}^{\infty}u_n$ converges and the series $\sum_{n=1}^{\infty}v_n$ diverges, then the series $\sum_{n=1}^{\infty}(u_n\pm v_n)$ must diverge. This is because if the series $\sum_{n=1}^{\infty}(u_n\pm v_n)$ converges, then the series $\sum_{n=1}^{\infty}[(u_n\pm v_n)-u_n]=\sum_{n=1}^{\infty}(\pm v_n)$ also converges, which contradicts the divergence of the series $\sum_{n=1}^{\infty}v_n$.

(2) If both series $\sum_{n=1}^{\infty}u_n$ and $\sum_{n=1}^{\infty}v_n$ diverge, then the series $\sum_{n=1}^{\infty}(u_n\pm v_n)$ may converge or diverge. This conclusion is left for the reader to consider.

Example 7.1.5 Determine the convergence of the series $\sum_{n=1}^{\infty}\left(\frac{1}{2^{n-1}}+\frac{2^n}{3^{n-1}}\right)$. If it converges, find its sum.

Solution Since the geometric series $\sum_{n=1}^{\infty}\frac{1}{2^{n-1}}$ $\left(\text{with the common ratio } q=\frac{1}{2}\right)$ converges to $\frac{1}{1-\frac{1}{2}}=2$, and $\sum_{n=1}^{\infty}\frac{2^n}{3^{n-1}}=\sum_{n=1}^{\infty}2\cdot\left(\frac{2}{3}\right)^{n-1}$ converges to $2\cdot\frac{1}{1-\frac{2}{3}}=6$, thus the series converges, and

$$\sum_{n=1}^{\infty}\left(\frac{1}{2^{n-1}}+\frac{2^n}{3^{n-1}}\right)=\sum_{n=1}^{\infty}\frac{1}{2^{n-1}}+\sum_{n=1}^{\infty}\frac{2^n}{3^{n-1}}=2+6=8.$$

Example 7.1.6 Discuss the convergence of the series $\sum_{n=1}^{\infty}\left[\frac{1}{4^n}+\cos(n\pi)\right]$.

Solution Since the geometric series $\sum_{n=1}^{\infty}\frac{1}{4^n}$ converges and the series $\sum_{n=1}^{\infty}\cos(n\pi)=\sum_{n=1}^{\infty}(-1)^n$ diverges, the original series diverges.

Property 7.1.4 Removing, adding, or changing a finite number of terms in a series does not change its convergence.

It suffices to prove that "removing or adding a finite number of terms at the beginning of a series does not alter the convergence of the series." Because other cases (i. e., removing, adding, or altering a finite number of terms anywhere in the series) can be seen as first removing a finite number of terms at the beginning and then adding a finite number of terms.

Proof Suppose the first k terms of the series $u_1+u_2+\cdots+u_k+u_{k+1}+\cdots+u_{k+n}+\cdots$ are removed, resulting in the series $u_{k+1}+u_{k+2}+\cdots+u_{k+n}+\cdots$. The partial sum of the new series is then

$$\sigma_n=u_{k+1}+u_{k+2}+\cdots+u_{k+n}=s_{k+n}-s_k,$$

where: s_{k+n} is the sum of the first $k+n$ terms of the original series. Since s_k is a constant, as $n\to\infty$ approaches infinity, σ_n and s_{k+n} either both have limits or both do not have limits.

Similarly, it can be proven that adding a finite number of terms at the beginning of a series does not change the convergence of the series.

Property 7.1.5 If the series $\sum_{n=1}^{\infty}u_n$ converges, then the series formed by arbitrarily grouping the terms of this series still converges, and its sum remains unchanged.

Proof Let the series $\sum_{n=1}^{\infty}u_n$ formed by arbitrarily grouping its terms be

$$(u_1+\cdots+u_{n_1})+(u_{n_1+1}+\cdots+u_{n_2})+\cdots+(u_{n_{k-1}+1}+\cdots+u_{n_k})+\cdots.$$

The partial sum of the series $\sum_{n=1}^{\infty}u_n$ (corresponding to the first n terms) is s_n, and the partial sum of the series (corresponding to the first k terms) is A_k, then

$$A_1=u_1+\cdots+u_{n_1}=s_{n_1},$$

$$A_2=(u_1+\cdots+u_{n_1})+(u_{n_1+1}+\cdots+u_{n_2})=s_{n_2},\cdots,$$

$$A_k=(u_1+\cdots+u_{n_1})+(u_{n_1+1}+\cdots+u_{n_2})+\cdots+(u_{n_{k-1}+1}+\cdots+u_{n_k})=s_{n_k},\cdots.$$

The visible sequence $\{A_k\}$ is a subsequence of the sequence $\{s_n\}$. Given the convergence of the sequence $\{s_n\}$ and the relationship between convergent sequences and their subsequences, it follows that the sequence $\{A_k\}$ must converge, and we have $\lim\limits_{k\to\infty}A_k=\lim\limits_{n\to\infty}s_n$, meaning that the series formed by grouping terms converges with the same sum.

Note: The converse of this proposition does not hold, meaning that even if a series converges when grouped, it does not necessarily converge without grouping. For example, the series $(1-1)+(1-1)+\cdots$ converges to zero, but the series $1-1+1-1+\cdots$ without grouping diverges. From Property 7.1.5, we can deduce the following corollary: If the series formed by grouping terms diverges, then the original series also diverges. This is because if the original series converges, then according to Property 7.1.5, the series formed by grouping terms should converge.

Exercise 7.1

1. Answer the following questions:

(1) Is the general term $a_n \to 0 (n \to \infty)$ a sufficient condition for the convergence of the series $\sum_{n=1}^{\infty} a_n$? If not, provide an example.

(2) If $\sum_{n=1}^{\infty} a_n$ converges and $\sum_{n=1}^{\infty} b_n$ diverges, is $\sum_{n=1}^{\infty} (a_n + b_n)$ convergent or divergent?

(3) If $\sum_{n=1}^{\infty} a_n$ and $\sum_{n=1}^{\infty} b_n$ both diverge, is $\sum_{n=1}^{\infty} (a_n + b_n)$ convergent or divergent?

(4) If $\sum_{n=1}^{\infty} u_n$ converges, is $\sum_{n=1}^{\infty} u_{n+50}$, $\sum_{n=1}^{\infty} \frac{1}{u_n}$ convergent or divergent?

(5) If $\sum_{n=1}^{\infty} u_n$ diverges, is $\sum_{n=1}^{\infty} u_{n+50}$, $\sum_{n=1}^{\infty} \frac{1}{u_n}$ convergent or divergent?

2. Write the first five terms of the following series:

(1) $\sum_{n=1}^{\infty} (-1)^{n-1} \frac{1}{n}$; (2) $\sum_{n=1}^{\infty} \frac{n!}{n^n}$; (3) $\sum_{n=1}^{\infty} \frac{1 \cdot 3 \cdot 5 \cdot \cdots \cdot (2n-1)}{2 \cdot 4 \cdot 6 \cdot \cdots \cdot (2n)}$.

3. Write the general term of the following series:

(1) $\frac{2}{1 \cdot 2} + \frac{3}{2 \cdot 3} + \frac{4}{3 \cdot 4} + \cdots$;

(2) $1 + \frac{6}{2^2} + \frac{12}{2^3} + \frac{20}{2^4} + \frac{30}{2^5} + \cdots$;

(3) $\frac{2}{1} - \frac{3}{2} + \frac{4}{3} - \frac{5}{4} + \frac{6}{5} - \cdots$;

(4) $\frac{\sqrt{x}}{2} + \frac{x}{2 \cdot 4} + \frac{x\sqrt{x}}{2 \cdot 4 \cdot 6} + \frac{x^2}{2 \cdot 4 \cdot 6 \cdot 8} + \cdots$.

4. Determine the convergence of the following series based on the definition of convergence and divergence:

(1) $\frac{1}{1 \cdot 6} + \frac{1}{6 \cdot 11} + \cdots + \frac{1}{(5n-4)(5n+1)} + \cdots$;

(2) $\sum_{n=1}^{\infty} (\sqrt{n+1} - \sqrt{n})$;

(3) $\sum_{n=1}^{\infty} (\sqrt{n+2} - 2\sqrt{n+1} + \sqrt{n})$;

(4) $\sum_{n=1}^{\infty} \sin \frac{n\pi}{6}$.

5. Determine the convergence of the following series:

(1) $-\frac{8}{9} + \frac{8^2}{9^2} - \frac{8^3}{9^3} + \cdots + (-1)^n \frac{8^n}{9^n} + \cdots$;

(2) $\left(\frac{1}{2} + \frac{1}{3}\right) + \left(\frac{1}{2^2} + \frac{1}{3^2}\right) + \left(\frac{1}{2^3} + \frac{1}{3^3}\right) + \cdots + \left(\frac{1}{2^n} + \frac{1}{3^n}\right) + \cdots$;

(3) $\sum_{n=1}^{\infty} \left(\frac{1}{3n} + \frac{1}{2^n}\right)$;

(4) $1! + 2! + 3! + 4! + \cdots + n! + \cdots$.

7.2 Positive Series

For the series $\sum_{n=1}^{\infty} u_n$, if all terms have the same sign, whether all positive or all negative, it is referred to as a series with terms of the same sign. The study of the convergence of such series is reduced to the study of the convergence of positive series. A positive series $\sum_{n=1}^{\infty} u_n$ refers to $u_n \geqslant 0$, $\forall n \in \mathbf{N}^+$. A series with all negative terms, when each term is multiplied by -1, is transformed into a positive series.

The reason for specifically studying positive series is that they are relatively simple and possess many good properties. Additionally, the convergence of many other types of series can be studied by transforming them into positive series.

7.2.1 Comparison Test

Let the series of positive terms be

$$u_1 + u_2 + \cdots + u_n + \cdots = \sum_{n=1}^{\infty} u_n,$$

s_n is its partial sum, it is obvious that $s_1 \leqslant s_2 \leqslant \cdots \leqslant s_n \leqslant \cdots$, i.e., $\{s_n\}$ is a monotonically increasing sequence, hence the necessary and sufficient condition for $\lim_{n \to \infty} s_n$ to exist is that $\{s_n\}$ is bounded, thus the following theorem holds:

Theorem 7.2.1 A series of positive terms $\sum_{n=1}^{\infty} u_n$ converges if and only if its sequence of partial sums $\{s_n\}$ is bounded above.

If the sequence of partial sums of a series of positive terms is unbounded, then it must diverge to $+\infty$, i.e., $\sum_{n=1}^{\infty} u_n = +\infty$.

Although the expression for s_n is difficult to derive, and determining its boundedness is not straightforward, this does not diminish the theoretical importance of Theorem 7.2.1.

Theorem 7.2.2 (Comparison Test) Let $\sum_{n=1}^{\infty} u_n$ and $\sum_{n=1}^{\infty} v_n$ be series of positive terms, with $\forall n \in \mathbf{N}^+$, such that $u_n \leqslant v_n$, then

(1) When $\sum_{n=1}^{\infty} v_n$ converges, $\sum_{n=1}^{\infty} u_n$ also converges;

(2) When $\sum_{n=1}^{\infty} u_n$ diverges, $\sum_{n=1}^{\infty} v_n$ also diverges.

Proof Let $s_n = \sum_{i=1}^{n} u_i$, $\sigma_n = \sum_{i=1}^{n} v_i$, since $\forall n \in \mathbf{N}^+$, has $u_n \leqslant v_n$, hence $s_n \leqslant \sigma_n$.

(1) When $\sum_{n=1}^{\infty} v_n$ converges, by Theorem 7.2.1, the sequence $\{\sigma_n\}$ is bounded, and thus the sequence $\{s_n\}$ is bounded. Similarly, by Theorem 7.2.1, $\sum_{n=1}^{\infty} u_n$ converges.

(2) Since $\sum_{n=1}^{\infty} u_n$ diverges, $s_n \to +\infty\ (n\to\infty)$, and thus $\sigma_n \to +\infty\ (n\to\infty)$, i.e., the sequence $\{\sigma_n\}$ is unbounded. By Theorem 7.2.1, $\sum_{n=1}^{\infty} v_n$ diverges.

Since multiplying each term of a series by a non-zero constant and removing a finite number of initial terms does not change the convergence of the series, the following corollary of Theorem 7.2.2 can be obtained.

Corollary 7.2.1 Let $\sum_{n=1}^{\infty} u_n$ and $\sum_{n=1}^{\infty} v_n$ be positive series,

(1) If $\sum_{n=1}^{\infty} v_n$ converges, and there exists a positive integer N, such that when $n \geqslant N$, $u_n \leqslant kv_n\ (k > 0)$, then $\sum_{n=1}^{\infty} u_n$ also converges.

(2) If $\sum_{n=1}^{\infty} v_n$ diverges, and there exists a positive integer N, such that when $n \geqslant N$, $u_n \geqslant kv_n\ (k > 0)$, then $\sum_{n=1}^{\infty} u_n$ also diverges.

Example 7.2.1 Discuss the convergence of the series p

$$1+\frac{1}{2^p}+\frac{1}{3^p}+\cdots+\frac{1}{n^p}+\cdots,$$

where the constant $p > 0$.

Solution When $p \leqslant 1$, $\frac{1}{n^p} \geqslant \frac{1}{n}$ $(\forall n \geqslant 1)$, and the series $\sum_{n=1}^{\infty}\frac{1}{n}$ is the harmonic series, which is divergent.

According to the comparison test, the series p diverges. Let $p > 1$, since when $n-1 \leqslant x \leqslant n$, there is $\frac{1}{n^p} \leqslant \frac{1}{x^p}$, hence

$$\frac{1}{n^p}=\int_{n-1}^{n}\frac{1}{n^p}\mathrm{d}x \leqslant \int_{n-1}^{n}\frac{1}{x^p}\mathrm{d}x=\frac{1}{p-1}\left[\frac{1}{(n-1)^{p-1}}-\frac{1}{n^{p-1}}\right] \quad (n=2,3,\cdots).$$

Consider the partial sums of the series $\sum_{n=2}^{\infty}\left[\frac{1}{(n-1)^{p-1}}-\frac{1}{n^{p-1}}\right]$,

$$s_n=\left(1-\frac{1}{2^{p-1}}\right)+\left(\frac{1}{2^{p-1}}-\frac{1}{3^{p-1}}\right)+\cdots+\left[\frac{1}{n^{p-1}}-\frac{1}{(n+1)^{p-1}}\right]=1-\frac{1}{(n+1)^{p-1}}.$$

Since

$$\lim_{n\to\infty} s_n=\lim_{n\to\infty}\left[1-\frac{1}{(n+1)^{p-1}}\right]=1,$$

therefore, the series p converges.

In summary, the series p converges when $p>1$, it diverges when $p\leqslant 1$.

Example 7.2.2 Determine the convergence of the series $\sum_{n=1}^{\infty}\frac{\ln n}{\sqrt{n}}$.

Solution When $n\geqslant 3$, $\ln n>1$, thus obtaining $\frac{\ln n}{\sqrt{n}}>\frac{1}{\sqrt{n}}$, since the series $\sum_{n=1}^{\infty}\frac{1}{\sqrt{n}}$ diverges, by the corollary of the comparison test, the original series diverges.

Example 7.2.3 Prove:

(1) The series $\sum_{n=2}^{\infty}\frac{1}{n\ln n}$ is divergent;

(2) The series $\sum_{n=2}^{\infty}\frac{1}{n\,(\ln n)^2}$ is convergent.

Proof

(1) $\frac{1}{n\ln n}=\int_n^{n+1}\frac{dx}{n\ln n}>\int_n^{n+1}\frac{dx}{x\ln x}=u_n\quad(n=2,3,\cdots)$,

obviously, $u_n>0$, and

$$u_2+u_3+\cdots+u_{n+1}=\int_2^{n+2}\frac{dx}{x\ln x}=\ln[\ln(n+2)]-\ln(\ln 2)\to+\infty\quad(n\to+\infty),$$

therefore, $\sum_{n=2}^{\infty}u_n$ diverges, and thus $\sum_{n=2}^{\infty}\frac{1}{n\ln n}$ diverges.

(2) $0<\frac{1}{n\,(\ln n)^2}=\int_{n-1}^{n}\frac{dx}{n\,(\ln n)^2}<\int_{n-1}^{n}\frac{dx}{x\,(\ln x)^2}=v_n\quad(n=3,4,\cdots)$,

while

$$v_3+v_4+\cdots+v_{n+2}=\int_2^{n+2}\frac{dx}{x\,(\ln x)^2}=\frac{1}{\ln 2}-\frac{1}{\ln(n+2)}\to\frac{1}{\ln 2}\quad(n\to\infty).$$

The series $\sum_{n=3}^{\infty}v_n$ converges, so the series $\sum_{n=3}^{\infty}\frac{1}{n\,(\ln n)^2}$ converges, and thus the original series converges.

Theorem 7.2.3 (Limit Comparison Test) Let $\sum_{n=1}^{\infty}u_n$ and $\sum_{n=1}^{\infty}v_n$ be positive series, if $\lim_{n\to\infty}\frac{u_n}{v_n}=l\ (v_n>0)$, then

(1) When $0<l<+\infty$, $\sum_{n=1}^{\infty}u_n$ and $\sum_{n=1}^{\infty}v_n$ either both converge or both diverge;

(2) When $l=0$ and $\sum_{n=1}^{\infty}v_n$ converges, $\sum_{n=1}^{\infty}u_n$ converges;

(3) When $l=+\infty$ and $\sum_{n=1}^{\infty}v_n$ diverges, $\sum_{n=1}^{\infty}u_n$ diverges.

Proof

(1) Let $0<l<+\infty$, for $\varepsilon=\frac{l}{2}>0$, from $\lim\limits_{n\to\infty}\frac{u_n}{v_n}=l$, there exists a positive integer N, such that when $n>N$, the inequality $l-\frac{l}{2}<\frac{u_n}{v_n}<l+\frac{l}{2}$ or $\frac{l}{2}<\frac{u_n}{v_n}<\frac{3}{2}l$ holds, i.e., $\frac{l}{2}v_n<u_n<\frac{3l}{2}v_n$. According to the Comparison Test, the series $\sum\limits_{n=1}^{\infty}u_n$ and $\sum\limits_{n=1}^{\infty}v_n$ either both converge or both diverge.

(2) Let $l=0$, for $\varepsilon=1$, then there exists a positive integer N, such that when $n>N$, $\frac{u_n}{v_n}<1$ holds, i.e., $u_n<v_n$. According to the comparison test, since $\sum\limits_{n=1}^{\infty}v_n$ converges, $\sum\limits_{n=1}^{\infty}u_n$ must also converge.

(3) Let $l=+\infty$, then there is a positive integer N, when $n>N$, there is $\frac{u_n}{v_n}>1$, i.e., $u_n>v_n$. According to the comparison test, since $\sum\limits_{n=1}^{\infty}v_n$ diverges, $\sum\limits_{n=1}^{\infty}u_n$ also diverges.

The limit form of the comparison test, under the condition that the general terms of two positive series both tend to zero, actually compares the orders of these general terms as infinitesimal quantities. That is, when u_n and v_n are infinitesimals of the same order (including u_n and v_n being equivalent infinitesimals), the series $\sum\limits_{n=1}^{\infty}u_n$ and $\sum\limits_{n=1}^{\infty}v_n$ converge or diverge simultaneously. When u_n is a higher-order infinitesimal of v_n, if $\sum\limits_{n=1}^{\infty}v_n$ converges, then $\sum\limits_{n=1}^{\infty}u_n$ must converge. When u_n is a lower-order infinitesimal of v_n, if $\sum\limits_{n=1}^{\infty}v_n$ diverges, then $\sum\limits_{n=1}^{\infty}u_n$ must diverge.

Example 7.2.4 Discuss the convergence of the series $\sum\limits_{n=1}^{\infty}\sin\frac{1}{n}$.

Solution Since $0<\frac{1}{n}<\frac{\pi}{2}$, hence $\sin\frac{1}{n}>0$, and $\sin\frac{1}{n}\sim\frac{1}{n}\ (n\to\infty)$, i.e., $\lim\limits_{n\to\infty}\frac{\sin\frac{1}{n}}{\frac{1}{n}}=1.$ Since $\sum\limits_{n=1}^{\infty}\frac{1}{n}$ diverges, the original series also diverges.

Example 7.2.5 Determine the convergence of the series $\sum\limits_{n=1}^{\infty}\frac{1}{\ln(n+1)}$.

Solution Because

$$\lim_{n\to\infty}\frac{\frac{1}{\ln(n+1)}}{\frac{1}{n}}=\lim_{n\to\infty}\frac{n}{\ln(n+1)}=\lim_{x\to+\infty}\frac{x}{\ln(x+1)}=\lim_{x\to+\infty}\frac{1}{\frac{1}{1+x}}=+\infty,$$

since $\sum_{n=1}^{\infty}\frac{1}{n}$ diverges, the original series diverges.

Theorem 7.2.4 (Limit Comparison Test or p Test)　Let $\sum_{n=1}^{\infty}u_n$ be a series of positive terms.

(1) If $\lim_{n\to\infty}nu_n=l>0$ (or $\lim_{n\to\infty}nu_n=+\infty$), then the series $\sum_{n=1}^{\infty}u_n$ diverges;

(2) If $p>1$, and $\lim_{n\to\infty}n^p u_n=l$ $(0\leqslant l<+\infty)$, then the series $\sum_{n=1}^{\infty}u_n$ converges.

Proof　(1) In the limit comparison test, take $v_n=\frac{1}{n}$, since the harmonic series $\sum_{n=1}^{\infty}\frac{1}{n}$ diverges, the conclusion holds.

(2) In the limit comparison test, take $v_n=\frac{1}{n^p}$, when $p>1$, the series p converges, hence the conclusion holds.

Example 7.2.6　Determine the convergence of the series $\sum_{n=1}^{\infty}\sqrt{n+1}\left(1-\cos\frac{\pi}{n}\right)$.

Solution　Because

$$\lim_{n\to\infty}n^{\frac{3}{2}}u_n=\lim_{n\to\infty}n^{\frac{3}{2}}\sqrt{n+1}\left(1-\cos\frac{\pi}{n}\right)=\lim_{n\to\infty}n^2\sqrt{\frac{n+1}{n}}\cdot\frac{1}{2}\left(\frac{\pi}{n}\right)^2=\frac{1}{2}\pi^2,$$

according to the p test, the original series converges.

It is worth noting that when using the comparison test or the limit comparison test, it is necessary to construct a series $\sum_{n=1}^{\infty}v_n$ whose convergence is known in advance, as a benchmark for comparison with the known series. The most commonly used benchmark series are the geometric series and the p series.

7.2.2　Ratio Test and Root Test

Comparing the given positive series with a geometric series yields a very practical ratio test.

Theorem 7.2.5 (D'Alembert's Test, Ratio Test)　Let $\sum_{n=1}^{\infty}u_n$ be a positive series, if $\lim_{n\to\infty}\frac{u_{n+1}}{u_n}=\rho$ (where ρ can be $+\infty$), then

(1) When $\rho<1$, $\sum_{n=1}^{\infty}u_n$ converges;

(2) When $1<\rho\leqslant+\infty$, $\sum_{n=1}^{\infty}u_n$ diverges;

(3) When $\rho=1$, $\sum_{n=1}^{\infty}u_n$ may converge or diverge.

Proof (1) When $\rho<1$, take a sufficiently small positive number ε, such that $\rho+\varepsilon=r<1$, then by the definition of the limit, there exists a positive integer m, such that when $n\geqslant m$, $\frac{u_{n+1}}{u_n}<\rho+\varepsilon=r$, therefore

$$u_{m+1}<ru_m, u_{m+2}<ru_{m+1}<r^2u_m, \cdots, u_{m+k}<r^ku_m, \cdots,$$

the series $\sum_{k=1}^{\infty} r^k u_m$ converges (with common ratio $r<1$), according to the corollary of the comparison test, the series $\sum_{n=1}^{\infty} u_n$ converges.

(2) When $\rho>1$, take a sufficiently small $\varepsilon>0$, such that $\rho-\varepsilon>1$, by the definition of the limit, when $n\geqslant m$, $\frac{u_{n+1}}{u_n}>\rho-\varepsilon>1$, i.e., $u_{n+1}>u_n$. Therefore, when $n\geqslant m$, the general term of the series u_n gradually increases, thus $\lim_{n\to\infty} u_n\neq 0$. According to the necessary condition for the convergence of a series, the series $\sum_{n=1}^{\infty} u_n$ diverges.

Similarly, it can be proven that when $\lim_{n\to\infty}\frac{u_{n+1}}{u_n}=+\infty$, the series $\sum_{n=1}^{\infty} u_n$ diverges.

(3) When $\rho=1$, the series may converge or diverge. For example, the series $\sum_{n=1}^{\infty}\frac{1}{n^p}$ with p, regardless of the value of p, always has $\lim_{n\to\infty}\frac{u_{n+1}}{u_n}=1$. However, in practice, when $p>1$, the series p converges; when $p\leqslant 1$, the series p diverges. Therefore, the convergence of the series cannot be determined based solely on $\rho=1$.

Example 7.2.7 Prove that the series $1+\frac{1}{1}+\frac{1}{1\cdot 2}+\frac{1}{1\cdot 2\cdot 3}+\cdots+\frac{1}{(n-1)!}+\cdots$ converges and estimate the error resulting from replacing the sum s with the partial sum s_n of the series.

Solution Since $\lim_{n\to\infty}\frac{u_{n+1}}{u_n}=\lim_{n\to\infty}\frac{(n-1)!}{n!}=\lim_{n\to\infty}\frac{1}{n}=0<1$, according to the ratio test, the original series converges. And

$$|r_n|=\frac{1}{n!}+\frac{1}{(n+1)!}+\frac{1}{(n+2)!}+\cdots=\frac{1}{n!}\left[1+\frac{1}{n+1}+\frac{1}{(n+1)(n+2)}+\cdots\right]$$

$$<\frac{1}{n!}\left(1+\frac{1}{n}+\frac{1}{n^2}+\cdots\right)=\frac{1}{n!}\cdot\frac{1}{1-\frac{1}{n}}=\frac{1}{(n-1)(n-1)!}.$$

Example 7.2.8 Determine the convergence of the series $\sum_{n=1}^{\infty}\frac{n\cos^2\frac{n}{3}\pi}{2^n}$.

Solution Since $\frac{n\cos^2\frac{n\pi}{3}}{2^n}\leqslant\frac{n}{2^n}\triangleq u_n$, first consider the convergence of the series $\sum_{n=1}^{\infty}\frac{n}{2^n}$.

$$\lim_{n\to\infty}\frac{u_{n+1}}{u_n}=\lim_{n\to\infty}\frac{\dfrac{n+1}{2^{n+1}}}{\dfrac{n}{2^n}}=\lim_{n\to\infty}\frac{1}{2}\cdot\frac{n+1}{n}=\frac{1}{2}<1.$$

By the ratio test, the series $\sum_{n=1}^{\infty}\frac{n}{2^n}$ converges, and then by the comparison test, it is known that the original series converges.

Example 7.2.9 Given $x>0$, discuss the convergence of the series $\sum_{n=1}^{\infty}n!\left(\frac{x}{n}\right)^n$ and find the limit $\lim_{n\to\infty}n!\left(\frac{2}{n}\right)^n$.

Solution Let $u_n=n!\left(\frac{x}{n}\right)^n$, because

$$\lim_{n\to\infty}\frac{u_{n+1}}{u_n}=\lim_{n\to\infty}\frac{(n+1)!\left(\dfrac{x}{n+1}\right)^{n+1}}{n!\left(\dfrac{x}{n}\right)^n}=\lim_{n\to\infty}\frac{x}{\left(1+\dfrac{1}{n}\right)^n}=\frac{x}{e},$$

when $0<\frac{x}{e}<1$, i.e., $0<x<e$, the original series converges; when $\frac{x}{e}>1$, i.e., $x>e$, the original series diverges; when $\frac{x}{e}=1$, i.e., $x=e$, the ratio test cannot determine convergence, but $\left(1+\frac{1}{n}\right)^n$ increases monotonically to $e\,(n>3)$, thus $\frac{u_{n+1}}{u_n}$ decreases monotonically to 1, i.e., for any positive integer N, when $n\geqslant N$, there is $u_n\geqslant u_{n-1}\geqslant\cdots\geqslant u_N>0$. Therefore, $\lim_{n\to\infty}u_n\neq 0$, so when $x=e$, the original series diverges. For the limit part, it is immediately obtained based on the necessary condition for the convergence of a series.

$$\lim_{n\to\infty}n!\left(\frac{2}{n}\right)^n=0\quad(0<2<e).$$

* **Theorem 7.2.6 (Cauchy's Root Test)** Let $\sum_{n=1}^{\infty}u_n$ be a positive series, if $\lim_{n\to\infty}\sqrt[n]{u_n}=\rho$ (where ρ can be $+\infty$), then

(1) When $\rho<1$, the series converges;

(2) When $1<\rho\leqslant+\infty$, the series diverges;

(3) When $\rho=1$, the series may converge or diverge.

The proof of Theorem 7.2.6 is similar to that of Theorem 7.2.5, and is omitted here.

Example 7.2.10 Determine the convergence of the series $\sum_{n=1}^{\infty}\frac{2+(-1)^n}{2^n}$.

Solution Because

$$\lim_{n\to\infty}\sqrt[n]{u_n}=\lim_{n\to\infty}\frac{1}{2}\sqrt[n]{2+(-1)^n}=\frac{1}{2}<1,$$

therefore, according to the root test, the original series converges.

From the ratio test, root test, and their examples, it is evident that both tests have their limitations (when $\rho=1$). Generally, when the general term of the series contains factorials, the ratio test is convenient for simplifying and finding the limit, while when the general term contains a power of n, the root test can be considered.

Exercise 7.2

1. Determine the convergence of the following series using the comparison test:

(1) $\sum_{n=1}^{\infty}\frac{2}{(n+1)(n+4)}$; (2) $\sum_{n=1}^{\infty}\frac{1}{\sqrt{4n^2+n}}$;

(3) $\sum_{n=1}^{\infty}\frac{1}{n}\frac{a}{1+a^n}\ (a>0)$; (4) $\sum_{n=1}^{\infty}\frac{1}{(2n-1)^2}$;

(5) $\sum_{n=1}^{\infty}\frac{\ln n}{n^{\frac{4}{3}}}$; (6) $\sum_{n=1}^{\infty}\frac{4n-1}{n^2+n}$;

(7) $\sum_{n=1}^{\infty}\left(1-\cos\frac{\pi}{n}\right)$; (8) $\sum_{n=1}^{\infty}\ln\left(1+\frac{a}{n}\right)\ (a>0)$;

(9) $\sum_{n=2}^{\infty}\frac{1}{\sqrt{n}}\ln\frac{n+1}{n-1}$; (10) $\sum_{n=1}^{\infty}\left(\frac{1+n^2}{1+n^3}\right)^2$.

2. Determine the convergence of the following series using the ratio test:

(1) $\sum_{n=1}^{\infty}\frac{2^n}{\sqrt{n^n}}$; (2) $\sum_{n=1}^{\infty}\frac{3^n\cdot n!}{n^n}$;

(3) $\sum_{n=1}^{\infty}n^2\sin\frac{\pi}{3^n}$; (4) $\sum_{n=1}^{\infty}n\tan\frac{\pi}{2^{n+1}}$;

(5) $\sum_{n=1}^{\infty}\frac{(2n-1)!!}{3^n\cdot n!}$; (6) $\sum_{n=1}^{\infty}\frac{1\cdot 5\cdot 9\cdot\cdots\cdot(4n-3)}{2\cdot 5\cdot 8\cdot\cdots\cdot(3n-1)}$;

(7) $\sum_{n=1}^{\infty}\frac{(n!)^2}{(2n)!}$.

3. Determine the convergence of the following series using the root test:

(1) $\sum_{n=1}^{\infty}\left(\frac{2n+1}{3n-2}\right)^n$; (2) $\sum_{n=1}^{\infty}\frac{1}{[\ln(n+1)]^n}$;

(3) $\sum_{n=1}^{\infty}n^n\left(\sin\frac{\pi}{n}\right)^n$; (4) $\sum_{n=1}^{\infty}\left(\frac{n}{3n-1}\right)^{2n-1}$;

(5) $\sum_{n=1}^{\infty}\left(\frac{b}{a_n}\right)^n$, where $\lim_{n\to\infty}a_n=a$, and both $a\neq b$, a, b and a_n are positive numbers.

4. Determine the convergence of the following series:

(1) $\sum_{n=1}^{\infty}\frac{1}{(an^2+bn+c)^{\beta}}$ ($a>0$, $b>0$, $c>0$, β are all constant numbers);

(2) $\sum_{n=1}^{\infty}\frac{n^4}{n!}$; (3) $\sum_{n=1}^{\infty}\left(\frac{n}{n+1}\right)^n$;

(4) $\sum_{n=1}^{\infty} 2^n \sin\frac{\pi}{3^n}$; (5) $\sum_{n=1}^{\infty} \frac{a^n}{n^s}$ $(a>0, s>0)$.

5. Suppose $u_n \geqslant 0$ $(n=1,2,3,\cdots)$, $\sum_{n=1}^{\infty} u_n$ converges, prove that $\sum_{n=1}^{\infty} u_n^2$ also converges. Conversely, if $\sum_{n=1}^{\infty} u_n^2$ converges, does $\sum_{n=1}^{\infty} u_n$ also converge? Why?

6. Prove the following limit equalities:

(1) $\lim\limits_{n\to\infty}\frac{n^n}{(n!)^2}=0$; (2) $\lim\limits_{n\to\infty}\frac{n^k}{a^n}=0$ $(a>1)$.

7. Suppose both series $\sum_{n=1}^{\infty} a_n$, $\sum_{n=1}^{\infty} b_n$ converge, and $a_n \leqslant c_n \leqslant b_n$, prove that series $\sum_{n=1}^{\infty} c_n$ also converges.

7.3 Series with Arbitrary Terms

The convergence of positive series was discussed in the previous section, while the determination of the convergence of arbitrary series (whose terms can be positive or negative, also known as general series or alternating series) is much more complex than that of series with the same sign. This section will introduce methods for determining the convergence of arbitrary series, mainly focusing on the convergence of a special type of arbitrary series (alternating series) and transforming arbitrary series into positive series to study their convergence.

7.3.1 Alternating Series

Definition 7.3.1 Let $u_n > 0$, $\forall n \in \mathbf{N}^+$, the series is called

$$u_1 - u_2 + u_3 - u_4 + \cdots + (-1)^{n-1}u_n + \cdots = \sum_{n=1}^{\infty} (-1)^{n-1}u_n$$

or the series

$$-u_1 + u_2 - u_3 + \cdots + (-1)^n u_n + \cdots = \sum_{n=1}^{\infty} (-1)^n u_n$$

is called an alternating series.

Clearly, the series can be transformed into each other by multiplying by -1, so we only discuss the convergence of first series.

Theorem 7.3.1 (Leibniz's Theorem) If the alternating series $\sum_{n=1}^{\infty} (-1)^{n-1}u_n$ satisfies:

(1) $u_n \geqslant u_{n+1}$ $(n=1,2,\cdots)$;

(2) $\lim\limits_{n\to\infty} u_n = 0$.

Then the series $u_1-u_2+u_3-u_4+\cdots+(-1)^{n-1}u_n+\cdots=\sum_{n=1}^{\infty}(-1)^{n-1}u_n$ converges, and the sum of the convergence is $s\leqslant u_1$, the absolute value of the remainder r_n is $|r_n|\leqslant u_{n+1}$.

Proof First, we prove that the limit of the sum of the first $2n$ terms exists. For this purpose, we write s_{2n} in the following two forms:

$$s_{2n}=(u_1-u_2)+(u_3-u_4)+\cdots+(u_{2n-1}-u_{2n})$$

and

$$s_{2n}=u_1-(u_2-u_3)-(u_4-u_5)-\cdots-(u_{2n-2}-u_{2n-1})-u_{2n}.$$

According to condition (1), the difference within all brackets is positive. From the first form, it is evident that the sequence $\{s_{2n}\}$ is monotonically increasing, and from the second form, it is evident that $s_{2n}<u_1$. Therefore, according to the criterion that a monotonically bounded sequence must have a limit, as n increases indefinitely, s_{2n} approaches a limit s, and s is not greater than u_1, i.e., $\lim_{n\to\infty}s_{2n}=s\leqslant u_1$.

Next, we prove that the limit of the sum of the first $2n+1$ terms s_{2n+1} is also s. In fact, we have $s_{2n+1}=s_{2n}+u_{2n+1}$. And from condition (2), it follows that $\lim_{n\to\infty}u_{2n+1}=0$, thus $\lim_{n\to\infty}s_{2n+1}=\lim_{n\to\infty}(s_{2n}+u_{2n+1})=s$.

In summary, for all $\forall n\in\mathbf{N}^+$, we have $\lim_{n\to\infty}s_n=s$, i.e., the series converges to s, and $s\leqslant u_1$.

Additionally, the remainder term $r_n=\pm(u_{n+1}-u_{n+2}+\cdots)$, whose absolute value $|r_n|=u_{n+1}-u_{n+2}+\cdots$, the right-hand side of the above equation is also an alternating series, which also satisfies the two conditions for convergence, so its sum is less than the first term of the series, that is, $|r_n|\leqslant u_{n+1}$. This completes the proof.

Note: The conditions in Theorem 7.3.1 are sufficient but not necessary for the convergence of an alternating series. Therefore, when an alternating series does not satisfy the conditions of the theorem, it cannot be concluded that the series diverges.

Example 7.3.1 Prove that the series $1-\frac{1}{2}+\frac{1}{3}-\frac{1}{4}+\cdots+(-1)^{n-1}\frac{1}{n}+\cdots$ converges and estimate the remainder.

Proof Because

$$u_n=\frac{1}{n}>\frac{1}{n+1}=u_{n+1}\quad(n=1,2,3,\cdots),$$

since $\lim_{n\to\infty}u_n=\lim_{n\to\infty}\frac{1}{n}=0$, the original series converges, and the sum of convergence is $s<1$. If we take $s_n=1-\frac{1}{2}+\frac{1}{3}-\cdots+(-1)^{n-1}\frac{1}{n}$ as an approximation of s, the resulting error is $|r_n|\leqslant\frac{1}{n+1}$.

Example 7.3.2 Prove that the series $\sum_{n=2}^{\infty} \frac{(-1)^n \ln n}{n}$ converges.

Proof Let $f(x)=\frac{\ln x}{x}$, then $f'(x)=\frac{1-\ln x}{x^2}<0\ (x \geqslant 3)$, that is, $f(x)$ is monotonically decreasing, so when $n \geqslant 3, f(n)>f(n+1)$, that is, $u_n > u_{n+1}$. And $\lim_{n \to \infty} u_n = \lim_{n \to \infty} \frac{\ln n}{n}=0$, by Theorem 7.3.1, the series $\sum_{n=3}^{\infty} \frac{(-1)^{n-1} \ln n}{n}$ converges, therefore, the series $\sum_{n=2}^{\infty} \frac{(-1)^n \ln n}{n}$ converges.

7.3.2 Absolute Convergence and Conditional Convergence

For an arbitrary term series $\sum_{n=1}^{\infty} u_n$, taking the absolute value of each term results in a positive term series $\sum_{n=1}^{\infty} |u_n|$. The convergence problem of arbitrary term series is often transformed into the convergence problem of positive term series for discussion. Below, we first introduce the concepts of absolute convergence and conditional convergence.

Definition 7.3.2 If the series of absolute values of the terms of the arbitrary term series $\sum_{n=1}^{\infty} u_n = u_1 + u_2 + \cdots + u_n + \cdots$ converges, then the series $\sum_{n=1}^{\infty} u_n$ is said to be absolutely convergent. If the arbitrary term series $\sum_{n=1}^{\infty} u_n$ converges while $\sum_{n=1}^{\infty} |u_n|$ diverges, then the series $\sum_{n=1}^{\infty} u_n$ is said to be conditionally convergent.

It is easy to verify from Definition 7.3.2 that the series $\sum_{n=1}^{\infty} (-1)^n \frac{1}{n^2}$ is absolutely convergent, while the series $\sum_{n=1}^{\infty} (-1)^{n-1} \frac{1}{n}$ is a conditionally convergent series.

The following important relationship exists between the absolute convergence of a series and its convergence:

Theorem 7.3.2 For an arbitrary term series $\sum_{n=1}^{\infty} u_n$, if the series $\sum_{n=1}^{\infty} |u_n|$ converges, then the arbitrary term series $\sum_{n=1}^{\infty} u_n$ converges.

Theorem 7.3.2 can also be stated as "If an arbitrary term series is absolutely convergent, then it must converge."

Proof Let $v_n = \frac{1}{2}(u_n + |u_n|)(n=1,2,\cdots)$, it is obvious that $v_n \geqslant 0$, and $v_n \leqslant |u_n|(n=1,2,\cdots)$, since the series $\sum_{n=1}^{\infty}|u_n|$ converges, by the comparison test, the series $\sum_{n=1}^{\infty}v_n$ converges, thus, the series $\sum_{n=1}^{\infty}2v_n$ also converges. And $u_n = 2v_n - |u_n|$, from the basic properties of convergent series, it follows that the series $\sum_{n=1}^{\infty}u_n = \sum_{n=1}^{\infty}2v_n - \sum_{n=1}^{\infty}|u_n|$ must converge, thus completing the proof.

Note the series $\sum_{n=1}^{\infty}v_n$ introduced in the proof, whose general term

$$v_n = \frac{1}{2}(u_n + |u_n|) = \begin{cases} u_n, & u_n > 0 \\ 0, & u_n \leqslant 0 \end{cases}.$$

The visible series $\sum_{n=1}^{\infty}v_n$ is obtained by replacing the negative terms in the series $\sum_{n=1}^{\infty}u_n$ with 0. Similarly, if we let $w_n = \frac{1}{2}(|u_n| - u_n)$, then the series $\sum_{n=1}^{\infty}w_n$ is the series consisting of the absolute values of all negative terms in the series $\sum_{n=1}^{\infty}u_n$. If the series $\sum_{n=1}^{\infty}u_n$ converges absolutely, then both series $\sum_{n=1}^{\infty}v_n$ and $\sum_{n=1}^{\infty}w_n$ converge. If the series $\sum_{n=1}^{\infty}u_n$ converges conditionally, then both series $\sum_{n=1}^{\infty}v_n$ and $\sum_{n=1}^{\infty}w_n$ diverge.

Theorem 7.3.2 states that for any series $\sum_{n=1}^{\infty}u_n$, if the convergence test for positive series determines that the series $\sum_{n=1}^{\infty}|u_n|$ converges, then the original series $\sum_{n=1}^{\infty}u_n$ must converge. This transforms the problem of determining the convergence of a large class of series into the problem of determining the convergence of positive series.

Example 7.3.3 Discuss the convergence of the series $\sum_{n=1}^{\infty}\frac{\sin(n\alpha)}{(\ln 10)^n}$ (α is a constant).

Solution Since $\left|\frac{\sin(n\alpha)}{(\ln 10)^n}\right| \leqslant \frac{1}{(\ln 10)^n} < \frac{1}{2^n}$, and because $\sum_{n=1}^{\infty}\frac{1}{2^n}$ converges, by the comparison test for positive series, the series $\sum_{n=1}^{\infty}\left|\frac{\sin(n\alpha)}{(\ln 10)^n}\right|$ converges. Therefore, the original series converges absolutely.

Example 7.3.4 Discuss the convergence of the series $\sum_{n=1}^{\infty}(-1)^{n-1}\frac{1}{n\cdot 2^n}$.

Solution From $|u_n| = \frac{1}{n\cdot 2^n}$, we have

$$\lim_{n\to\infty}\frac{|u_{n+1}|}{|u_n|}=\lim_{n\to\infty}\frac{\dfrac{1}{(n+1)\,2^{n+1}}}{\dfrac{1}{n\cdot 2^n}}=\frac{1}{2}<1.$$

Hence, the series $\sum_{n=1}^{\infty}|u_n|$ converges, and thus the original series converges absolutely.

It must be noted, generally speaking, that if the series $\sum_{n=1}^{\infty}|u_n|$ diverges, it does not necessarily imply that the series $\sum_{n=1}^{\infty}u_n$ also diverges. Let's analyze an example, the series

$$1-1+\frac{1}{2}-\frac{1}{2}+\frac{1}{3}-\frac{1}{3}+\cdots+\frac{1}{n}-\frac{1}{n}+\cdots,$$

obtained by taking the absolute values of its terms is

$$1+1+\frac{1}{2}+\frac{1}{2}+\frac{1}{3}+\frac{1}{3}+\cdots+\frac{1}{n}+\frac{1}{n}+\cdots,$$

it is divergent. In fact, the sum of the first $2n$ terms of this series is

$$\sigma_{2n}=1+1+\frac{1}{2}+\frac{1}{2}+\frac{1}{3}+\frac{1}{3}+\cdots+\frac{1}{n}+\frac{1}{n}=2\sum_{i=1}^{n}\frac{1}{i},$$

therefore,

$$\lim_{n\to\infty}\sigma_{2n}=2\lim_{n\to\infty}\sum_{i=1}^{n}\frac{1}{i}=+\infty.$$

That is, the series with absolute values of terms is divergent. However, the original series is convergent. Because the sum of the first $2n$ terms of the original series $s_{2n}=0$, the sum of the first $2n+1$ terms $s_{2n+1}=\dfrac{1}{n+1}\to 0\ (n\to\infty)$, hence $\lim_{n\to\infty}s_n=0$, indicating that the original series converges.

If the ratio test or the root test determines that the series $\sum_{n=1}^{\infty}|u_n|$ diverges, then it can be concluded that the series $\sum_{n=1}^{\infty}u_n$ must also diverge. This is because when $\rho>1$, it can be deduced that $|u_n|\to 0\ (n\to\infty)$, and thus $u_n\to 0$, so the series $\sum_{n=1}^{\infty}u_n$ diverges.

Example 7.3.5 Discuss the convergence of the series $\sum_{n=1}^{\infty}(-1)^{n-1}\dfrac{1}{n^p}\ (p>0)$.

Solution Since $\sum_{n=1}^{\infty}\left|(-1)^{n-1}\dfrac{1}{n^p}\right|=\sum_{n=1}^{\infty}\dfrac{1}{n^p}$, by the convergence conclusion of the p series, when $p>1$, the series $\sum_{n=1}^{\infty}\left|(-1)^{n-1}\dfrac{1}{n^p}\right|$ converges, and thus, the original series converges absolutely. When $p\leqslant 1$, the series $\sum_{n=1}^{\infty}\left|(-1)^{n-1}\dfrac{1}{n^p}\right|$ diverges. That is, the original series is not absolutely convergent, but in the original series, $u_n=\frac{1}{n^p}\to 0\ (n\to\infty)$, and $u_{n+1}=\frac{1}{(n+1)^p}<\frac{1}{n^p}\ (n=1,2,3,\cdots)$, which satisfies the two conditions of Leibniz's Theorem, therefore, the original series converges and is conditionally convergent.

Example 7.3.6 Determine the convergence of the series $\sum_{n=1}^{\infty}(-1)^n\frac{1}{2^n}\left(1+\frac{1}{n}\right)^{n^2}$.

Solution Let $|u_n|=\frac{1}{2^n}\left(1+\frac{1}{n}\right)^{n^2}$, then

$$\sqrt[n]{|u_n|}=\frac{1}{2}\left(1+\frac{1}{n}\right)^n\to\frac{e}{2}\quad(n\to\infty).$$

Since $\frac{e}{2}>1$, it follows that $|u_n|\nrightarrow 0\ (n\to\infty)$ and thus $u_n\nrightarrow 0$, therefore, the original series diverges.

Absolutely convergent series have many properties that conditionally convergent series do not have. For example, the addition of absolutely convergent series satisfies the commutative law, whereas for conditionally convergent series, changing the order of its terms may change its sum. For instance, the alternating series $\sum_{n=1}^{\infty}\frac{(-1)^{n+1}}{n}$ is conditionally convergent, and its sum

$$s=1-\frac{1}{2}+\frac{1}{3}-\frac{1}{4}+\frac{1}{5}-\frac{1}{6}+\frac{1}{7}-\frac{1}{8}+\frac{1}{9}-\cdots.$$

Multiplying both sides by $\frac{1}{2}$ gives

$$\frac{s}{2}=\frac{1}{2}-\frac{1}{4}+\frac{1}{6}-\frac{1}{8}+\frac{1}{10}-\frac{1}{12}+\cdots,$$

rewriting it in the following form

$$\frac{s}{2}=0+\frac{1}{2}+0-\frac{1}{4}+0+\frac{1}{6}+0-\frac{1}{8}+0+\frac{1}{10}+0-\frac{1}{12}+\cdots,$$

adding it to the first equation yields

$$\frac{3}{2}s=1+\frac{1}{3}-\frac{1}{2}+\frac{1}{5}+\frac{1}{7}-\frac{1}{4}+\frac{1}{9}+\frac{1}{11}-\frac{1}{6}+\cdots.$$

Clearly, its sum has now become $\frac{3}{2}s$.

Exercise 7.3

1. Determine the convergence of the following alternating series:

(1) $\frac{1}{3}\cdot\frac{1}{2}-\frac{1}{3}\cdot\frac{1}{2^2}+\frac{1}{3}\cdot\frac{1}{2^3}-\frac{1}{3}\cdot\frac{1}{2^4}+\cdots$;

(2) $\sum_{n=1}^{\infty} (-1)^{n-1} \frac{n}{3^{n-1}}$; (3) $\sum_{n=2}^{\infty} \frac{(-1)^{n-1} n}{n^2-1}$;

(4) $\sum_{n=1}^{\infty} \frac{(-1)^n}{\sqrt{n^2+2}}$; (5) $\sum_{n=1}^{\infty} \frac{(-1)^{n+1} n!}{2^{n^2}}$.

2. Determine whether the following series converges? If it converges, determine whether it is absolutely convergent or conditionally convergent.

(1) $\sum_{n=1}^{\infty} (-1)^{n-1} \frac{1}{\sqrt[n]{n}}$; (2) $\sum_{n=1}^{\infty} (-1)^n \frac{n^2}{2^n}$;

(3) $\sum_{n=1}^{\infty} (-1)^{n-1} \frac{2 \cdot 4 \cdot 6 \cdot \cdots \cdot (2n)}{1 \cdot 3 \cdot 5 \cdot \cdots \cdot (2n-1)}$; (4) $\sum_{n=1}^{\infty} (-1)^{n-1} (\sqrt{n+1} - \sqrt{n})$;

(5) $\sum_{n=1}^{\infty} (-1)^n \frac{\ln(1+n)}{1+n}$; (6) $\sum_{n=1}^{\infty} (-1)^{n-1} \frac{2+(-1)^n}{n^{5/4}}$.

3. Multiple choice.

(1) Let the constant $k > 0$, then the series $\sum_{n=1}^{\infty} (-1)^n \frac{k+n}{n^2}$ is ( ).

A. divergent B. absolutely convergent

C. conditionally convergent D. convergence depends on k

(2) The series $\sum_{n=1}^{\infty} (-1)^n \left(1 - \cos\frac{\alpha}{n}\right)$ (constant $\alpha>0$) is ( ).

A. divergent B. conditionally convergent

C. absolutely convergent D. convergence depends on α

(3) Let the constant $\lambda > 0$, and the series $\sum_{n=1}^{\infty} a_n^2$ converges, then the series $\sum_{n=1}^{\infty} (-1)^n \frac{|a_n|}{\sqrt{n^2+\lambda}}$ is ( ).

A. divergent B. conditionally convergent

C. absolutely convergent D. convergence depends on λ

4. Study the convergence of the series $\sum_{n=1}^{\infty} \frac{a^n}{n^p}$ ($p > 0$, a is a constant).

5. Suppose the series $\sum_{n=1}^{\infty} a_n^2$, $\sum_{n=1}^{\infty} b_n^2$ all converge, prove that the series $\sum_{n=1}^{\infty} \frac{a_n}{n}$ and the series $\sum_{n=1}^{\infty} a_n b_n$ both converge absolutely.

6. If the series $\sum_{n=1}^{\infty} a_n$ converges absolutely, prove that the series $\sum_{n=1}^{\infty} a_n^2$, $\sum_{n=1}^{\infty} \frac{a_n}{1+a_n}$, $\sum_{n=1}^{\infty} \frac{a_n^2}{1+a_n^2}$ also converges absolutely.

7. For the alternating series $\sum_{n=1}^{\infty} (-1)^{n-1} a_n$ ($a_n > 0, n = 1, 2, \cdots$), if the condition

$a_n \geqslant a_{n+1}$ is not satisfied, does the series $\sum_{n=1}^{\infty}(-1)^{n-1}a_n$ necessarily diverge? Discuss the series $\sum_{n=2}^{\infty}\frac{(-1)^n}{\sqrt{n+(-1)^n}}$.

7.4 Power Series

The Taylor's Mean Value Theorem discusses the problem of approximating a function $f(x)$ with a polynomial. If $f(x)$ has $n+1$ derivatives at point x_0, then with n some neighborhood of point x_0,

$$f(x)=P_n(x)+R_n(x)=f(x_0)+f'(x_0)(x-x_0)+\cdots+\frac{f^{(n)}(x_0)}{n!}(x-x_0)^n+R_n(x).$$

Imagine when $f(x)$ has derivatives of any order at point x_0, the result is

$$f(x)=a_0+a_1(x-x_0)+a_2(x-x_0)^2+\cdots+a_n(x-x_0)^n+\cdots,$$

in the form, and $R_n(x)$ approaches zero as n increases. The right-hand side of the above expression is a power series.

7.4.1 Concept of Function Series

Definition 7.4.1 Let $u_1(x), u_2(x), \cdots, u_n(x), \cdots$ be a sequence of functions defined on the interval I, then the expression

$$u_1(x)+u_2(x)+\cdots+u_n(x)+\cdots=\sum_{n=1}^{\infty}u_n(x)$$

is called an (infinite) series of functions on the interval I. When $x_0 \in I$, the series

$$u_1(x_0)+u_2(x_0)+\cdots+u_n(x_0)+\cdots=\sum_{n=1}^{\infty}u_n(x_0).$$

For a series of terms. If the series converges, then the point x_0 is called a convergence point of the series. The set of all convergence points of the series is called the convergence domain of the first one. The points where the series does not converge are called its divergence points. The set of all divergence points of the series is called its divergence domain.

If we assume that I is the convergence domain of the series, $\forall x \in I$, then $u_1(x)+u_2(x)+\cdots+u_n(x)+\cdots=\sum_{n=1}^{\infty}u_n(x)$ converges to a unique number $s(x)$. Thus, on the convergence domain I, the sum of the series of functions is a function of x $s(x)$, which is commonly referred to as the sum function of the series of functions, and the domain of definition of $s(x)$ is exactly the convergence domain of the series, and it is written as

$$s(x)=u_1(x)+u_2(x)+\cdots+u_n(x)+\cdots.$$

If the partial sum of the first n terms of the series is denoted as $s_n(x)$, then on the convergence domain, $\lim_{n\to\infty} s_n(x)=s(x)$. The term $r_n(x)=s(x)-s_n(x)$ is still referred to as the remainder of the series of functions, and thus we have $\lim_{n\to\infty} r_n(x)=0$.

7.4.2 Power Series and Their Convergence

$$a_0+a_1(t-t_0)+a_2(t-t_0)^2+\cdots+a_n(t-t_0)^n+\cdots,$$

a series of the form is called a power series, where the constant a_n $(n=0,1,2,3,\cdots)$ is called the coefficient of the power series. If we let $x=t-t_0$, we obtain

$$\sum_{n=0}^{\infty} a_n x^n = a_0 + a_1 x + a_2 x^2 + \cdots + a_n x^4 + \cdots.$$

Transforming equation using $x=t-t_0$ results in equation. Therefore, the primary focus is on the power series represented by equation.

The power series is a simpler type of series of functions, where each term is a constant multiplied by then power of x. It is also noted that power series are strictly ordered in ascending powers. For example, the series $1+x+x^2+\cdots+x^n+\cdots$ and the series $1-\frac{x^2}{2!}+\frac{x^4}{4!}-\frac{x^6}{6!}+\cdots+\frac{(-1)^k}{(2k)!}x^{2k}+\cdots$ are both power series, with the second series lacking odd power terms, which can be considered to have coefficients $a_{2k+1}=0$.

For power series, the concern is about what the convergence and divergence domains of the power series are. That is, for which points on the number axis x does the power series converge or diverge? This is the convergence problem of power series.

First, consider examples of the convergence of three power series:

Example 7.4.1 Discuss the convergence of the power series $1+x+x^2+\cdots+x^n+\cdots$.

Solution This is a power series with a common ratio of x. Therefore, when $|x|<1$, the series converges to the sum $\frac{1}{1-x}$, when $|x|\geqslant 1$, the series diverges. Hence, the convergence domain of this power series is the open interval $(-1,1)$, and the divergence domain is $(-\infty,-1]$ and $[1,+\infty)$. If x takes values within the interval $(-1,1)$, then

$$\frac{1}{1-x}=1+x+x^2+\cdots+x^n+\cdots.$$

Example 7.4.2 Discuss the convergence of the power series $\sum_{n=1}^{\infty}\frac{x^n}{n!}$.

Solution Given any $x\neq 0$, then $\lim_{n\to\infty}\frac{\left|\frac{x^{n+1}}{(n+1)!}\right|}{\left|\frac{x^n}{n!}\right|}=\lim_{n\to\infty}\frac{|x|}{n+1}=0$, so when $x\neq 0$, the original series converges absolutely, and when $x=0$, the original series is obviously also absolutely convergent.

Example 7.4.3 Discuss the convergence of the power series $\sum_{n=1}^{\infty} n!x^n$.

Solution For any $x \neq 0$, there is

$$\lim_{n\to\infty} \frac{|(n+1)!x^{n+1}|}{|n!x^n|} = \lim_{n\to\infty}(n+1)|x| = +\infty.$$

The series diverges at the point $x \neq 0$, but converges at the point $x=0$.

From the above examples, it is analyzed that the power series always converges at the point $x=0$. The convergence domain of the power series is an interval symmetric about the origin, such as the open interval $(-1,1)$ in Example 7.4.1. It can also be the entire number line, as in Example 7.4.2 with $(-\infty,+\infty)$. In fact, this conclusion holds for general power series as well, as stated in the following theorem:

Theorem 7.4.1(Abel's Theorem) If the series $\sum_{n=0}^{\infty} a_n x^n$ converges when $x=x_0$ $(x_0 \neq 0)$, then for all x satisfying the inequality $|x|<|x_0|$, the power series converges absolutely. Conversely, if the series $\sum_{n=0}^{\infty} a_n x^n$ diverges when $x=x_0$, then for all x satisfying the inequality $|x|>|x_0|$, the power series diverges.

Proof First, assume that x_0 is a point of convergence for the power series $\sum_{n=0}^{\infty} a_n x^n$, i.e., the series $a_0+a_1x_0+a_2x_0^2+\cdots+a_nx_0^n+\cdots$ converges. By the necessary condition for the convergence of a series, we have $\lim_{n\to\infty} a_n x_0^n = 0$. Consequently, there exists a constant $M>0$, such that $|a_n x_0^n| \leqslant M(n=0,1,2,\cdots)$. Thus,

$$|a_n x^n| = \left| a_n x_0^n \cdot \frac{x^n}{x_0^n} \right| = |a_0 x_0^n| \left| \frac{x}{x_0} \right|^n \leqslant M \left| \frac{x}{x_0} \right|^n.$$

Since, when $|x|<|x_0|$, the series $\sum_{n=0}^{\infty} M \left| \frac{x}{x_0} \right|^n$ converges (with common ratio $\left| \frac{x}{x_0} \right| < 1$), the series $\sum_{n=0}^{\infty} |a_n x^n|$ converges, which means that the series $\sum_{n=0}^{\infty} a_n x^n$ converges absolutely.

The second part of the theorem can be proven by contradiction. If the power series diverges when $x=x_0$, and there exists a point x_1 satisfying $|x_1|>|x_0|$, such that the series converges, then according to the first part of this theorem, the series should converge when $x=x_0$, which contradicts the given condition. Hence, the theorem is proven.

Theorem 7.4.1 states that: If the power series converges at a point $x=x_0$ $(x_0 \neq 0)$, then it absolutely converges for any point x within the open interval $(-|x_0|,|x_0|)$. If the power series diverges at a point $x=x_1$, then it diverges for any x outside the closed interval $[-|x_1|,|x_1|]$. As shown in Figure 7-2, all convergence points are concentrated within an interval centered at the origin, and the convergence points are contiguous without intermixing with divergence points. Additionally, any point on the number line is either a convergence point or a divergence point. Thus, there exist two completely determined points $x=R$ and $x=-R$ $(R>0)$ on either side of the origin, which are the boundary points between the convergence domain and the divergence domain of the power series $\sum_{n=0}^{\infty} a_n x^n$. Based on this, the following conclusion is drawn:

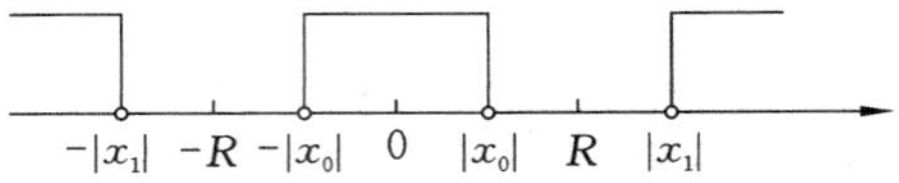

Figure 7-2

Inference If the power series $\sum_{n=0}^{\infty} a_n x^n$ does not converge only at $x=0$ or converges throughout the entire number line, there must exist a certain positive number R such that the power series converges absolutely when $|x|<R$; diverges when $|x|>R$; and may converge or diverge when $x=R$ and $x=-R$.

The positive number R mentioned in the inference is commonly referred to as the convergence radius of the power series $\sum_{n=0}^{\infty} a_n x^n$, and the open interval $(-R, R)$ is called the convergence interval of the power series $\sum_{n=0}^{\infty} a_n x^n$. At the endpoints of the convergence interval, the power series may converge or diverge. Therefore, the convergence domain of the power series is convergence domain$=(-R,R)\cup\{$ convergent endpoints $\}$.

If the power series $\sum_{n=0}^{\infty} a_n x^n$ converges only at the point $x=0$, i.e., the convergence domain consists of only the point $x=0$, then the convergence radius is defined as convergence radius $R=0$; if it converges for all x, then $R=+\infty$ is defined, and the convergence domain is $(-\infty,+\infty)$.

From the above analysis, it is clear that the problem of studying the convergence of power series $\sum_{n=0}^{\infty} a_n x^n$ is to determine its convergence domain, and the problem of finding the convergence domain is essentially to find the convergence radius and discuss the convergence at the endpoints of the convergence interval.

The following theorem provides a method for finding the convergence radius of a power series:

Theorem 7.4.2 If $\lim_{n\to\infty}\left|\frac{a_{n+1}}{a_n}\right|=\rho$, where a_n, a_{n+1} are the coefficients of two adjacent terms in the power series, then the convergence radius of this power series is

$$R=\begin{cases}\frac{1}{\rho}, & 0<\rho<+\infty \\ +\infty, & \rho=0 \\ 0, & \rho=+\infty\end{cases}.$$

Proof First, consider the series formed by taking the absolute values of each term in the power series

$$\sum_{n=0}^{\infty} |a_n x^n|.$$

The ratio of two adjacent terms in this series is $\left|\frac{a_{n+1}x^{n+1}}{a_n x^n}\right| = \left|\frac{a_{n+1}}{a_n}\right| |x|$.

If $\lim_{n\to\infty}\left|\frac{a_{n+1}}{a_n}\right| = \rho\ (\rho \neq 0)$, then by the ratio test, when $\rho|x|<1$, i.e., $|x|<\frac{1}{\rho}$, the series $\sum_{n=0}^{\infty}|a_n x^n|$ converges, and thus the series converges absolutely. When $\rho|x|>1$, i.e., $|x|>\frac{1}{\rho}$, the series diverges, which implies $a_n x^n \to 0$, and thus the series diverges. Therefore, the convergence radius is $R=\frac{1}{\rho}$.

If $\rho=0$, then for any $x\neq 0$, $\lim_{n\to\infty}\frac{|a_{n+1}x^{n+1}|}{|a_n x^n|}=0$, it follows that the series converges, and thus the series converges absolutely, hence $R=+\infty$.

If $\rho=+\infty$, then for all values of x except the point $x=0$, the series must diverge. This is because $\lim_{n\to\infty}\frac{|a_{n+1}x^{n+1}|}{|a_n x^n|}=+\infty$. The series diverges. Suppose there is some value of $x_1\neq 0$ that makes the series converge, then by Theorem 7.4.1, the series converges absolutely for all $x=0$, i.e., the series converges, which leads to a contradiction.

By Theorem 7.4.1, for all x that satisfy $|x|<|x_1|$, the series converges absolutely, i.e., the series converges, which leads to a contradiction. Therefore $R=0$.

Using the Cauchy convergence test, we can also derive the following theorem for finding the radius of convergence of a power series:

Theorem 7.4.3 If the coefficients of the power series satisfy $\lim_{n\to\infty}\sqrt[n]{|a_n|}=\rho$, then the radius of convergence of the power series is

$$R=\begin{cases}\frac{1}{\rho}, & \rho\neq 0\\ +\infty, & \rho=0\\ 0, & \rho=+\infty\end{cases}.$$

The proof of Theorem 7.4.3 is similar to that of Theorem 7.4.2 and is left as an exercise for the reader.

Example 7.4.4 Find the radius of convergence and the convergence domain of the power series $\sum_{n=1}^{\infty}\frac{(-1)^{n-1}}{\sqrt{n}}x^n$.

Solution Since $\rho=\lim_{n\to\infty}\left|\frac{a_{n+1}}{a_n}\right|=\lim_{n\to\infty}\frac{\sqrt{n}}{\sqrt{n+1}}=1$, the radius of convergence is $R=\frac{1}{\rho}=$

1.At the endpoint $x=1$, the series becomes an alternating series $\sum_{n=1}^{\infty}\frac{(-1)^{n-1}}{\sqrt{n}}$, which converges by the Leibniz theorem. At the endpoint $x=-1$, the series becomes $-\sum_{n=1}^{\infty}\frac{1}{\sqrt{n}}$, which diverges. Therefore, the convergence domain of the original series is $(-1,1]$.

Example 7.4.5 Find the radius of convergence and the convergence domain of the power series $\sum_{n=1}^{\infty}\left(\frac{n}{2n+1}\right)^n x^n$.

Solution Since $\rho=\lim_{n\to\infty}\sqrt[n]{|a_n|}=\lim_{n\to\infty}\frac{n}{2n+1}=\frac{1}{2}$, it follows that $R=2$. Also, when $|x|=2$, because

$$\lim_{n\to\infty}\left(\frac{n}{2n+1}\right)^n 2^n=\lim_{n\to\infty}\left(\frac{2n}{2n+1}\right)^n=\mathrm{e}^{-\frac{1}{2}}\neq 0.$$

Therefore, the convergence domain of the original series is $(-2,2)$.

Example 7.4.6 Find the convergence domain of the power series $\sum_{n=1}^{\infty}\frac{x^{2n}}{n^2}$.

Solution The series

$$\sum_{n=1}^{\infty}\frac{x^{2n}}{n^2}=x^2+\frac{x^4}{4}+\frac{x^6}{9}+\cdots+\frac{x^{2n}}{n^2}+\cdots.$$

The coefficients are $a_1=0, a_2=1, a_3=0, a_4=\frac{1}{4}, a_5=0, a_6=\frac{1}{9}, \cdots$, and in general, $a_{2k-1}=0, a_{2k}\neq 0$. Therefore, the series lacks all odd power terms and cannot be directly evaluated using Theorem 7.4.2, but the ratio test can be used to find the radius of convergence:

$$\lim_{n\to\infty}\left|\frac{\frac{x^{2n+2}}{(n+1)^2}}{\frac{x^{2n}}{n^2}}\right|=\lim_{n\to\infty}\left(\frac{n}{n+1}\right)^n\cdot|x|^2=|x|^2.$$

When $|x|^2<1$, i.e., $|x|<1$, the original series converges; when $|x|^2>1$, i.e., $|x|>1$, the original series diverges. Therefore, the radius of convergence is $R=1$.

When $x=\pm 1$, the original series becomes $\sum_{n=1}^{\infty}\frac{1}{n^2}$, which is obviously convergent. Therefore, the convergence domain of the series is $[-1,1]$.

Note that the series can also be written in the following form:

$$\sum_{n=1}^{\infty}\frac{x^{2n}}{n^2}=(x^2)+\frac{(x^2)^2}{4}+\frac{(x^2)^3}{9}+\cdots.$$

For the power series of x^2, the coefficients

$$a_1=1, a_2=\frac{1}{4}, a_3=\frac{1}{9}, \cdots, a_n=\frac{1}{n^2}, \cdots.$$

At this point,

$$\lim_{n\to\infty}\left|\frac{a_{n+1}}{a_n}\right|=\lim_{n\to\infty}\frac{n^2}{(n+1)^2}=1.$$

Therefore, for the power series of x^2, the radius of convergence is $R=1$, meaning that the original series converges when $|x^2|=|x|^2<1$, i.e., $|x|<1$, and diverges when $|x^2|=|x|^2>1$, i.e., $|x|>1$. Hence, the radius of convergence of the original series is $R=1$.

Example 7.4.7 Find the convergence domain of the power series $\sum_{n=1}^{\infty}(-1)^{n-1}\frac{(x-1)^n}{n}$.

Solution Let $t=x-1$, then the series becomes

$$\sum_{n=1}^{\infty}(-1)^{n-1}\frac{t^n}{n}=t-\frac{t^2}{2}+\frac{t^3}{3}-\cdots+(-1)^{n-1}\frac{t^n}{n}-\cdots.$$

For this power series,

$$\lim_{n\to\infty}\left|\frac{a_{n+1}}{a_n}\right|=\lim_{n\to\infty}\frac{n}{n+1}=1.$$

Hence, the radius of convergence is $R=1$. That is, for the power series of t, the convergence interval is $(-1,1)$, and when $t=1$, the series $\sum_{n=1}^{\infty}(-1)^{n-1}\frac{1}{n}$ is a convergent alternating series, when $t=-1$, the series becomes

$$\sum_{n=1}^{\infty}(-1)^{n-1}\frac{(-1)^n}{n}=-1-\frac{1}{2}-\frac{1}{3}-\cdots-\frac{1}{n}-\cdots,$$

it is a divergent series. Therefore, the convergence domain of the power series of t is $(-1,1]$, and thus the convergence domain of the original series is $-1<x-1\leqslant 1$, i.e., $(0,2]$.

7.4.3 Operations on Power Series

Let the power series $\sum_{n=0}^{\infty}a_nx^n$ and $\sum_{n=0}^{\infty}b_nx^n$ converge within the intervals $(-R_1,R_1)$ and $(-R_2,R_2)$ respectively. Below are the four arithmetic operations for power series.

1. Addition and Subtraction

$$\sum_{n=0}^{\infty}a_nx^n\pm\sum_{n=0}^{\infty}b_nx^n=\sum_{n=0}^{\infty}(a_n\pm b_n)x^n,$$

according to the basic properties of convergent series in Section 7.1, Property 7.1.3, the above equation holds within the interval $(-R_1,R_1)\cap(-R_2,R_2)$.

2. Cauchy Product

$$\left(\sum_{n=0}^{\infty}a_nx^n\right)\left(\sum_{n=0}^{\infty}b_nx^n\right)=a_0b_0+(a_0b_1+a_1b_0)x+(a_0b_2+a_1b_1+a_2b_0)x^2$$
$$+\cdots+(a_0b_n+a_1b_{n-1}+\cdots+a_nb_0)x^n+\cdots$$
$$=\sum_{n=0}^{\infty}\Big(\sum_{i+j=n}a_ib_j\Big)x^n.$$

After integration, the resulting new power series still converges within the original convergence interval, but at the endpoints of the convergence interval, its convergence should be reconsidered.

3. Division

When $b_0 \neq 0$ and $|x|$ are sufficiently small, the two power series can be divided, and their quotient is also a power series:

$$\frac{\sum_{n=0}^{\infty} a_n x^n}{\sum_{n=0}^{\infty} b_n x^n} = c_0 + c_1 x + \cdots + c_n x^n + \cdots,$$

where: the coefficient $c_0, c_1, \cdots, c_n, \cdots$ can be determined by the relation

$$\left(\sum_{n=0}^{\infty} b_n x^n\right)\left(\sum_{n=0}^{\infty} c_n x^n\right) = \sum_{n=0}^{\infty} a_n x^n.$$

The series of equations derived from this relation are

$$b_0 c_0 = a_0, b_1 c_0 + b_0 c_1 = a_1, \cdots, b_n c_0 + b_{n-1} c_1 + \cdots + b_0 c_n = a_n, \cdots.$$

The convergence radius of the quotient series determined in this way is difficult to determine and is generally much smaller than R_1, R_2.

Regarding the analytical operations on power series (continuity, integrability, differentiability of power series), it is reduced to the properties of the sum function of the power series, and the proof is omitted.

Property 7.4.1 The sum function $s(x)$ of the power series $\sum_{n=0}^{\infty} a_n x^n$ is continuous on its convergence domain I.

Property 7.4.2 The sum function $s(x)$ of the power series $\sum_{n=0}^{\infty} a_n x^n$ is integrable on its convergence domain I, and there is a term-by-term integration formula:

$$\int_0^x s(x) \mathrm{d}x = \int_0^x \left(\sum_{n=0}^{\infty} a_n x^n\right) \mathrm{d}x = \sum_{n=0}^{\infty} \int_0^x a_n x^n \mathrm{d}x = \sum_{n=0}^{\infty} \frac{a_n}{n+1} x^{n+1} \quad (x \in I).$$

After term-by-term integration, the resulting power series has the same convergence radius as the original series.

Property 7.4.3 The sum function $s(x)$ of the power series $\sum_{n=0}^{\infty} a_n x^n$ is differentiable within its convergence interval $(-R, R)$, the term-by-term differentiation formula holds:

$$s'(x) = \left(\sum_{n=0}^{\infty} a_n x^n\right)' = \sum_{n=0}^{\infty} (a_n x^n)' = \sum_{n=0}^{\infty} n a_n x^{n-1} \quad (|x| < R).$$

After term-by-term differentiation, the resulting power series has the same radius of convergence as the original series.

By repeatedly applying the above conclusion, it follows that the sum function $s(x)$ of the power series $\sum_{n=0}^{\infty} a_n x^n$ has derivatives of all orders within its convergence interval $(-R, R)$.

It is particularly emphasized here that when using Property 7.4.2 or Property 7.4.3 to differentiate or integrate a power series term by term, the new power series obtained still converges within the original convergence interval, but at the endpoints of the interval, its convergence may change. Therefore, the convergence at the endpoints of the interval needs to be re-evaluated.

Using the properties of power series, one can find the sum functions of some power series.

Example 7.4.8 Find the sum function of the power series $\sum_{n=0}^{\infty} \frac{x^n}{n+1}$.

Solution First, find the convergence domain. From $\lim_{n\to\infty} \left|\frac{a_{n+1}}{a_n}\right| = \lim_{n\to\infty} \frac{n+1}{n+2} = 1$, the radius of convergence is $R=1$. At the point $x=-1$, the power series becomes $\sum_{n=0}^{\infty} \frac{(-1)^n}{n+1}$, which is a convergent alternating series. At the point $x=1$, the power series becomes $\sum_{n=0}^{\infty} \frac{1}{n+1}$, which is divergent. Therefore, the convergence domain is $I=[-1,1)$.

Given the function $s(x)=\sum_{n=0}^{\infty} \frac{x^n}{n+1}$, $x \in [-1,1)$, then $xs(x)=\sum_{n=0}^{\infty} \frac{x^{n+1}}{n+1}$ is differentiated term by term, and using

$$\frac{1}{1-x}=1+x+x^2+\cdots+x^n+\cdots \quad (-1<x<1),$$

we obtain $[xs(x)]'=\sum_{n=0}^{\infty} \left(\frac{x^{n+1}}{n+1}\right)'=\sum_{n=0}^{\infty} x^n=\frac{1}{1-x} \quad (-1<x<1)$.

Integrating the above equation from 0 to x, we get

$$xs(x)=\int_0^x \frac{1}{1-x}\mathrm{d}x=-\ln(1-x) \quad (-1 \leqslant x<1).$$

Thus, when $x \neq 0$, we have $s(x)=-\frac{1}{x}\ln(1-x)$, and $s(0)=a_0=1$. Therefore,

$$s(x)=\begin{cases} -\frac{1}{x}\ln(1-x), & x \in [-1,0) \cup (0,1) \\ 1, & x=0 \end{cases}.$$

It is easy to see that $s(0)$ can also be obtained based on the continuity of the sum function, i.e.,

$$s(0)=\lim_{x\to 0} s(x)=\lim_{x\to 0}\left[-\frac{1}{x}\ln(1-x)\right]=1.$$

Example 7.4.9 Prove: $\frac{\pi}{4}=1-\frac{1}{3}+\frac{1}{5}-\frac{1}{7}+\cdots+(-1)^n\frac{1}{2n+1}+\cdots$.

Proof　The problem can be seen as finding the sum of the numerical series $1-\frac{1}{3}+\frac{1}{5}-\frac{1}{7}+\cdots+(-1)^n\frac{1}{2n+1}+\cdots$. First, construct an auxiliary power series

$$\sum_{n=0}^{\infty}(-1)^n\frac{x^{2n+1}}{2n+1}=x-\frac{x^3}{3}+\frac{x^5}{5}-\frac{x^7}{7}+\cdots.$$

Thus, the problem can also be seen as finding the sum function $s(x)$ of this auxiliary power series at the point $x=1$. It is easy to find that the convergence domain of the auxiliary power series is $(-1,1]$, so $\forall x\in(-1,1)$, we have

$$s(x)=\sum_{n=0}^{\infty}(-1)^n\frac{x^{2n+1}}{2n+1}=\sum_{n=0}^{\infty}(-1)^n\int_0^x t^{2n}\,\mathrm{d}t=\int_0^x\left[\sum_{n=0}^{\infty}(-1)^n t^{2n}\right]\mathrm{d}t.$$

While

$$\sum_{n=0}^{\infty}(-1)^n t^{2n}=\sum_{n=0}^{\infty}(-1)^n(t^2)^n=\frac{1}{1+t^2}\quad(|t|<1),$$

hence $s(x)=\int_0^x\frac{\mathrm{d}t}{1+t^2}=\arctan x$, and thus we obtain

$$1-\frac{1}{3}+\frac{1}{5}-\frac{1}{7}+\cdots+(-1)^n\frac{1}{2n+1}+\cdots=\arctan 1=\frac{\pi}{4}.$$

Exercise 7.4

1. Find the convergence domain of the following power series:

(1) $\sum_{n=1}^{\infty}(-1)^n\frac{2^n}{\sqrt{n}}x^n$;　(2) $\sum_{n=1}^{\infty}\frac{1}{n!}x^n$;

(3) $\sum_{n=1}^{\infty}\frac{1}{(2n)!!}x^n$;　(4) $\sum_{n=1}^{\infty}\frac{2^n}{n^2+1}x^n$;

(5) $\sum_{n=1}^{\infty}\frac{2^n}{n}x^{2n}$;　(6) $\sum_{n=1}^{\infty}(-1)^n\frac{x^{2n+1}}{2n+1}$;

(7) $\sum_{n=1}^{\infty}\frac{(-1)^n}{n}\left(\frac{x}{2x+1}\right)^n$;　(8) $\sum_{n=1}^{\infty}\frac{2^n+(-1)^n}{n}(x-1)^n$.

2. Find the sum function of the following series:

(1) $\sum_{n=1}^{\infty}nx^{n-1}$;　(2) $\sum_{n=1}^{\infty}\frac{x^{2n-1}}{2n-1}$;

(3) $\sum_{n=1}^{\infty}\frac{n}{n+1}x^n$;　(4) $\sum_{n=1}^{\infty}\frac{n(n+1)}{2}x^{n-1}$;

(5) $\sum_{n=1}^{\infty}\frac{1}{n(n+1)}x^n$.

3. Multiple choice questions.

(1) If the series $\sum_{n=0}^{\infty}a_n(x-3)^n$ converges at point $x=-1$, then the series at point $x=6$ is (　　).

A. divergent B. conditionally convergent

C. absolutely convergent D. cannot be determined

(2) Suppose the power series $\sum_{n=1}^{\infty}\frac{(x-a)^n}{n}$ converges at point $x=2$, then the range of values for the real number a is ( ).

A. $1<a\leqslant 3$ B. $1\leqslant a<3$ C. $1<a<3$ D. $1\leqslant a\leqslant 3$

4. Suppose the power series $\sum_{n=1}^{\infty}a_n(x-1)^n$ converges at point $x=0$ and diverges at point $x=2$, determine the convergence domain of the power series and explain the reason.

7.5 Expansion of Functions into Power Series

This section aims to determine whether a given function can be represented by a power series, thereby using the power series as a tool to study the function. An important application of power series is their use in representing functions.

7.5.1 Taylor Series

If a given function $f(x)$ can be found to converge within a certain interval, and the sum of the converging series is exactly the given function $f(x)$, then it is said that the function $f(x)$ can be expanded into a power series within that interval, and this power series expresses the function $f(x)$ within that interval.

If $f(x)$ has derivatives up to the $(n+1)$th order in a neighborhood of the point x_0, then the nth order Taylor formula for $f(x)$ in that neighborhood is

$$f(x)=f(x_0)+f'(x_0)(x-x_0)+\frac{f''(x_0)}{2!}(x-x_0)^2+\cdots+\frac{f^{(n)}(x_0)}{n!}(x-x_0)^n+R_n(x),$$

where: $R_n(x)$ is the Lagrange form of the remainder:

$$R_n(x)=\frac{f^{(n+1)}(\xi)}{(n+1)!}(x-x_0)^{n+1},$$

ξ is some value between x and x_0.

Note:

$$P_n(x)=f(x_0)+f'(x_0)(x-x_0)+\frac{f''(x_0)}{2!}(x-x_0)^2+\cdots+\frac{f^{(n)}(x_0)}{n!}(x-x_0)^n,$$

from equation, we have $f(x)=P_n(x)+R_n(x)$. Thus, using $P_n(x)$ as an approximate expression for $f(x)$, the error is $|R_n(x)|$. If $|R_n(x)|$ decreases as n increases, then, let's consider: By increasing the number of terms in the polynomial and approximating $f(x)$ with $P_n(x)$, the accuracy is bound to improve.

Instead of $f(x)$ having derivatives up to order $(n+1)$ in a neighborhood of point x_0, let $f(x)$ have derivatives of all orders $f'(x), f''(x), \cdots, f^{(n)}(x), \cdots$ in a neighborhood of point x_0. Thus, envisioning that the number of terms in equation $P_n(x)$ tends to infinity and becomes a power series.

$$f(x_0)+f'(x_0)(x-x_0)+\frac{f''(x_0)}{2!}(x-x_0)^2+\cdots+\frac{f^{(n)}(x_0)}{n!}(x-x_0)^n+\cdots,$$

the equation is called the Taylor series of $f(x)$, and the Taylor series of $f(x)$ converges to $f(x_0)$ at point $x=x_0$. The question is, at other points in the neighborhood of point x_0, does the Taylor series necessarily converge? If it converges, does it necessarily converge to $f(x)$? The following theorem can answer these questions.

Theorem 7.5.1 Let the function $f(x)$ have derivatives of all orders in a neighborhood $U(x_0)$ of the point x_0. Then the necessary and sufficient condition for $f(x)$ to be expanded into a Taylor series in this neighborhood is that the remainder term $R_n(x)$ in the Taylor formula of $f(x)$ tends to zero as $n\to\infty$, i.e., $\lim\limits_{n\to\infty}R_n(x)=0$ $(x\in U(x_0))$.

Proof First, we prove the necessity. Suppose $f(x)$ can be expanded into a Taylor series in $U(x_0)$, i.e.,

$$f(x)=f(x_0)+f'(x_0)(x-x_0)+\frac{f''(x_0)}{2!}(x-x_0)^2+\cdots+\frac{f^{(n)}(x_0)}{n!}(x-x_0)^n+\cdots,$$

this holds for all $x\in U(x_0)$. Writing the nth order Taylor formula of $f(x)$ as $f(x)=s_{n+1}(x)+R_n(x)$, where $s_{n+1}(x)$ is the sum of the first $(n+1)$ terms of the Taylor series of $f(x)$. Since from equation we have $\lim\limits_{n\to\infty}s_{n+1}(x)=f(x)$, it follows that

$$\lim_{n\to\infty}R_n(x)=\lim_{n\to\infty}[f(x)-s_{n+1}(x)]=f(x)-f(x)=0,$$

this proves that the condition is necessary.

Next, we prove the sufficiency. Since $\lim\limits_{n\to\infty}R_n(x)=0$ holds for all $x\in U(x_0)$, from $s_{n+1}(x)=f(x)-R_n(x)$, we get

$$\lim_{n\to\infty}s_{n+1}(x)=\lim_{n\to\infty}[f(x)-R_n(x)]=f(x),$$

that is, the Taylor series of $f(x)$ converges in $U(x_0)$ and converges to $f(x)$. Therefore, the condition is sufficient. The theorem is proved.

Letting $x_0=0$ in equation, we obtain

$$f(0)+f'(0)x+\frac{f''(0)}{2!}x^2+\cdots+\frac{f^{(n)}(x_0)}{n!}x^2+\cdots,$$

the series is called the Maclaurin series of the function $f(x)$.

Theorem 7.5.2 (Uniqueness Theorem)

(1) If the function $f(x)$ can be expanded into a power series of x within a neighborhood $(-R,R)$ of the point $x_0=0$:

$$f(x)=\sum_{n=0}^{\infty}a_nx^n=a_0+a_1x+a_2x^2+\cdots+a_nx^n+\cdots \quad (x\in(-R,R)),$$

then, it must be that $a_n=\frac{1}{n!}f^{(n)}(0)$, denoted as $f^{(0)}(0)=f(0)$.

(2) If the function $f(x)$ can be expanded into a power series of $(x-x_0)$ within a

neighborhood $U(x_0)$ of the point $x=x_0$:

$$f(x)=\sum_{n=0}^{\infty}a_n(x-x_0)^n=a_0+a_1(x-x_0)+a_2(x-x_0)^2+\cdots$$
$$+a_n(x-x_0)^n+\cdots\quad(x\in U(x_0)),$$

then, it must be that $a_n=\frac{1}{n!}f^{(n)}(x_0)$, denoted as $f^{(0)}(x_0)=f(x_0)$.

Proof (1) If

$$f(x)=a_0+a_1x+a_2x^2+\cdots+a_nx^n+\cdots\quad(x\in(-R,R)),$$

Let $x=0$, we get $a_0=f(0)$, and by differentiating term by term, we obtain

$$f'(x)=a_1+2a_2x+3a_3x^2+\cdots+na_nx^{n-1}+\cdots,$$
$$f''(x)=2!\ a_2+3\cdot 2a_3x+\cdots+n(n-1)a_nx^{n-2}+\cdots,$$
$$f'''(x)=3!\ a_3+\cdots+n(n-1)(n-2)a_nx^{n-3}+\cdots,$$
$$\cdots$$
$$f^{(n)}(x)=n!\ a_n+(a+1)n(n-1)+\cdots+2a_{n+1}x+\cdots,$$
$$\cdots.$$

Substituting $x=0$ into the above equations yields

$$a_0=f(0),a_1=f'(0),a_2=\frac{f''(0)}{2!},\cdots,a_n=\frac{f^{(n)}(0)}{n!},\cdots.$$

Thus, (1) is proven.

The proof of (2) is entirely analogous to that of (1). The reader can follow the same approach.

According to Theorem 7.5.2, if the function $f(x)$ can be expanded into a power series of x within $(-R,R)$, then this power series must be the Maclaurin series of $f(x)$. If the function $f(x)$ can be expanded into a power series of $(x-x_0)$ within $U(x_0)$, then this power series must be the Taylor series of $f(x)$ at the point x_0. Hence, Theorem 7.5.2 is called the Uniqueness Theorem.

As mentioned earlier, if the function $f(x)$ has derivatives of all orders at the point $x_0=0$, then the Maclaurin series of $f(x)$ can be constructed, but it is not guaranteed that the Maclaurin series will converge, and even if it does, it is not guaranteed to converge to $f(x)$. The issue of convergence to $f(x)$ depends on Theorem 7.5.1, which requires checking whether $\lim_{n\to\infty}R_n(x)=0$ holds.

7.5.2 Methods for Expanding Functions into Power Series

1. Direct Expansion Method

To expand a function into a power series of x, follow these steps:

Step 1: Find the derivatives $f'(x),f''(x),\cdots,f^{(n)}(x),\cdots$ of the function $f(x)$ of all orders. If a derivative of some order does not exist at the point $x=0$, it indicates that the function cannot be expanded into a power series of x.

Step 2: Determine $f(0), f'(0), f''(0), \cdots, f^{(n)}(0), \cdots$.

Step 3: Construct the power series $f(0)+f'(0)x+\frac{f''(0)}{2!}x^2+\cdots+\frac{f^{(n)}(0)}{n!}x^2+\cdots$, and find the radius of convergence R.

Step 4: Check the limit of the remainder $R_n(x)$ when x is within $(-R, R)$, i.e.,

$$\lim_{n\to\infty}R_n(x)=\lim_{n\to\infty}\frac{f^{(n+1)}(\xi)}{(n+1)!}x^{n+1} \quad (\xi \text{ is between } 0 \text{ and } x),$$

whether it is zero. If it is zero, then the expansion holds.

$$f(x)=f(0)+f'(0)x+\frac{f''(0)}{2!}x^2+\cdots+\frac{f^{(n)}(0)}{n!}x^n+\cdots,$$

where: $x\in(-R, R)$.

The method described above, which directly calculates the coefficients of the power series using formula $a_n=\frac{f^{(n)}(0)}{n!}$, then verifies $\lim_{n\to\infty}R_n(x)=0$, and finally substitutes into the Maclaurin expansion to obtain the power series of x, is commonly referred to as the direct expansion method.

Example 7.5.1 Expand the function $f(x)=e^x$ into a power series of x.

Solution Since $f^{(n)}(x)=e^x$ $(n=1,2,\cdots)$, therefore, $f^{(n)}(0)=1$ $(n=1,2,\cdots)$, thus obtaining the series $1+x+\frac{x^2}{2!}+\cdots+\frac{x^n}{n!}+\cdots$ $[x\in(-\infty,+\infty)]$. It is evident that for any x, ξ (ξ between 0 and x) holds.

$$|R_n(x)|=\left|\frac{e^\xi}{(n+1)!}x^{n+1}\right|<e^{|x|}\cdot\frac{|x|^{n+1}}{(n+1)!},$$

here, $e^{|x|}$ is a finite number independent of n. Considering that the series $\sum_{n=1}^{\infty}\frac{|x|^{n+1}}{(n+1)!}$ is a convergent series, it follows that

$$\lim_{n\to\infty}e^{|x|}\cdot\frac{|x|^{n+1}}{(n+1)!}=0.$$

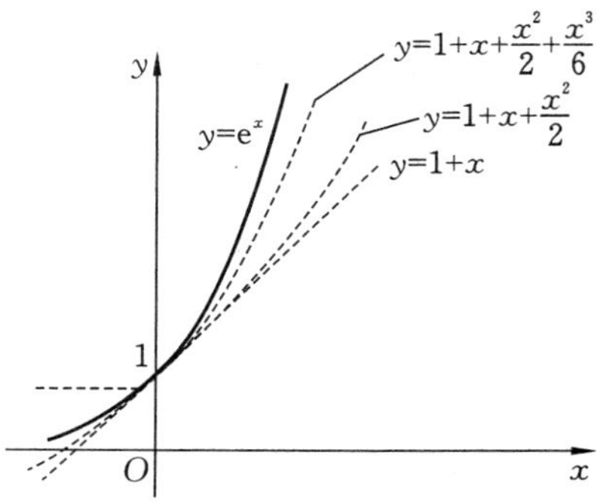

Figure 7-3

That is, $\lim_{n\to\infty}R_n(x)=0$. Thus, $f(x)=e^x$ can be expanded into a power series of x, namely

$$e^x=1+x+\frac{x^2}{2!}+\cdots+\frac{x^n}{n!}+\cdots \quad [x\in(-\infty,+\infty)].$$

As shown in Figure 7-3, near the point $x=0$, approximating e^x with partial sums of the series (i.e., polynomials), as the number of polynomial terms increases, they increasingly approximate e^x.

Example 7.5.2 Expand the function $f(x)=\sin x$ into a power series of x.

Solution Since

$$f^{(n)}(x)=\sin\left(x+n\cdot\frac{\pi}{2}\right)\quad(n=0,1,2,\cdots),$$

$f^{(n)}(0)$ sequentially cycles through $0,1,0,-1,\cdots(n=0,1,2,\cdots)$, thus obtaining the series

$$x-\frac{x^3}{3!}+\frac{x^5}{5!}+\cdots+(-1)^{n-1}\frac{x^{2n-1}}{(2n-1)!}+\cdots\quad(x\in(-\infty,+\infty)),$$

for any finite number x, ξ (ξ between 0 and x), there exists

$$|R_n(x)|=\left|\frac{\sin\left[\xi+\left(\frac{n+1}{2}\right)\pi\right]}{(n+1)!}x^{n+1}\right|\leqslant\frac{|x|^{n+1}}{(n+1)!}\to 0\quad(n\to\infty).$$

Hence, the expansion is

$$\sin x=x-\frac{x^3}{3!}+\frac{x^5}{5!}+\cdots+(-1)^{n-1}\frac{x^{2n-1}}{(2n-1)!}+\cdots\quad(x\in(-\infty,+\infty)).$$

From the above two examples, we can see that when using the direct expansion method to expand a function into a power series of x, it is necessary to calculate the nth derivative of the function and to check whether the limit of the remainder term tends to zero. Both of these tasks are cumbersome. The method introduced below (indirect expansion method) can avoid these two major difficulties.

2. Indirect Expansion Method

The so-called indirect expansion method refers to the method of expanding a given function into a power series by utilizing known function expansions, arithmetic operations of power series, term by term integration and differentiation, and variable substitution.

Example 7.5.3 Expand the function $f(x)=\cos x$ into a power series of x.

Solution Differentiating term by term the expansion yields

$$\cos x=1-\frac{x^2}{2!}+\frac{x^4}{4!}-\cdots+(-1)^n\frac{x^{2n}}{(2n)!}+\cdots\quad(-\infty<x<+\infty).$$

Example 7.5.4 Expand the function $f(x)=\ln(1+x)$ into a power series of x.

Solution Since $f'(x)=\frac{1}{1+x}$, and $\frac{1}{1+x}$ is the sum function of the convergent geometric series $\sum_{n=0}^{\infty}(-1)^n x^n$, i.e.,

$$\frac{1}{1+x}=1-x+x^2-x^3+\cdots+(-1)^n x^n+\cdots\quad(-1<x<1).$$

Integrating term by term from 0 to x, we obtain

$$\ln(1+x)=x-\frac{x^2}{2}+\frac{x^3}{3}-\frac{x^4}{4}+\cdots+(-1)^n\frac{x^{n+1}}{n+1}+\cdots\quad(-1<x\leqslant 1).$$

Noticing that the expansion holds for $x=1$ because the power series on the right converges when $x=1$, and the sum function $\ln(1+x)$ is defined and continuous at the point $x=1$.

Example 7.5.5 expands the function $f(x)=(1+x)^m$ into a power series of x, where m is an arbitrary constant.

Solution Because

$$f'(x)=m(1+x)^{m-1},$$
$$f''(x)=m(m-1)(1+x)^{m-2},$$
$$\cdots$$
$$f^{(n)}(x)=m(m-1)(m-2)\cdots(m-n+1)(1+x)^{m-n},$$

therefore

$$f(0)=1, f'(0)=m, f''(0)=m(m-1), \cdots,$$
$$f^{(n)}(0)=m(m-1)\cdots(m-n+1), \cdots.$$

Thus, we obtain the series

$$1+mx+\frac{m(m-1)}{2!}x^2+\cdots+\frac{m(m-1)\cdots(m-n+1)}{n!}x^n+\cdots,$$

also

$$\lim_{n\to\infty}\left|\frac{a_{n+1}}{a_n}\right|=\lim_{n\to\infty}\left|\frac{m-n}{n+1}\right|=1,$$

hence, for any constant m, the above series converges within the interval $(-1,1)$.

To avoid direct study of the remainder term, assume that the series converges within $(-1,1)$ to the sum function $F(x)$, i.e.,

$$F(x)=1+mx+\frac{m(m-1)}{2!}x^2+\cdots+\frac{m(m-1)\cdots(m-n+1)}{n!}x^n+\cdots \quad (x\in(-1,1)).$$

Next, we need to prove $F(x)=(1+x)^m \quad (x\in(-1,1))$.

Differentiating term by term yields

$$F'(x)=m\left[1+\frac{m-1}{1}x+\cdots+\frac{(m-1)\cdots(m-n+1)}{(n-1)!}x^{n-1}+\cdots\right].$$

Multiplying both sides by $(1+x)$ and combining the two terms containing x^n $(n=1,2,\cdots)$, while utilizing the identity

$$\frac{(m-1)\cdots(m-n+1)}{(n-1)!}+\frac{(m-1)\cdots(m-n)}{n!}=\frac{m(m-1)\cdots(m-n+1)}{n!} \quad (n=1,2,\cdots),$$

thus, we have

$$(1+x)F'(x)=m\left[1+mx+\frac{m(m-1)}{2!}x^2+\cdots+\frac{m(m-1)\cdots(m-n+1)}{n!}x^n+\cdots\right]$$
$$=mF(x) \quad (-1<x<1).$$

Let $\varphi(x)=\dfrac{F(x)}{(1+x)^m}$, then $\varphi(0)=F(0)=1$, and

$$\varphi'(x)=\frac{(1+x)^mF'(x)-m(1+x)^{m-1}F(x)}{(1+x)^{2m}}=\frac{(1+x)^{m-1}[(1+x)F'(x)-mF(x)]}{(1+x)^{2m}}=0,$$

therefore $\varphi(x)=C$ (constant), but due to $\varphi(0)=1$, thus $\varphi(x)=1$, i.e., $F(x)=(1+x)^m$, which proves the expansion within the interval $(-1,1)$:

$$(1+x)^m=1+mx+\frac{m(m-1)}{2!}x^2+\cdots+\frac{m(m-1)\cdots(m-n+1)}{n!}x^n+\cdots \quad (-1<x<1).$$

Whether the expansion holds at the endpoints depends on the value of m. Equation is called the binomial expansion. Specifically, when m is a positive integer, the series becomes a polynomial of degree m in x, which is the binomial theorem in algebra. The binomial expansions of $m=\frac{1}{2},-\frac{1}{2}$ are respectively

$$\sqrt{1+x}=1+\frac{1}{2}x-\frac{1}{2\cdot 4}x^2+\frac{1\cdot 3}{2\cdot 4\cdot 6}x^3-\frac{1\cdot 3\cdot 5}{2\cdot 4\cdot 6\cdot 8}x^4+\cdots \quad (-1\leqslant x\leqslant 1),$$

$$\frac{1}{\sqrt{1+x}}=1-\frac{1}{2}x+\frac{1\cdot 3}{2\cdot 4}x^2-\frac{1\cdot 3\cdot 5}{2\cdot 4\cdot 6}x^3+\frac{1\cdot 3\cdot 5\cdot 7}{2\cdot 4\cdot 6\cdot 8}x^4-\cdots \quad (-1<x\leqslant 1).$$

Example 7.5.6 Expand the function $f(x)=\arctan x$ into a power series of x.

Solution Since $f'(x)=\frac{1}{1+x^2}$, and

$$\frac{1}{1+x^2}=\frac{1}{1-(-x^2)}=1+(-x^2)+(-x^2)^2+(-x^2)^3+\cdots+(-x^2)^n+\cdots$$

$$=1-x^2+x^4-x^6+\cdots+(-1)^n x^{2n}+\cdots=\sum_{n=0}^{\infty}(-1)^n x^{2n} \quad (-1<x<1).$$

Integrating the above equation term by term from 0 to x, and noting that $f(0)=\arctan 0=0$, we get

$$\arctan x=\sum_{n=0}^{\infty}(-1)^n\frac{x^{2n+1}}{2n+1} \quad (-1<x<1).$$

When $x=\pm 1$, the series on the right-hand side of the above equation becomes $\pm\sum_{n=0}^{\infty}\frac{(-1)^n}{2n+1}$, which is obviously convergent, and since $f(x)=\arctan x$ is continuous at $x=\pm 1$, we obtain

$$\arctan x=x-\frac{x^3}{3}+\frac{x^5}{5}-\cdots+(-1)^n\frac{x^{2n+1}}{2n+1}+\cdots \quad (-1\leqslant x\leqslant 1).$$

Example 7.5.7 Expand the function $f(x)=\cos^2 x$ into a power series of x.

Solution Since $f(x)=\cos^2 x=\frac{1}{2}+\frac{1}{2}\cos 2x$, and

$$\cos 2x=\sum_{n=0}^{\infty}(-1)^n\frac{2^{2n}}{(2n)!}x^{2n} \quad (x\in(-\infty,+\infty)),$$

therefore

$$\cos^2 x=\frac{1}{2}+\frac{1}{2}\sum_{n=0}^{\infty}\frac{(-1)^n 2^{2n}}{(2n)!}x^{2n} \quad (x\in(-\infty,+\infty)).$$

Example 7.5.8 Expand the function $f(x)=\frac{x}{x^2-x-2}$ into a power series of x.

Solution Since

$$f(x)=\frac{x}{x^2-x-2}=\frac{x}{(x-2)(x+1)}=\frac{1}{3}\left(\frac{1}{x+1}+\frac{2}{x-2}\right)=\frac{1}{3}\left(\frac{1}{1+x}-\frac{1}{1-\frac{x}{2}}\right),$$

while

$$\frac{1}{1+x}=\sum_{n=0}^{\infty}(-1)^n x^n \quad (-1<x<1),$$

$$\frac{1}{1-\frac{x}{2}}=\sum_{n=0}^{\infty}\left(\frac{x}{2}\right)^n=\sum_{n=0}^{\infty}\frac{1}{2^n}x^n \quad (-2<x<2),$$

thus

$$\frac{x}{x^2-x-2}=\frac{1}{3}\left[\sum_{n=0}^{\infty}(-1)^n x^n-\sum_{n=0}^{\infty}\frac{1}{2^n}x^n\right]=\frac{1}{3}\sum_{n=0}^{\infty}\left[(-1)^n-\frac{1}{2^n}\right]x^n \quad (-1<x<1).$$

Below is an example of expanding a function into a power series of $(x-x_0)$.

Example 7.5.9 Expand the following functions into Taylor series at the specified points.

(1) $f(x)=\ln x, x=2$; (2) $f(x)=\cos x, x=-\frac{\pi}{3}$.

Solution (1) Because

$$f(x)=\ln[2+(x-2)]=\ln 2+\ln\left[1+\left(\frac{x-2}{2}\right)\right],$$

while

$$\ln\left[1+\left(\frac{x-2}{2}\right)\right]=\sum_{n=1}^{\infty}(-1)^{n+1}\frac{1}{n\cdot 2^n}(x-2)^n \quad (x\in[0,4]),$$

therefore

$$\ln x=\ln 2+\sum_{n=1}^{\infty}(-1)^{n+1}\frac{1}{n\cdot 2^n}(x-2)^n \quad (x\in(0,4]).$$

(2)

$$\cos x=\cos\left[-\frac{\pi}{3}+\left(x+\frac{\pi}{3}\right)\right]=\cos\left(-\frac{\pi}{3}\right)\cos\left(x+\frac{\pi}{3}\right)-\sin\left(-\frac{\pi}{3}\right)\sin\left(x+\frac{\pi}{3}\right)$$

$$=\cos\frac{\pi}{3}\cos\left(x+\frac{\pi}{3}\right)+\sin\frac{\pi}{3}\sin\left(x+\frac{\pi}{3}\right)=\frac{1}{2}\sum_{n=0}^{\infty}\frac{(-1)^n}{(2n)!}\left(x+\frac{\pi}{3}\right)^{2n}$$

$$+\frac{\sqrt{3}}{2}\sum_{n=0}^{\infty}\frac{(-1)^n}{(2n+1)!}\left(x+\frac{\pi}{3}\right)^{2n+1} \quad (-\infty<x<+\infty).$$

7.5.3 Applications of Power Series Expansions of Functions

Using the power series expansion of a function, one can perform approximate calculations, i. e., within the valid interval of the expansion, the function value can be approximated by calculating this series to the required accuracy.

Example 7.5.10 Calculate the approximate value of $\sqrt[5]{240}$, requiring the error to be no more than 0.0001.

Solution Since $\sqrt[5]{240}=\sqrt[5]{243-3}=3\left(1-\frac{1}{3^4}\right)^{\frac{1}{5}}$, therefore, in the binomial expansion, taking $m=\frac{1}{5}$, $x=-\frac{1}{3^4}$, we obtain

$$\sqrt[5]{240}=3\left(1-\frac{1}{5}\cdot\frac{1}{3^4}-\frac{1\cdot 4}{5^2\cdot 2!}\cdot\frac{1}{3^8}-\frac{1\cdot 4\cdot 9}{5^3\cdot 3!}\cdot\frac{1}{3^{12}}-\cdots\right).$$

This series converges rapidly. Taking the sum of the first two terms as an approximation for $\sqrt[5]{240}$, the error (also known as the truncation error) is

$$|r_n|=3\left(\frac{1\cdot 4}{5^2\cdot 2!}\cdot\frac{1}{3^8}+\frac{1\cdot 4\cdot 9}{5^3\cdot 3!}\cdot\frac{1}{3^{12}}+\frac{1\cdot 4\cdot 9\cdot 14}{5^4\cdot 4!}\cdot\frac{1}{3^{16}}+\cdots\right)$$

$$<3\cdot\frac{1\cdot 4}{5^2\cdot 2!}\cdot\frac{1}{3^8}\left[1+\frac{1}{81}+\left(\frac{1}{81}\right)^2+\cdots\right]=\frac{6}{25}\cdot\frac{1}{3^8}\cdot\frac{1}{1-\frac{1}{81}}=\frac{1}{25\cdot 27\cdot 40}<\frac{1}{20000},$$

thus, the approximation is $\sqrt[5]{240}\approx 3\left(1-\frac{1}{5}\cdot\frac{1}{3^4}\right)$.

To ensure that the error caused by rounding combined with the truncation error does not exceed 10^{-4}, calculations should be performed to five decimal places, and then rounded off, resulting in $\sqrt[5]{240}\approx 2.9926$.

Example 7.5.11 Calculate the approximate value of $\frac{1}{\sqrt{e}}$, requiring the error to be no more than 10^{-4}.

Solution In the expansion of e^x, let $x=-\frac{1}{2}$ to obtain

$$\frac{1}{\sqrt{e}}=e^{-\frac{1}{2}}=1-\frac{1}{2}+\frac{\left(-\frac{1}{2}\right)^2}{2!}+\cdots+\frac{\left(-\frac{1}{2}\right)^n}{n!}+\cdots.$$

This is a Leibniz-type alternating series, where $|r_n|\leqslant u_{n+1}$. Here,

$$u_1=1, u_2=\frac{1}{2}, u_3=\frac{1}{8}, u_4=\frac{1}{48}, u_5=\frac{1}{384}, u_6=\frac{1}{3840}, u_7=\frac{1}{6!}\left(\frac{1}{2}\right)^6<10^{-4}.$$

Therefore,

$$\frac{1}{\sqrt{e}}\approx 1-\frac{1}{2}+\frac{1}{8}-\frac{1}{48}+\frac{1}{384}-\frac{1}{3840}\approx 0.6065.$$

Example 7.5.12 Calculate the approximate value of the integral $\int_0^1\frac{\sin x}{x}dx$, requiring the error to be no more than 0.0001.

Solution Since $\lim\limits_{x\to 0}\frac{\sin x}{x}=1$, the given integral is not an improper integral. If the value of the integrand function at point $x=0$ is defined as 1, it is continuous over the integral interval $[0,1]$.

Expanding the integrand function yields

$$\frac{\sin x}{x}=1-\frac{x^2}{3!}+\frac{x^4}{5!}-\frac{x^6}{7!}+\cdots \quad (-\infty<x<+\infty).$$

Integrating term by term over the interval $[0,1]$, we get

$$\int_0^1 \frac{\sin x}{x}\mathrm{d}x=1-\frac{1}{3\cdot 3!}+\frac{1}{5\cdot 5!}-\frac{1}{7\cdot 7!}+\cdots.$$

Because the absolute value of the fourth term $\frac{1}{7\cdot 7!}<\frac{1}{30000}$, taking the sum of the first three terms as the approximate value of the integral:

$$\int_0^1 \frac{\sin x}{x}\mathrm{d}x \approx 1-\frac{1}{3\cdot 3!}+\frac{1}{5\cdot 5!}\approx 0.9461.$$

Exercise 7.5

1. Expand the following function into a power series of x, and determine the interval on which the expansion is valid.

(1) a^x;

(2) $\ln(a+x)$ $(a>0)$;

(3) $\operatorname{sh}x=\frac{e^x-e^{-x}}{2}$;

(4) $\sin^2 x$;

(5) $(1+e^x)^3$;

(6) $\frac{x}{\sqrt{1+x^2}}$;

(7) $(4-x^2)^{-\frac{1}{2}}$.

2. Expand the following function into a power series of $(x-1)$, and determine the interval on which the expansion is valid.

(1) $\frac{1}{3-x}$;

(2) $\sqrt{x^3}$;

(3) $\lg x$;

(4) $\frac{1}{x^2+4x+3}$.

3. Expand the function $f(x)=\frac{1}{x}$ into a power series of $(x-3)$.

4. Find the sum function of the power series $\sum_{n=2}^{\infty}\frac{x^{2n+1}}{n!}$.

5. Find the sum function of the power series $\frac{1}{2}\sum_{n=0}^{\infty}(-1)^n\frac{1-3^{2n}}{(2n)!}x^{2n}$.

6. Find the approximate value of $\cos 18°$, accurate to 10^{-3}.

7. Find the approximate value of $\int_0^{\frac{1}{2}}\frac{\arctan x}{x}\mathrm{d}x$, accurate to 10^{-3}.

8. Find the approximate value of $\sqrt[3]{130}$, accurate to 10^{-3}.

Comprehensive Review 7

1.Fill in the blanks.

(1) For the series $\sum_{n=1}^{\infty} u_n$, $\lim_{n\to\infty} u_n = 0$, it is a ________ condition for convergence, not a ________ condition for convergence.

(2) The boundedness of the sequence of partial sums $\{s_n\}$ is a ________ condition for the convergence of the positive series $\sum_{n=1}^{\infty} u_n$.

(3) If the series $\sum_{n=1}^{\infty} u_n$ is absolutely convergent, then the series $\sum_{n=1}^{\infty} u_n$ must be ________.

(4) If the series $\sum_{n=1}^{\infty} u_n$ diverges, then the series $\sum_{n=1}^{\infty} |u_n|$ must ________.

(5) If the positive series $\sum_{n=1}^{\infty} u_n$ converges, then the series $\sum_{n=1}^{\infty} \sqrt{u_n u_{n+1}}$ ________.

(6) If $\lim_{n\to\infty} u_n \neq 0$, then the series $\sum_{n=1}^{\infty} u_n$ ________.

(7) If the series $\sum_{n=1}^{\infty} u_n$ converges to A, then the series $\sum_{n=1}^{\infty} (u_n + u_{n+1})$ converges to ________.

(8) If $\lim_{n\to\infty} \left| \frac{a_n}{a_{n+1}} \right| = \frac{1}{3}$, then the convergence radius of the series $\sum_{n=0}^{\infty} a_n \left(\frac{x+1}{2} \right)^n$ equals ________.

(9) Given that the convergence radius of the power series $\sum_{n=0}^{\infty} a_n x^n$ is 3, the convergence interval of the power series $\sum_{n=1}^{\infty} n a_n (x-1)^{n+1}$ is ________.

(10) If the series $\sum_{n=1}^{\infty} a_n (x-1)^n$ converges at point $x=-1$, then the series converges at point $x=2$ ________.

2. Multiple choice.

(1) Let $a_n > 0 (n=1,2,3,\cdots)$, and $\sum_{n=1}^{\infty} a_n$ converges, constant $\lambda \in \left(0, \frac{\pi}{2}\right)$, then the series $\sum_{n=1}^{\infty} (-1)^n \left(n \tan \frac{\lambda}{n} \right) a_{2n}$ is (　　).

A. absolutely convergent　　B. conditionally convergent

C. divergent　　D. convergence depends on λ

(2) Let a be a constant, then the series $\sum_{n=1}^{\infty}\left[\frac{\sin(na)}{n^2}-\frac{1}{\sqrt{n}}\right]$ is ( ).

A. absolutely convergent B. conditionally convergent

C. divergent D. convergence depends on the value of a

(3) Let $0 \leqslant a_n < \frac{1}{n}$ $(n=1,2,3,\cdots)$, then the series that definitely converges is ( ).

A. $\sum_{n=1}^{\infty} a_n$ B. $\sum_{n=1}^{\infty} (-1)^n a_n$ C. $\sum_{n=1}^{\infty} \sqrt{a_n}$ D. $\sum_{n=1}^{\infty} a_n^2$

(4) Given $\sum_{n=1}^{\infty} (-1)^{n+1} a_n = 2$, $\sum_{n=1}^{\infty} a_{2n-1} = 5$, then $\sum_{n=1}^{\infty} a_n =$ ( ).

A. 3 B. 7 C. 8 D. 9

(5) Which of the following options is correct? ( )

A. If both $\sum_{n=1}^{\infty} u_n^2$ and $\sum_{n=1}^{\infty} v_n^2$ converge, then $\sum_{n=1}^{\infty} (u_n+v_n)^2$ converges.

B. If $\sum_{n=1}^{\infty} |u_n v_n|$ converges, then both $\sum_{n=1}^{\infty} u_n^2$ and $\sum_{n=1}^{\infty} v_n^2$ converge.

C. If the positive series $\sum_{n=1}^{\infty} u_n$ diverges, then $u_n \geqslant \frac{1}{n}$.

D. If the series $\sum_{n=1}^{\infty} u_n$ converges and $u_n \geqslant v_n$ $(n=1,2,3,\cdots)$, then the series $\sum_{n=1}^{\infty} v_n$ also converges.

3. Determine the convergence of the following series:

(1) $\sum_{n=1}^{\infty} \frac{(n+1)!}{n^{n+1}}$; (2) $\sum_{n=1}^{\infty} \frac{n\cos^2\frac{n\pi}{3}}{2^n}$;

(3) $\sum_{n=1}^{\infty} \frac{a^n}{1+a^{2n}}$ $(a>0)$; (4) $\sum_{n=1}^{\infty} \frac{1}{\sqrt{n^2+n}}$;

(5) $\sum_{n=1}^{\infty} \frac{1!+2!+\cdots+n!}{(2n)!}$.

4. Discuss for what values of $x>0$, the following series converges:

(1) $\sum_{n=1}^{\infty} \frac{x^n}{3^n}$; (2) $\sum_{n=1}^{\infty} n!\left(\frac{x}{n}\right)^n$.

5. Discuss the absolute and conditional convergence of the following series:

(1) $\sum_{n=1}^{\infty} (-1)^{\frac{n(n-1)}{2}} \frac{n^{10}}{2^n}$; (2) $\sum_{n=1}^{\infty} (-1)^{n+1} \frac{\sin\frac{\pi}{n+1}}{\pi^{n+1}}$;

(3) $\sum_{n=1}^{\infty} (-1)^{n+1} \frac{1}{2n-1}$; (4) $\frac{1}{\pi^2}\sin\frac{\pi}{2}-\frac{1}{\pi^3}\sin\frac{\pi}{3}+\frac{1}{\pi^4}\sin\frac{\pi}{4}-\cdots$.

6. Discuss for what values of x, the series $\sum_{n=1}^{\infty} \frac{x^n}{1+x^{2n}}$ converges, converges absolutely, and converges conditionally.

7. Find the limit $\lim_{n\to\infty} [2^{\frac{1}{3}} \cdot 4^{\frac{1}{9}} \cdot 8^{\frac{1}{27}} \cdot \cdots \cdot (2^n)^{\frac{1}{n}}]$.

8. Find the convergence domain of the following power series:

(1) $\sum_{n=1}^{\infty} \frac{1}{3^n+(-2)^n} \cdot \frac{x^n}{n}$; (2) $\sum_{n=1}^{\infty} \frac{(x-2)^{2n}}{n\,4^n}$;

(3) $\sum_{n=1}^{\infty} \frac{3^n+(-2)^n}{n}(x-2)^n$; (4) $\sum_{n=1}^{\infty} \frac{[3+(-1)^n]^n}{n}x^n$.

9. Find the sum function of the following power series:

(1) $\sum_{n=1}^{\infty} n(n+1)x^n$; (2) $\sum_{n=1}^{\infty} \frac{(-1)^{n-1}}{n(2n-1)}x^{2n}$;

(3) $\sum_{n=1}^{\infty} n(x-1)^n$.

10. Find the sum of the series $\sum_{n=0}^{\infty} (-1)^n \frac{n+1}{(2n+1)!}$.

11. Expand the following functions into Maclaurin series:

(1) $\int_0^x t\cos t\,dt$; (2) $x\arctan x - \ln\sqrt{1+x^2}$.

12. Suppose $f(x)$ has continuous second derivatives in a neighborhood of $x=0$, and $\lim_{x\to 0}\frac{f(x)}{x}=0$, prove that $\sum_{n=1}^{\infty} f\left(\frac{1}{n}\right)$ converges absolutely.

13 .If $\sum_{n=0}^{\infty} a_n(x-b)^n$, converges when $x=0$ and diverges when $x=2b$, indicate the convergence radius of $\sum_{n=0}^{\infty} a_n x^n$, and prove it.

14. Suppose the series $\sum_{n=1}^{\infty} u_n$ converges, and $\lim_{n\to\infty}\frac{v_n}{u_n}=1$, ask whether the series $\sum_{n=1}^{\infty} v_n$ also converges? Explain the reason.

15. Suppose the positive sequence $\{a_n\}$ is monotonically decreasing, and $\sum_{n=1}^{\infty}(-1)^n a_n$ diverges, ask whether the series $\sum_{n=1}^{\infty}\left(\frac{1}{a_n+1}\right)^n$ converges? Explain the reason.

16. Suppose $a_n \neq 0(n=1,2,\cdots)$, and $\lim_{n\to\infty} a_n = a\,(a\neq 0)$. Prove: The series $\sum_{n=1}^{\infty}|a_{n+1}-a_n|$ and $\sum_{n=1}^{\infty}\left|\frac{1}{a_{n+1}}-\frac{1}{a_n}\right|$ converge or diverge simultaneously.

References

[1] STEWART J,CLEGG D,WATSON S.Calculus:Early Transcendentals[M].9th ed.Farmington:Cengage Learning,2020.

[2] HASS J,HEIL C,WEIR M D.Thomas' Calculus[M].14th ed.New York:Pearson-Education Inc.,2018.

[3] APOSTOL T M.Calculus(Volume Ⅰ & Ⅱ)[M].2nd ed.New York:John Wiley & Sons, Inc.,1991.

[4] 阮正顺,张忠诚,刘雁鸣.高等数学及其应用(上)[M].北京:科学出版社,2018.

[5] 刘为凯,王志宏,杨建华.高等数学及其应用(下)[M].北京:科学出版社,2018.